INTERNATIONAL DEVELOPMENT CENTRE
LIBRARY

Please return this book to the Library on or before the

Natural Resources Management in African Agriculture
Understanding and Improving Current Practices

Natural Resources Management in African Agriculture
Understanding and Improving Current Practices

Edited by

Christopher B. Barrett

Cornell University, USA

Frank Place

International Centre for Research in Agroforestry (ICRAF), Kenya

and

Abdillahi A. Aboud

Egerton University, Kenya

CABI *Publishing*
in association with the
International Centre for Research in Agroforestry

CABI *Publishing* is a division of CAB *International*

CABI Publishing
CAB International
Wallingford
Oxon OX10 8DE
UK

Tel: +44 (0)1491 832111
Fax: +44 (0)1491 833508
Email: cabi@cabi.org
Web site: www.cabi-publishing.org

CABI Publishing
10 E 40th Street
Suite 3203
New York, NY 10016
USA

Tel: +1 212 481 7018
Fax: +1 212 686 7993
Email: cabi-nao@cabi.org

A catalogue record for this book is available from the British Library, London, UK.

Library of Congress Cataloging-in-Publication Data
Natural resources management in African agriculture : understanding and improving current practices / edited by Christopher B. Barrett, Frank Place, and Abdillahi A. Aboud.
 p. cm.
 Includes bibliographical references (p.).
 ISBN 0-85199-584-5 (alk. paper)
  1. Agricultural resources--Africa--Management. I. Barrett, Christopher B. (Christopher Brendan) II. Place, Frank. III. Aboud, Abdillahi A.

 S472.A1 N38 2002
 333.76′096--dc21   2001043833

ISBN 0 85199 584 5

Typeset by AMA DataSet Ltd, UK.
Printed and bound in the UK by Cromwell Press, Trowbridge.

Contents

Contributors

Abdillahi A. Aboud *is Dean of Faculty, Environmental Sciences and Natural Resources, Egerton University, PO Box 536, Njoro, Kenya.*

Akinwumi A. Adesina *is Associate Director (Food Security) and Resident Representative for Southern Africa at the Rockefeller Foundation, Southern Africa Office, 7th Floor Kopje Plaza, PO Box MP 172, Harare, Zimbabwe.*

Christopher B. Barrett *is Associate Professor in the Department of Applied Economics and Management, Cornell University, 315 Warren Hall, Ithaca, NY 14853, USA.*

Douglas R. Brown *is a PhD candidate in resource economics in the Department of Applied Economics and Management, Cornell University, 315 Warren Hall, Ithaca, NY 14853, USA.*

Louise Buck *is a Senior Extension Associate in the Department of Natural Resources and Senior Associate Scientist with the Center for International Forestry Research, Cornell University, 10B Fernow Hall, Ithaca, NY 14853, USA.*

Jonas Chianu *is a Research Associate at the International Institute of Tropical Agriculture (IITA), Ibadan, Nigeria, c/o IITA, L.W. Lambourn & Co., Carolyn House, 26 Dingwall Road, Croydon CR9 3EE, UK.*

Daniel C. Clay *is Director of the Institute of International Agriculture, Office of International Programs, College of Agriculture and Natural Resources, Michigan State University, 324 Agriculture Hall, East Lansing, MI 48824-1039, USA.*

Richard Coe *is the Principal Statistician and Head of Research Support Unit at the International Centre for Research in Agroforestry, PO Box 30677, Nairobi, Kenya.*

David Colman *is Professor of Agricultural Economics in the School of Economic Studies, University of Manchester, Manchester M13 9PL, UK.*

N.C. de Haan *is Rural Sociologist at the International Institute of Tropical Agriculture, Ibadan, Nigeria, c/o IITA, L.W. Lambourn & Co., Carolyn House, 26 Dingwall Road, Croydon CR9 3EE, UK.*

Joris DeWolf *works with the International Centre for Research in Agroforestry (ICRAF), PO Box 30677, Nairobi, Kenya.*

B. Douthwaite *works with the International Institute of Tropical Agriculture (IITA), Ibadan, Nigeria, c/o IITA, L.W. Lambourn & Co., Carolyn House, 26 Dingwall Road, Croydon CR9 3EE, UK.*

Steven Franzel *is Principal Agricultural Economist at the International Centre for Research in Agroforestry, PO Box 30677, Nairobi, Kenya.*

H. Ade Freeman *is an Economist at the International Crops Research Institute for the Semi-Arid Tropics, PO Box 39063, Nairobi, Kenya.*

Karen S. Freudenberger *is an Independent Consultant in natural resource management issues, c/o Chemonics International, 1133 20th Street, NW, Washington, DC 20036, USA.*

Mark S. Freudenberger *is Regional Director, Landscape Development Interventions Program, c/o Chemonics International, 1133 20th Street, NW, Washington, DC 20036, USA.*

Marcel Galiba *is Country Director for Mali and Burkina Faso, Sasakawa Global 2000, Bamako, Mali.*

Berhanu Gebremedhin *is a Post-Doctoral Scientist at the International Livestock Research Institute, PO Box 5689, Addis Ababa, Ethiopia.*

Christina H. Gladwin *is Professor at the Department of Food and Resource Economics, University of Florida, Box 110240 IFAS, Gainesville, FL 32611-0500, USA.*

Nulu Hatibu *is Associate Professor and Leader of the Soil-Water Management Research Group, Sokoine University of Agriculture, PO Box 3003, Morogoro, Tanzania.*

Stein T. Holden *is Associate Professor in the Department of Economics and Social Sciences, Agricultural University of Norway, PO Box 5033, N-1432 Ås, Norway.*

M.A. Jabbar *is an Agricultural Economist at the International Livestock Research Institute (ILRI), PO Box 5689, Addis Ababa, Ethiopia.*

Bashir Jama *is Senior Scientist at the International Centre for Research in Agroforestry (ICRAF), PO Box 30677, Nairobi, Kenya.*

Valerie Kelly *is Visiting Associate Professor in the Department of Agricultural Economics, Michigan State University, Agriculture Hall, East Lansing, MI 48824-1039, USA.*

Patti Kristjanson *is an Agricultural Economist at the International Livestock Research Institute (ILRI), PO Box 30709, Nairobi, Kenya.*

R. Kruska *is the Geographical Information Systems Specialist at the International Livestock Research Institute (ILRI), PO Box 30709, Nairobi, Kenya.*

Freddie Kwesiga *works with the International Centre for Research in Agroforestry (ICRAF), PO Box 30677, Nairobi, Kenya.*

Evelyne A. Lazaro *is Senior Research Fellow in agricultural economics and member of the Soil-Water Management Research Group, Sokoine University of Agriculture, PO Box 3003, Morogoro, Tanzania.*

John Lynam *is Director, the Rockefeller Foundation, PO Box 47543, Nairobi, Kenya.*

H.F. Mahoo *is Senior Lecturer in irrigation and member of the Soil-Water Management Research Group, Sokoine University of Agriculture, PO Box 3003, Morogoro, Tanzania.*

V.M. Manyong *is an Agricultural Economist at the International Institute for Tropical Agriculture (IITA), Ibadan, Nigeria, c/o IITA, L.W. Lambourn & Co., Carolyn House, 26 Dingwall Road, Croydon CR9 3EE, UK.*

Mulugetta Mekuria *is Senior Regional Economist at the Economics and Maize Programmes, CIMMYT-Zimbabwe, PO Box MP163, Mount Pleasant, Harare, Zimbabwe.*

Edson Mpyisi *is In-country Coordinator at the Food Security Research Project, Rwanda Ministry of Agriculture, Animal Resources and Forests, c/o USAID, BP 2848, Kigali, Rwanda.*

Priscah H. Mugabe *is Lecturer in the Department of Animal Science, PO Box MP 167, University of Zimbabwe, Mount Pleasant, Harare, Zimbabwe.*

Lindela R. Ndlovu *is Pro-Vice Chancellor (Academic & Research) at the National University of Science & Technology, Corner Gwanda Road/Cecil Avenue, PO Box AC939, Ascot, Bulawayo, Zimbabwe.*

Amadou Niang *works with the International Centre for Research in Agroforestry (ICRAF), PO Box 30677, Nairobi, Kenya.*

I. Okike *is an Agricultural Economist working as a consultant for the International Livestock Research Institute (ILRI), Oyo Road, PMB 5320, Ibadan, Nigeria.*

Bernard N. Okumu *is Research Associate in the Department of Applied Economics and Management, Cornell University, 444 Warren Hall, Ithaca, NY 14850, USA.*

Pauline E. Peters *is Lecturer on Public Policy at the John F. Kennedy School of Government, Harvard University, 79 John F. Kennedy Street, Cambridge, MA 02138, USA.*

Jennifer S. Peterson *is a Consultant in Niamey, Niger.*

Donald Phiri *is at World Vision, Zambia.*

Frank Place *is an Agricultural Economist at the International Centre for Research in Agroforestry (ICRAF), PO Box 30677, Nairobi, Kenya.*

Jules Pretty *is Professor of Environment and Society, University of Essex, Wivenhoe Park, Colchester CO4 3SQ, UK.*

Thomas Reardon *is Professor in the Department of Agricultural Economics, Michigan State University, 211F Agriculture Hall, East Lansing, MI 48824-1039, USA.*

Ralph Rommelse *works with the International Centre for Research in Agroforestry (ICRAF), PO Box 30677, Nairobi, Kenya.*

Noel Russell *is Senior Lecturer in agricultural economics, University of Manchester, Manchester M13 9PL, UK.*

Filbert B.R. Rwehumbiza *is Senior Lecturer in soil physics and member of the Soil-Water Management Research Group, Sokoine University of Agriculture, PO Box 3003, Morogoro, Tanzania.*

John H. Sanders *is Professor in the Department of Agricultural Economics, Purdue University, 1145 Krannert Building, West Lafayette, IN 47907-1145, USA.*

Barry I. Shapiro *is Program Director at the International Crops Research Institute for the Semi-Arid Tropics (ICRISAT), BP 320, Bamako, Mali.*

Bekele Shiferaw *works with the Department of Economics and Social Sciences, Agricultural University of Norway, PO Box 5033, N-1432 Ås, Norway.*

B.B. Singh *is an Agronomist at the International Institute for Tropical Agriculture (IITA), Ibadan, Nigeria, c/o IITA, L.W. Lambourn & Co., Carolyn House, 26 Dingwall Road, Croydon CR9 3EE, UK.*

Brent M. Swallow *is Programme Leader, Natural Resource Strategies and Policy, International Centre for Research in Agroforestry (ICRAF), PO Box 30677, Nairobi, Kenya.*

Scott M. Swinton *is Associate Professor in the Department of Agricultural Economics, Michigan State University, East Lansing, MI 48824-1039, USA.*

Mamadou Lamine Sylla *is Director of the Natural Resource Management Programme at Office de la Haute Vallée du Niger, BP 178, Bamako, Mali.*

Gbassay Tarawali *is Consultant Agronomist at the International Institute of Tropical Agriculture (IITA), Ibadan, Nigeria, c/o IITA, L.W. Lambourn & Co., Carolyn House, 26 Dingwall Road, Croydon CR9 3EE, UK.*

Shirley A. Tarawali *is an Agronomist at the International Livestock Research Institute (ILRI), Oyo Road, PMB 5320, Ibadan, Nigeria, and the International Institute for Tropical Agriculture (IITA), Ibadan, Nigeria, c/o IITA, L.W. Lambourn & Co., Carolyn House, 26 Dingwall Road, Croydon CR9 3EE, UK.*

Robert Uttaro *is a former graduate student in the Department of Political Science, University of Florida, Gainesville, FL 32611, USA.*

Stephen R. Waddington *is Soil Fert Net Coordinator/Maize Agronomist, CIMMYT-Zimbabwe, PO Box MP163, Mount Pleasant, Harare, Zimbabwe.*

Justine Wangila *is a Research Assistant at the International Centre for Research in Agroforestry (ICRAF), PO Box 30677, Nairobi, Kenya.*

David Weight *is Coordinator of the Partnership for Enhancing Agriculture in Rwanda through Linkages (PEARL), Institute of International Agriculture, Michigan State University, East Lansing, MI 48824, USA.*

T J Wyatt *is an Environmental Protection Specialist in the Economic Analysis Branch, Office of Pesticide Programs for the United States Environmental Protection Agency, Mail Code 7503C, 1200 Pennsylvania Ave, NW, Washington, DC 20460, USA.*

Preface

It is no secret that both the land and the people of rural Africa are suffering. In recent years, researchers, practitioners and policy-makers concerned with persistently high rates of rural poverty and food insecurity and declining per capita agricultural productivity in sub-Saharan Africa have begun to attend seriously to the formidable natural resource management problems that are both cause and consequence of these ills. Much has been written about both the vicious circle in which poverty leads to natural resource degradation, which in turn leads to low resource productivity and renewed poverty, as well as about the need for agricultural intensification on existing cultivated and grazed lands. Researchers have dedicated considerable time and resources over the past decade to developing, often in collaboration with farmers, farming technologies and natural resource management practices to break the vicious circle, to facilitate intensification and thereby to increase agricultural productivity, food security and rural incomes across the continent.

Unfortunately, rates of adoption and diffusion of improved natural resource management practices have generally fallen short of expectations. There are no simple answers to the questions of why many African farmers unsustainably exploit soils and water and why many do not adopt or adapt other, seemingly superior technologies already available. A clear understanding of these processes is none the less urgently needed. Any such understanding must also adequately explain important examples of farm- and community-level innovation and careful natural resource stewardship across the continent, or else it will provide a poor platform on which to base future policy and research.

The chapters that follow cultivate such an understanding, developed from detailed reports on both failures and success stories from across the full range of agroecosystems and economic and institutional conditions found on the continent. This volume thereby breaks new ground in identifying important regularities regarding core determinants of and constraints on natural resource management adoption patterns. Perhaps more importantly, the volume's breadth and depth make clear the key policy and research priorities on which new initiatives need to focus in order to foster substantive improvements. Understanding and improving current practices remain a core challenge in the important task of eliminating poverty and malnutrition in rural Africa over the course of the 21st century.

Early versions of most of these chapters were presented at an international conference on 'Understanding Adoption Processes for Natural Resources Management for Sustainable Agricultural Production in Sub-Saharan Africa', held at the headquarters of the International Centre for Research in Agroforestry (ICRAF) in Nairobi, Kenya, 3–5 July 2000. Conference participants included a broad range of social scientists, biophysical scientists, development practitioners and representatives of international agencies, private foundations and conservation and development organizations. In addition to the presentation and review of earlier versions of these chapters, the conference devoted much time to discussion of the policy and research implications stemming from the papers' results. The editors thank all of the conference participants, especially those whose work is incorporated in this volume, for their contributions both to the conference and to this book.

We are particularly grateful to the Rockefeller Foundation, which sponsored the conference and the subsequent publication of this volume, to ICRAF, which not only hosted the event with its usual grace but also contributed considerable amounts of staff time and resources to the conference and to subsequent editorial work, and to the Danish Agency for Development Assistance (DANIDA), which provided valuable financial support. In this connection, we especially wish to thank Akin Adesina, John Lynam and Ruben Puentes of the Rockefeller Foundation and Pedro Sanchez and Brent Swallow of ICRAF. For unstinting assistance in conference planning and support, we express our sincere thanks to ICRAF staff members Marion Kihori, Oscar Ochieng, Antonia Okono, Justine Wangila and Kijo Waruhiu and to Joyce Knuutila at Cornell. Quinn Avery did a truly extraordinary job organizing and copy-editing the final volume. Joy Learman at Cornell skilfully saw the final product through to publication.

In addition to the authors and co-authors of the chapters included in this volume, we thank many others who provided input and guidance in the planning and execution of the earlier conference and who, directly or indirectly, contributed in important ways to this volume. These include Jane Alumira, Erick Fernandes, Jim Gockowski, Susan Kaaria, Wilberforce Kisamba-Mugerwa, Paul Laizer, Chris Moser, Fridah Mugo, Lincoln Mwarasomba, Brima Ngombi, Jemimah Njuki, Tom Reardon, Kimsey Savadogo and Chris Wien.

The high quality of the revised chapters owes much to the constructive criticism provided by talented colleagues who generously lent their expertise and scarce time to this project as reviewers of draft chapters. We heartily thank Suresh Babu, Bruno Barbier, Brad Barham, Larry Becker, Sara Berry, Hugo DeGroote, Cheryl Doss, Merle Faminow, Jeremy Foltz, George Frisvold, Sarah Gavian, Doug Gollin, Garth Holloway, John Kerr, Arie Kuyvenhoven, Bruce Larson, Melissa Leach, David Lee, Peter Little, Ruth Meinzen-Dick, Bart Minten, John McPeak, Stefano Pagiola, Alice Pell, John Pender, Mark Powell, Tom Randolph, Tom Reardon, Mitch Renkow, Ruerd Ruben, Sara Scherr, Jerry Shively, Melinda Smale, Kevin Smith, Denise Stanley, Camilla Toulmin, Matt Turner, Norman Uphoff, Steve Vosti and Alex Winter-Nelson for their valuable service and insights.

Sincere thanks also go to Tim Hardwick, our development editor at CAB *International*, for his assistance and patience in the course of completing this book, and to Margaret C. Last, our technical editor, and Phyllis Muthee, our indexer.

Finally, we thank our wives and children from the bottom of our hearts for their patience with us as we spent many long evenings poring over draft chapters or sitting at the computer rather than with them. They inspire and sustain us, not just in this book project, but always.

Foreword

Sub-Saharan Africa's agricultural development contrasts markedly with that of Asia and Latin America during the last 40 years. A recent study presented by Hans Gregersen at the Consultative Group for International Agricultural Research (CGIAR) Mid-term Meeting in Durban, South Africa, May 2001, concluded: 'Sub-Saharan Africa stands out as the only region in the world where almost no progress has been made in raising average per capita food consumption or in the incidence of undernourishment.' In hindsight, the overarching reason is quite simple: the natural resource base on which agriculture depends is so depleted that genetic-improvement efforts have been able to bear only limited fruit. In Africa, most farmers are smallholders with 0.5–2.0 ha, earn less than US$1 day^{-1}, face 3–5 hunger months, are malnourished and have large families and 30% are human immunodeficiency virus (HIV)-positive. Women do most of the farming and collect fuel wood and water, while men do off-farm work. The way forward is integrated natural resource management (INRM) to tackle the loss of soil fertility and forested watersheds and to replenish other lost resources. Then, the full weight of genetic improvement and enabling government policies can come into play, as it has in the rest of the developing world.

This book focuses on how farmers, researchers and development workers are tackling these complex issues and presents evidence of substantial progress. Farmers operating under quite varied conditions in Ethiopia, Kenya, Madagascar, Zimbabwe, Mali, Nigeria and elsewhere on the continent are adopting improved fallows, terraces, tied ridges or other improvements that increase yields, conserve scarce soil and water, replenish soil fertility and generate increased profits. Often, improvements are spontaneous adaptations of technologies developed through national and international agricultural research systems. But the scale of these successes remains on the order of thousands of farm families. The pressing challenge is to scale up these promising practices and improved processes of technology adaptation to millions of farmers in order to eliminate this last bastion of hunger and malnutrition from our planet. The authors and editors are to be congratulated for a very useful contribution towards African agricultural development.

Pedro A. Sanchez
Director General
International Centre for Research in Agroforestry
21 June 2001

1 The Challenge of Stimulating Adoption of Improved Natural Resource Management Practices in African Agriculture

Christopher B. Barrett,[1] Frank Place,[2] Abdillahi Aboud[3] and Douglas R. Brown[1]

[1]*Department of Applied Economics and Management, Cornell University, 315 Warren Hall, Ithaca, NY 14853, USA;* [2]*International Centre for Research in Agroforestry (ICRAF), PO Box 30677, Nairobi, Kenya;* [3]*Faculty of Environmental Sciences and Natural Resources, Egerton University, PO Box 536, Njoro, Kenya*

Introduction

Sub-Saharan Africa (SSA) faces a formidable challenge. Most rural Africans remain poor and food-insecure in the aftermath of widespread macroeconomic, political and sectoral reforms that have largely failed to stimulate significant agricultural productivity improvements. Meanwhile, the extensive margin of new arable land available to bring into cultivation, so as to satisfy population-driven increases in food demand, is rapidly being exhausted across most of the continent. There are thus intense pressures for agricultural intensification so as to improve factor productivity without expanding the area under cultivation.

Sustainable agricultural intensification requires prudent long-term management of the natural resource base on which agriculture fundamentally depends. A wide range of traditional and modern techniques[1] exist for effective natural resource management (NRM). None the less, degradation of soils and other natural resources proceeds at a high rate in much of the continent, reflecting in large measure disturbingly low rates of use of sustainable NRM strategies, especially among the poorer subpopulations of smallholder producers.

The challenge of improving smallholder NRM practices lies at the heart of the broader imperative for sustainable agricultural intensification in Africa today, and thus of the universal objective of reducing poverty and vulnerability on a continent in which most people today are employed in agriculture and poverty remains most acute in the countryside. Improved NRM is every bit as much about increasing productivity and incomes for the current generation as it is about preserving the quality of resources to safeguard the livelihoods of future generations.

This volume provides an unprecedented synthesis of findings from across the continent. In drawing together lessons learned from the full range of African agroecosystems and economic and cultural contexts, several empirical regularities stand out. First, while some global principles of effective NRM exist and can be extended across the continent (indeed, probably more broadly), the extraordinary biophysical and economic microvariability of rural Africa makes it difficult to identify effective local solutions without early, active involvement of local farmers and communities. Development, extension and evaluation of NRM innovations must be more than farmer-focused.[2] The process must be farmer-centred, fully involving intended beneficiaries from the early, problem-identification stage onwards. Top-down processes have an undistinguished history, while farmer-driven processes can spur rapid and widespread adoption and adaptation, as several chapters in this volume document encouragingly.

Secondly, improved NRM practices are considerably more knowledge-intensive than are agricultural production technologies, which are typically embodied in inputs such as seed, ploughs or chemicals. Packaged NRM does not exist. Learning processes thus become central to the cycle of developing, disseminating and evaluating new methods. The knowledge intensity of NRM reinforces the necessity of farmer-centred development strategies, since different people and communities learn and communicate in different ways.

Thirdly, NRM is an investment choice. All of the social, economic and cultural factors that weigh on individuals' decisions to invest in financial, human or human-made capital affect choice with respect to natural capital as well, including prices, property rights, patience, opportunity costs, risk preferences, externalities and credit-market limitations. The adoption of improved NRM techniques occurs as a result of decisions made by a wide range of people, each influenced by the incentives and constraints they face. Although they may be farmers, rural households' objectives transcend farming and extend to their pursuit of improved livelihoods for themselves and their families. Scientists need to take these social and economic features of farm-level NRM more seriously than they have in the past. Similarly, economists need to consider

the ecological and geographical context in which farmers make choices.

From these regularities, three priorities emerge clearly for those committed to improved NRM, sustainable agricultural intensification and poverty reduction in rural Africa. First, NRM research and development methods must accelerate the replacement of top-down processes with farmer-centred approaches that take seriously the broader livelihood objectives of rural Africans. Secondly, improved information flow between and within the research community, rural communities, and individual farmers demands new practices and technologies, as well as general improvements in access to and quality of education in rural Africa. Thirdly, necessary public investments and policy reforms must be undertaken to reduce the structural impediments that discourage investment in improved NRM and thereby trap rural Africans in long-standing cycles of poverty and vulnerability. Such impediments include insecure property rights, severe gender inequities, the absence of reliable social security systems and limited access to financial credit, insurance or savings.

The ensuing 21 chapters make the case for the regularities and priorities claimed above. In the remainder of this introductory chapter, we present background – on the present situation, the historical context and current options for improved NRM in African agriculture – and a summary of the key issues and previous findings in the literature.

Background

Present situation

Roughly two-thirds of all Africans work in agriculture, most suffering poverty and associated food insecurity. Improving the productivity of assets – labour, soil, vegetation, water and livestock – used in agricultural production is therefore central to the objective of improving African livelihoods and well-being in the next few decades. Per capita agricultural production is down more than 16% from what it was in the early 1970s in SSA, having recovered only to the level of the early 1980s, immediately prior to the depths of the mid-1980s crisis (Fig. 1.1). Per capita livestock productivity has been in steady decline for the past two decades. Given projected increases in food demand (Pinstrup-Andersen *et al.*, 1997), the rapid exhaustion of unexploited lands onto which SSA agriculture can expand, the lack of comparative advantage in industry or services and the decline of foreign assistance that could finance significant increases in commercial food imports, the necessity of improved per capita agricultural productivity weighs heavily on contemporary policy-makers, development researchers and practitioners in SSA.

Declining SSA agricultural productivity is both a cause and a consequence of deterioration in the natural resource base on which agriculture depends. Oldeman (1998) estimates cumulative productivity losses of 8–14% in Africa since the Second World War as a result of human-induced soil degradation. Meanwhile, the need for increased output has fostered ecologically unsustainable agricultural intensification in many places, leading in particular to soil degradation (Reardon *et al.*, 1999). Scherr (1999) reports that 65% of agricultural crop land and 31% of permanent pasture in Africa

are estimated to be degraded, with 19% of all land 'seriously degraded'. Other widely respected estimates range higher still (Stoorvogel and Smaling, 1990; Dregne and Chou, 1992). Compaction, sealing, crusting, gulleying and other physical degradation pose serious problems, while scientists are only just beginning to explore the extent and impact of biological degradation associated with change in and loss of microbial populations. The primary and obvious problem appears to be chemical degradation of SSA soils due to nutrient depletion and erosion (Sanchez *et al.*, 2000). With precious few exceptions (e.g. Botswana, Mauritius, South Africa), SSA countries are currently losing 30 or more kilograms of nitrogen, phosphorus and potassium (NPK) ha^{-1} year^{-1}, with many countries losing 60–100 kg ha^{-1} annually (Henao and Baanante, 1999).

Soil degradation partly reflects the extraordinarily low use of mineral fertilizers in SSA (Gladwin *et al.*, 1997a; Heisey and Mwangi, 1997). Gross fertilizer use fell 9% between 1992 and 1998, and is largely unchanged from the 1.2 million metric tonnes applied in the mid-1980s, despite a roughly 10% expansion in cultivated area over the past 15 years. More broadly, soil-nutrient loss in SSA agriculture reflects generally low rates of managed nutrient deposition (e.g. through composting,

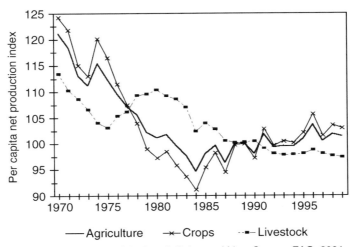

Fig. 1.1. Per capita agricultural productivity in sub-Saharan Africa. Source: FAO, 2001.

manuring and mulching) and fixation (e.g. through improved fallow rotations and intercropping) and high rates of water and wind erosion. Water availability is an equally important, interrelated issue. Seasonally variable or low rainfall can limit diversification through short growing seasons, thereby accentuating both the competition for land and other inputs and the seasonality of smallholder income. The impact of nutrient applications is highly conditioned on the availability of water as well. Only about 5.2 million ha are irrigated in SSA, only 3.3% of cultivated land, more than an order of magnitude lower than the irrigated share of crop lands in other continents. Moreover, the rate of growth of irrigated area has slowed from 2.5% annually in 1970–1990 to about 0.5% $year^{-1}$ in the past decade, with growth in irrigated area not accounting for depreciation of existing irrigation infrastructure. Poor water control also contributes to erosion, especially in hill and highland regions experiencing deforestation due to logging and agricultural expansion, and to soil-nutrient leaching in wetter areas. Investments in water control therefore become important, whether through conventional irrigation or through NRM practices to facilitate water harvesting, drainage or diversion, as appropriate to the biophysical context.

Many of the agricultural productivity and natural resource degradation problems in SSA agriculture derive from the broader economic, social and political environment. For years, colonial and postcolonial governments preyed on smallholder agriculture through land-use and marketing restrictions, price controls, trade and exchange-rate distortions and urban bias in the provision of public infrastructure and services. Had the initial post-independence impulse to draw surplus from agriculture to invest in industrialization not failed so badly, perhaps the deterioration in agricultural productivity and the natural resource base would be less tragic.

Since the early 1980s there has been a significant reduction in the macroeconomic and sectoral policy distortions that contributed to the crisis in African agriculture. However, they have not often been matched by complementary public investments (e.g. rural roads), with the result that these policy correctives have largely failed both to elicit the sort of robust agricultural supply response so many had expected and to induce improved natural resource conservation (Barrett and Carter, 1999; Reardon *et al.*, 1999). Attention has thus again returned to the farm- and landscape-level factors affecting adoption of agricultural technologies and NRM strategies that condition both agricultural factor productivity and agroecology dynamics.

African agriculture in historical context

Observers of African agriculture have often characterized it as inefficient, unproductive and backward. The statement by a Rhodesian administrator in 1926 that intercropping is nothing more than 'hit and miss planting in mixtures' typifies a derogatory view of indigenous African agriculture that persists today, including within national agricultural research and extension systems (NARES) and international agricultural research centres (IARCs) and among expatriate researchers and technicians (Jiggins, 1989; Peters, Chapter 3, this volume).

Much of the scientific research that has supposedly demonstrated the inefficiency of African agriculture has failed to take account of crucial differences in the biophysical and socio-economic contexts (Spencer, 1996). For example, the overwhelming mass of the published literature finds that African small-holders operate far inside their production-possibility frontier (Ali and Byerlee, 1991). Yet, once one controls appropriately for crucial differences in the extraordinarily variable biophysical, economic and institutional environment in which African farmers make production decisions, the limited available evidence suggests instead that African farmers are indeed producing on or near the frontier (Barrett, 1997a; Sherlund *et al.*, 2000). Biophysical conditions limit output and productivity more than most analysts acknowledge. Given stochastic natural inputs, financial constraints, etc., production levels are commonly at or near their constrained frontiers. The output gains from improved NRM may therefore be substantial.

Such findings are consistent with historical reviews of traditional African agriculture based on locally available inputs and emphasizing such practices as diversity within agricultural fields, staggered planting and intensive micro-management of the cropping pattern for resilience, productivity, security and sustainability (Dommen, 1989; Reij et al., 1996; Pretty and Shah, 1997). We are not suggesting that it is either desirable or feasible to turn back the clock or that contemporary agricultural research on improved germ-plasm, fertilizers or breeding stock is unhelpful. Quite the contrary. Traditional agricultural practices can, in many cases, be improved, e.g. by the use of inorganic fertilizers to complement biological methods of soil-nutrient management. The point is rather that traditional practices,

the evolutionary product of farmer experimentation, were often well adapted to the particular ecological, social and economic contexts within which they developed.

Dommen (1988, 1989) and Dupriez (1982) argue persuasively that resource conservation, which they each refer to as conservation of equilibrium biomass (CEB), is central to most traditional African agricultural systems. CEB was traditionally treated as a joint output with annual crop and livestock production. Soil fertility was maintained through CEB, which involved any of several activities – bush and forest fallow systems, integrated plant–animal systems, the addition of household waste or green manure, etc. – chosen to suit the characteristics of local soils, climate, insect, disease and microbial populations, institutions and economic and social incentives.

This view underscores the importance of dynamic concepts of agroecosystem equilibrium and resilience. The top right quadrant of Fig. 1.2 offers a heuristic

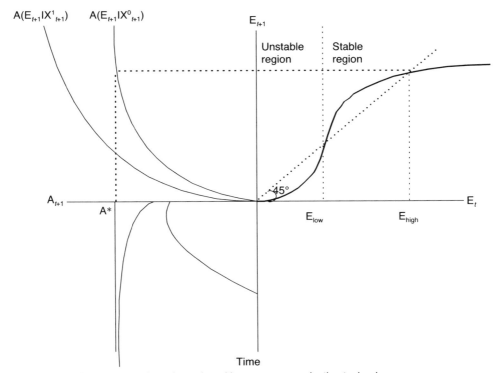

Fig. 1.2. Non-linear agroecology dynamics with a concave production technology.

depiction of the non-linear dynamics typical of many agroecosystems due to strategic complementarities and competition (Holling, 1978; Levin, 2000). The x and y axes represent the agroecosystem condition in successive periods, with the dashed 45° line representing dynamic equilibrium, in which the value is the same across time. Non-linear dynamics produce multiple equilibria, one of which, E_{high}, is stable and the other of which, E_{low}, is unstable. A system in the notional high-level, stable equilibrium, E_{high}, may be driven into disequilibrium – any spot off the 45° line – by external (human or natural) disturbances; but its resilience will in time return it to E_{high} in the wake of modest to significant perturbations, be they negative or positive. Such is not the case either for systems in the low-level, unstable equilibrium, E_{low}, that are hit with even a modest, adverse shock or for those in high-level, stable equilibrium that experience a catastrophic shock driving system health below E_{low}. In the absence of a countervailing (institutional, natural or policy) shock to restore the system to at least E_{low}, such systems may then only stabilize at very low levels of ecosystem productivity or even collapse over time. There is inherent variability in agroecosystem health and resilience, and not all disruptions are permanent or catastrophic (Behnke *et al.*, 1993; Izac and Swift, 1994; Leach and Mearns, 1996; Fairhead and Leach, 1998).

This useful, albeit oversimplified, model implies that acceptance of the claim that traditional systems once enjoyed stable agroecological equilibria is not tantamount to claiming that all such systems remain viable today. Over the past century, in some cases the past decade, traditional African agricultural systems have endured tremendous shocks due, for example, to increased human population densities, war, drought, monocropping and state-imposed revisions of land-tenure regimes or crop pricing. Changing circumstances sometimes render traditional systems inappropriate, and their continued use may result in further declines in productivity and agroecosystem health.

The inextricability of agroecological dynamics and agricultural productivity can be readily understood by defining agricultural production as an increasing function of agroecosystem (A) health (E) and inputs of labour, land and capital (X), as shown in the top left quadrant of Fig. 1.2. This generates the agricultural productivity dynamics shown in the bottom left quadrant of Fig. 1.2. As long as the agroecological state remains in the neighbourhood of E_{high} and inputs remain constant at X^0, agricultural productivity varies stably around A^*, as shown by the vertical line. When the agroecological state is just above E_{low}, agricultural output rises over time to A^* due to the salutary effects of natural regeneration, as shown by the centre curve. However, when the agroecological state is just below E_{low}, agricultural output collapses in time, as shown in the curve that moves to the vertical axis, mirroring the discouraging performance of African agriculture since the early 1970s (Fig. 1.1).

Traditional systems commonly attended to the need for CEB, thereby ensuring reasonably stationary output and ecological dynamics. Since the 1950s, greater emphasis has been put on increasing application of labour and other inputs ($X^1 > X^0$), thereby shifting the production function outward to $A(E|X^1)$ and a higher level of output. Such improvements persist, however, only so long as this change does not adversely affect the underlying agroecology. If, as a consequence, agroecosystem health falls below E_{low}, accelerating rates of input-application growth will be necessary to maintain output in the face of a deteriorating natural resource base. This view of the relationship between agricultural productivity and the health of the underlying ecosystem is consistent with definitions of sustainability that emphasize the need for stable or growing total factor productivity and the resiliency of the ecosystem in the face of inevitable shocks (Herdt and Steiner, 1995; Conway, 1997).

Current problems of stagnant or declining agricultural productivity may in some cases relate directly to insufficient past attention paid to the underlying agroecology, inattention that has sometimes led to inadvertent disruption and degradation (e.g. salinization and waterlogging of poorly irrigated soils, more rapid erosion of tilled former forest lands or disruption of soil biology through monocropping). The resulting disequilibria now demand attention and reversal lest per capita agricultural productivity should continue to decline even after one accounts for added use of land and other non-labour manual inputs.[3]

A wide range of NRM methods exists for CEB and enhanced agricultural productivity. The sometimes complementary options span three distinct groups: (i) traditional practices; (ii) traditional methods adapted to changed biophysical or socio-economic circumstances; and (iii) new NRM techniques. Too often, researchers' and policymakers' instincts are to try something novel, although sometimes the best approach may be to abandon inappropriate 'modern' practices that displaced effective traditional ones (Fernandes, 1998). Much of the recent emphasis on agroecological approaches to sustainable agricultural development arise from heightened researcher appreciation for the crucial details that drive smallholders' decisions to employ traditional NRM practices (Uphoff *et al.*, 2001). African agricultural history is rife with examples of well-intentioned initiatives that unsuccessfully tried to introduce 'modern' agricultural practices and, in so doing, displaced traditional systems more appropriate to the local context (Reij *et al.*, 1996; Pretty and Shah, 1997). Sectoral and macroeconomic policies that diminished returns to farming often created disincentives to continue traditional practices or to invest in agriculture more generally. In many places, researchers and farmers are finding that traditional methods resurface effectively when policy and science give them the space to do so.

In other cases, traditional methods must be adapted to a current, changed context. Newer adaptations, such as mulch farming (Aina *et al.*, 1991), improved green-manure cover crops (GMCC) (Tarawali *et al.*, 1999), fodder banks (Enoh *et al.*, 1999), composting, home gardening, hedgerows and improved short-rotational fallows (Fernandes, 1998), minimum tillage and tied ridging (Reij *et al.*, 1996; Sanders *et al.*, 1996)

are examples of NRM practices that can serve the objective of CEB within traditional cultivation systems. One key area for integrating modern and traditional methods concerns the use of inorganic fertilizers. Financially strapped smallholders typically cannot afford to rely entirely on purchased fertilizers for the nutrient amendments necessary to stimulate crop productivity. Improved fallow management and application of organic matter often provide lower-cost methods of meeting a reasonable portion of farmers' nitrogen needs, although it often pays to supplement biological methods with some inorganic fertilizer. But some conditions almost always require mineral fertilizer application (e.g. lime in the case of acid soils, phosphorus when soils are deficient in that crucial nutrient). The synergies between organic and inorganic inputs and between conservation and replenishment methods for soil-fertility improvement appear widespread and certainly deserve closer scrutiny.

New NRM practices are needed where changed circumstances prevent traditional systems from achieving the productivity and CEB required by farmers. Externally designed practices none the less need to fit the context. The adaptation of contour ridges imposed during the colonial era in Zimbabwe for the purpose of diverting water, in order to prevent erosion, towards water harvesting in drier areas provides a fine example in which the engineers' original design provided an excellent foundation for useful adaptation (Hagmann and Murwirwa, 1996). Adesina and Chianu (Chapter 4, this volume) and Tarawali *et al.* (Chapter 5, this volume) document similar successes of farmer adaptation of researcher-designed NRM innovations. Adoption of high-value, non-indigenous trees producing marketable fruit, medicine or timber has likewise been successful in places as diverse as the Kenyan or Rwandan highlands, the rain forests of Madagascar and the drylands of Morocco (Tiffen *et al.*, 1994; Reij *et al.*, 1996; Clay *et al.*, 1998; Sanchez *et al.*, 2000).

Certainly, the integration of modern science with indigenous knowledge offers much promise for mitigating poverty and natural resource degradation in rural SSA and thus deserves greater attention. However, although researchers have devised many improved NRM methods, both on station and in partnership with farmers, and a wide range of effective NRM strategies have evolved naturally within traditional agricultural systems in SSA, adoption rates none the less appear relatively low. We now turn our attention to understanding this challenge.

Understanding Improved NRM Practice Adoption Patterns

This section synthesizes findings from a wide range of studies of factors that condition adoption of improved NRM methods. The literature on NRM or technology adoption is too vast to survey here in depth, much less comprehensively. Our aim is far more modest: to highlight key issues related to the adoption of improved NRM practices. We adapt the framework introduced by Scherr and Hazell (1994) to consider both additional factors and, like Place and Dewees (1999), multiple scales of analysis. The subsequent 19 chapters are organized into four parts following this same structure. Although many inferences about causal relations remain to be demonstrated conclusively,[4] the evidence to date, combined with the findings reported in Parts I–IV, reveals empirical regularities carrying clear implications for policy and research. Part V concludes the book by exploring these implications in some detail.

Increasing population density and access to commercial markets tend to promote spontaneous intensification through investment in land improvements, including improved NRM practices, fostered in part by induced institutional and technological innovation (Boserup, 1981; Ruttan and Hayami, 1991). Although such patterns have been observed within SSA (Turner et al., 1993; Tiffen et al., 1994; Fairhead and Leach, 1996), the process is not automatic in its timing, extent or irreversibility (Scherr

and Hazell, 1994; Reardon and Vosti, 1995; Shiferaw and Holden, 1998).

That unpredictability arises largely because the adoption of innovations by individuals and their gradual diffusion through a particular societal group/area/region is a social process (Rogers, 1995). At one level, adoption essentially consists of a personal decision about what to do, with choice influenced by both the nature of the innovation itself and the individual's preferences, incentives and constraints. At another level, individual decisions are conditioned by the social context within which they are made. Therefore, although different people adopt, decide not to adopt or disadopt the same innovation at different times and for different purposes, it is not unusual to find social, economic and geographical clustering of particular NRM behaviours. Similar circumstances breed more similar behaviours.

At least from an economist's perspective, NRM practices are a form of investment behaviour because NRM sustains the natural capital on which agriculture depends and also entails real opportunity costs. It costs something to earn dividends. Investment choices have several important characteristics worth keeping in mind as the book progresses.

First, investments are multi-period decisions since some of the benefits, and some of the costs, are incurred well after the adoption decision is made. This necessarily implies that individuals' intertemporal preferences – their patience – and the intertemporal opportunity cost of assets invested in adoption (e.g. the interest rate on financial savings or credit) matter. It also means that decisions are subject to temporal uncertainty. People do not know everything about the future, but they learn over time. And individuals have discretion over the timing of a multi-period decision. This raises a thorny and more pervasive methodological issue in adoption studies. In many cases, research that sets out to measure the factors affecting adoption of an innovation may simply wind up determining the distinguishing characteristics common to individual members of distinct adopter groups – innovator, early adopter, early majority, late majority and laggard (Rogers, 1995) – rather than those factors that actually preclude or induce adoption of the innovation or practice in question.

Secondly, investments entail 'sunk costs', a fixed cost that is not recoverable through subsequent resale. Sunk costs reduce responsiveness to incentives; people are slower to undertake profitable investments or to divest from low-return strategies (Chavas, 1994; Dixit and Pindyck, 1994). In the case of NRM adoption, sunk costs include time spent learning about a new technique, the cost of necessary semi-permanent inputs (e.g. rocks for terracing), etc. The level of sunk costs in NRM adoption varies considerably by practice and depends in large measure on things like the extent of asset markets and land-tenure regimes, since improvements to land are rarely capitalized into marketable land values in SSA, with the occasional exception of tree crops.

Thirdly, investments commonly have limited divisibility, i.e. they are 'lumpy'. One cannot easily, for example, benefit from terracing, wind-breaks or drainage systems over very small units of land. Similarly, smallholders living below the poverty line find it difficult to afford a cow or a 50 kg sack of fertilizer. Lumpiness puts a premium on liquidity or on cooperative organization for sharing lumpy resources, as well as on marketing innovations, for example, smaller fertilizer packets, which Freeman and Coe (Chapter 11, this volume) report on from Kenya.

Fourthly, seemingly similar investments may differ significantly in terms of quality. For example, one may see a variety of terracing methods, ranging from trash lines to vegetative strips to bench terracing, with markedly different levels of quality within each, due to differences in materials used, spacing or the quality of workmanship. At some point, these cease being comparable. Those who lack the managerial skills or financial resources to undertake a high quality investment or to hire someone with the necessary skills to make it may find adoption of a similar but lower-quality innovation unattractive.

Fifthly, investments are not made in isolation. Choices with respect to a particular investment depend on the portfolio of other investments made or pending. Farmers who also have lucrative, stable non-farm income sources that demand their time (e.g. a store owner or a seasonal migrant to South African mines) may have a relatively high minimum required return to labour or capital investment in NRM practices. Conversely, farmers who have invested in valuable perennials for commercial production (e.g. of tea or fruit) may find complementary adoption of soil and water conservation (SWC) practices more attractive than in the absence of such prior investments. The primary criterion driving investment choice is relative net return, not gross net returns.

Sixthly and finally, investments affect social status by helping define an individual within his or her community. In some settings, an early adopter is seen as 'progressive', 'innovative' and a 'leader'. In others, the same behaviour can be branded 'reckless', 'brazen' or disrespectful of tradition or can violate religious taboos. Such influences breed herd behaviour extending beyond social learning processes (Moser, 2001). Rural Africans, like people throughout the world, largely choose investments based on what they feel is best for themselves and those they love. In order to understand NRM behaviour, one must therefore get a reasonably accurate understanding of: (i) individuals' objectives and learning processes; (ii) their willingness and capacity to make long-term investments; (iii) the incentives and constraints they face in making choices; and (iv) the biophysical, institutional and policy context within which choices get made. We now turn to these issues in sequence.

Farmers' Objectives and Learning Processes

Objectives and preferences

SSA farmers respond to material incentives. Since consumption depends on the income one has available to purchase goods and services, and agricultural output is a major – sometimes the only – source of income for farm households, farmers clearly prefer higher yields to lower yields and higher profits to lower ones, all else held equal. It is essential, however, to view rural households through a livelihoods perspective (Ellis, 2000). SSA farm households typically have non-farm interests as well and often prefer non-farm to farm investments (Barrett *et al.*, 2001d; Wyatt, Chapter 10, this volume).

Moreover, all else is rarely equal. Complications emerge because of heterogeneous farmer preferences with respect to time (i.e. patience) and risk. In so far as the poor tend to have higher discount rates and greater risk aversion, for any of several reasons, poorer smallholders may be less willing to undertake long-term NRM investments (Reardon and Vosti, 1995; Holden *et al.*, 1998; Holden and Shiferaw, Chapter 7, this volume).[5] Similarly, price and yield risk typically impede risk-averse farmers' adoption of yield-increasing technologies, while favouring cost-reducing technologies instead (Feder *et al.*, 1985; Kim *et al.*, 1992). Incomplete and imperfect rural markets in SSA also mean that one typically needs to look at household production and consumption decisions together, rather than a simpler, profit maximizing firm perspective (Singh *et al.*, 1986; DeJanvry *et al.*, 1991; Barrett, 1997a; Reardon and Vosti, 1997).

Perhaps the biggest challenge comes from the obvious importance of non-material factors to smallholder welfare. Rural land is often coveted as a retirement residence and as a cheaper location for the children to attend school while the father works in the urban sector. Pride and a desire for self-sufficiency or independence can induce resistance to recommendations for change, perhaps especially from outsiders. Appreciation for the aesthetics of farming can foster openness to some practices (e.g. terracing) but opposition to others (e.g. alley-cropping, cover crops) perceived as disruptive of the natural order. Similarly, a sense of stewardship of the land may accelerate identification of emerging degradation problems and motivate adoption of proved

NRM methods, but foster resistance to experimentation with unproved and yet promising ones.

Certain communities exhibit extraordinarily high or low adoption rates that just do not lend themselves easily to satisfactory explanation on the basis of biophysical and economic criteria alone. Individuals commonly conform to reasonably homogeneous standards of behaviour despite heterogeneous individual preferences, whether because they fear stigma or value tradition or conformity/solidarity or because deviance threatens the social insurance on which they depend (Bernheim, 1994; Moser, 2001).

Genuinely participatory approaches offer the best means for ascertaining farmer preferences, not least of which by creating space in which farmers can direct the development and adaptation of NRM methods themselves (Chambers, 1997; Adesina and Chianu, Chapter 4, this volume; Tarawali *et al.*, Chapter 5, this volume). Yet preferences are also mutable, especially in so far as smallholders' non-material values respond to moral suasion by community and national leaders. Sustained, sincere emphasis on the virtues and necessity of resource stewardship and rhetorical advocacy of what Nash (1991) calls the 'ecological virtues' can help motivate farmers to overcome the many tangible, material obstacles to improved NRM in SSA agriculture.

Learning and knowledge

Even receptive farmers need to learn about new methods in order to adopt them. Slow diffusion of improved NRM practices is often due not to the insufficient supply of methods appropriate to smallholder context, but rather to problems of weak information transfer (Napier, 1991; Pretty, 1995a). Where economists tend to focus on the incentives and constraints to adoption, taking information and beliefs as given (Feder *et al.*, 1985; Norris and Batie, 1987), other social science models of adoption and diffusion commonly emphasize the importance of knowledge and learning (Duff *et al.*, 1990; Pretty, 1995a; Rogers, 1995). The individual decision to invest in improved NRM practices depends fundamentally on the farmer's awareness of the need for improvement and his or her beliefs about the potential of the new practice.

Farmers form and update beliefs – i.e. learn – about the current and prospective situation differently depending on their information channels and the form(s) of knowledge involved (scientific, indigenous or otherwise). The available evidence, including several chapters that follow, suggests that formal mechanisms for disseminating information – schools, extension services, farmer cooperatives – exhibit a mixed performance in stimulating NRM practice adoption in SSA. If extension services are not trusted because of a chequered performance history or local personalities, recommendations may be discounted irrespective of the underlying worth of the recommended practice because farmers' perception of the characteristics of an innovation, not necessarily its demonstrable properties, matter to adoption decisions (Adesina and Zinnah, 1993; Adesina and Baidu-Forson, 1995; Shiferaw and Holden, 1998; Negatu and Parikh, 1999).[6] Extension appears most effective for introducing practices that are entirely new to a system (Adesina and Zinnah, 1993; Adesina and Baidu-Forson, 1995; Moser, 2001). Using Lindner *et al.*'s (1992) distinction between 'discovery', 'evaluation' and 'trial' phases of farmers learning about new technologies, extension services appear most important in the first, discovery phase, in which farmers become aware of innovations hitherto unknown in their community. Given the funding crisis gripping most extension services, their effectiveness in transferring knowledge and understanding of information-intensive NRM practices is expectedly limited. New approaches to farmer education are desperately needed in SSA (see Barrett *et al.*, Chapter 22, this volume).

Ultimately, the crucial phase in adoption decisions is typically that of innovation evaluation. Many farmers learn by experimenting themselves with new methods, even though experimentation is typically costly, implying both a willingness and an ability to invest to learn (Richards, 1985; Foster and Rosenzweig, 1995; Cameron, 1999).

Learning is mainly a social process, however. Interpersonal interactions within communities are typically the dominant channels for evaluating the potential and the best means of implementing innovations. The information shared among individuals appears to be less data than analysis. For example, very few Ghanaian farmers accurately report data on others' harvest of pineapples in an area of recent introduction; yet they share and exploit each others' analysis of their own individual experience (Goldstein and Udry, 1999). Seeing technologies work on fields managed by farmers of similar characteristics can instil confidence among learning farmers that would not occur from visiting research stations.

Learning from others can none the less create an incentive to delay adoption in order to let others incur the costs of experimentation while one observes the results relatively costlessly (Foster and Rosenzweig, 1995). Moreover, as Neill and Lee's (2001) study of the adoption and disadoption of GMCC in Honduras points out, when spontaneous adoption–diffusion occurs, those who adopt based on what they see their neighbours doing may not fully understand the essential management practices. As a result they may not employ the innovative technique as needed to ensure success and may therefore subsequently abandon the practice.

These factors underscore the importance and opportunities of social learning through local institutions, such as the various initiatives reported in Chapter 2 (this volume) by Pretty and Buck. Farmer field schools, which are based on experiential learning, and community groups can create and maintain social capital that complements human capital, particularly education, in acquiring, interpreting and applying valuable information.

Farmers' sharing of processed information, not just raw data, also underscores the place of indigenous knowledge in SSA agriculture and the importance of early farmer participation in problem definition and the design and evaluation of new NRM practices. But, as Peters (Chapter 3, this volume) cautions, one must not romanticize indigenous knowledge, for it too has its limits.

Furthermore, while some NRM problems (e.g. gulley erosion, waterlogging) are readily observable and well understood by SSA smallholders (Ndiaye and Sofranko, 1994; Place and Dewees, 1999; Boyd et al., 2000), others are not so immediately obvious (e.g. inefficient nitrogen fixation due to phosphorus deficiencies, semi-toxicity in forages that impede the functioning of microbial guilds in rumen and soils). So-called 'scientific' knowledge can help identify NRM problems and their scale and severity. People must perceive a problem before they become willing to incur costs to halt or reverse degradation (Gebremedhin and Swinton, Chapter 6, this volume).

Although farmers tend to know their production environments better than outsiders do, aptitude, experience and training clearly matter (Shiferaw and Holden, 1998). Younger, more inexperienced farmers and recent immigrants typically have a less keen understanding of local agroecosystem dynamics than do more established smallholders. This does not imply, however, that older, more established farmers are most likely to adopt improved NRM practices, since the value they place on respecting tradition is commonly greater than among the young and the mobile.

In principle, educational attainment – literacy, at least – should increase rates of adoption by accelerating information flow and improving farmers' capacity to follow printed instructions correctly and to respond productively to technology and policy shocks (Schultz, 1975; Barrett et al., 2001e). The empirical evidence with respect to the effect of education on improved NRM adoption, however, is relatively weak, possibly for statistical reasons.[7]

NRM methods are generally more knowledge-intensive than are production technologies based on improved seed, chemicals or mechanical implements embodying improvements. More complex NRM methods are also harder to learn. Partly for this

reason, Batz et al. (1999) and Moser (2001) independently find that the adoption of improved practices requiring simultaneous adoption of several new techniques tends to meet with some resistance, even when the characteristics and performance of the new methods are preferable from the farmers' standpoint.

Special mention is perhaps also due to the threats posed by acquired immune deficiency syndrome (AIDS) and continued violence. By decimating young adult populations in many SSA rural communities and forcing many families to withdraw children from school to conserve school expenses or replace lost labour at home, civil strife and AIDS pose unprecedented threats to the preservation of indigenous knowledge and the acquisition of literacy, numeracy and 'scientific' knowledge skills. While relatively little attention has been paid thus far to the prospective effects of AIDS or war on agricultural and NRM practices, the consequences could become profound in the absence of explicit efforts to record and codify indigenous knowledge at risk of being lost and to help parents keep children in school.

Willingness and Capacity to Make Long-term Investments

Since investments in improved NRM practices pay dividends over an extended period, farmers are more inclined to undertake such investments when they are more likely to reap the full stream of benefits over time and where that future stream of net returns is more predictably favourable. So incentives to adopt depend on security of usufructure rights in land, animals and other durable, productive assets whose returns vary with the choice of NRM regime and on temporal risk in prices and yields.

Rights in land and water are central to questions of improved NRM practice adoption. Secure rights[8] are associated with increased agricultural productivity, with farm investment and with higher rates of conservation practices or land improvements, though these are not entirely robust across locations (Migot-Adholla *et al.*, 1991; Place and Hazell, 1993; Besley, 1995; Gavian and Fafchamps, 1996; Reij *et al.*, 1996; Sjaastad and Bromley, 1997; Templeton and Scherr, 1999; Gebremedhin and Swinton, Chapter 6, this volume). One must be careful, however, not to misinterpret security of access as equivalent to statutory legal titling or even alienable, individualized property rights, since a wide range of customary tenurial institutions in SSA provide reasonably secure long-term usufructure and even transfer rights (Atwood, 1990; Ostrom, 1990; Baland and Platteau, 1996). Moreover, many tenurial institutions have evolved to serve multiple purposes in rural SSA, in particular pooling risks by offering others residual claims on crop residues, dead wood for fuel, grazing/browsing corridors or dry-season watering points, etc. (Van den Brink *et al.*, 1995; Baland and Platteau, 1996). Increasing the exclusivity of primary holders' rights will then have analytically ambiguous effects on investments in conservation structures to protect common-pool resources or other ecologically favourable practices.

Land tenurial systems do not treat all persons equally, so incentives to adopt improved NRM practices vary with some predictability across demographic groups. Recent immigrants and women typically have the least secure rights and are therefore less inclined to undertake long-term investments in improved NRM practices.[9] Moreover, if improved NRM increases the value of land or involves commercialization (e.g. by the introduction of high-value tree products), then women may risk losing their assets to their husbands, fathers or brothers (Agarwal, 1997; Meinzen-Dick *et al.*, 1997; Place and Swallow, 2000). So long as a degraded, low-productivity plot is better than none at all, gender differences in security of land access will contribute to gender differences in improved NRM adoption patterns.

The empirical relationship between tenure and adoption is also complicated by the real possibility of bidirectional causality. Conservation that adds value to the land may induce enclosure and privatization (Reij *et al.*, 1996). Moreover, many indigenous rights systems clearly allow for investments in trees or conservation structures that strengthen one's claims (Shepherd, 1991; Bruce and Migot-Adholla, 1994; Place and Swallow, 2000; Gray and Kevane, 2001).

Tenurial regimes are not the only factors determining the security of claims to the long-term stream of benefits generated by investments in NRM improvements. Places subject to sociopolitical upheaval tend to exhibit much lower rates of NRM investment. If people are unsure whether their lands, water rights or animals will be stolen or destroyed or if they will need to flee and abandon their assets, then they have little reason to undertake costly improvements. Boyd *et al.* (2000) found cattle rustling to be a serious obstacle to investment in improved NRM practices, because asset loss compromises households' ability to undertake necessary investments and because cattle are commonly a complementary input to many sustainable intensification methods.

The temporal risk SSA farmers face extends beyond asset security to physical productivity and prices. Most empirical evidence strongly supports the hypothesis of smallholder risk aversion, so it is natural that improved NRM methods that are perceived as risk-increasing are less commonly adopted than those methods that reduce risk. Batz *et al.* (1999) indeed found that greater relative uncertainty about a practice discouraged adoption and that risk-reducing technologies are more likely to be adopted than are income-increasing ones in uncertain environments (Wyatt, Chapter 10, this volume). Chavas *et al.* (1991) and Fafchamps (1993) similarly find that Burkinabe farmers choose production practices in part to maintain flexibility in the face of temporal uncertainty, even though this involves foregoing expected income.

Price risk also matters. When faced with price spikes, farmers may be inclined to mine a resource, either to generate greater current income or to self-insure against prospective shortages (Barrett, 1999). Since liberalized agricultural markets have tended to increase price variabiliity in SSA (Barrett and Carter, 1999), there is renewed, widespread interest on the continent in means of effective price stabilization of key commodities. In so far as price stabilization reduces the temporal uncertainty surrounding the returns to investments in improved NRM, it may help stimulate increased adoption and productivity.

Willingness to invest long-term applies not just to farmers, but also to communities and to national governments. Where leaders are committed to the long-term health of their jurisdictions and are held accountable for the conditions under which their subjects live, it is far easier to secure their attention than when they operate largely without accountability and with little concern for the legacy they will ultimately leave behind.

Even when willing to invest, poor smallholders often have limited capacity to mobilize labour, land or cash for investment in even effective and profitable NRM improvements (Reardon and Vosti, 1995; Vosti and Reardon, 1997; Shepherd and Soule, 1998; Holden and Shiferaw, 2000; Holden and Shiferaw, Chapter 7, and Clay *et al.*, Chapter 8, this volume). The nature and severity of these constraints varies markedly by NRM practice and across and within villages.

Some NRM investments have significant up-front, sunk costs (e.g. terracing, irrigation) that put them beyond the reach of households with limited cash or labour availability (Wyatt, Chapter 10, this volume). Other promising NRM techniques – such as commercial cultivation of high-value trees and leguminous vegetables – demand significant purchased inputs, including planting materials, labour and fertilizer. Still other practices, like improved fallows (Gladwin *et al.*, Chapter 9, and Place *et al.*, Chapter 12, this volume) or livestock production (Ndlovu and Mugabe, Chapter 19, this volume), require access to ample land. Because cash can alleviate household constraints on labour, land, equipment and the like so long as markets exist for these inputs, financing may be the most widespread limitation on smallholder capacity to invest in improved NRM practices. Rural SSA financial markets are plagued by structural problems of covariate risk and information asymmetries that induce credit rationing, while physical insecurity and inappropriate government financial policies have impeded the formation of safe deposit-taking institutions (Carter, 1988; Stiglitz, 1990; Yaron *et al.*, 1997; Zeller *et al.*, 1997). The consequence can be limited smallholder cash savings and credit access because informal finance (e.g. moneylenders, store credit, loans from relatives or neighbours) tends to be of too short a duration (i.e. months) to match the multi-year pay-out on NRM investments well (Zeller *et al.*, 1997). Moreover, even when credit is available, NRM investments are not often the most desired use of funds, so NRM investments may not increase at the same rate as credit access (Wyatt, Chapter 10, this volume).

None the less, SSA farmers have other sources of financing besides cash savings and credit. Where livestock markets function well, cattle and small stock serve as reasonably liquid stores of wealth. Trees are also known to have significant 'savings' values, such as with eucalyptus in Ethiopia (Gebremedhin and Swinton, Chapter 6, this volume). There is widespread evidence that African smallholders use non-farm and off-farm income to invest in perennials (e.g. trees) and variable inputs for other, complementary annuals (e.g. inorganic fertilizer for vegetables sold in urban markets), livestock or SWC structures that increase whole-farm productivity and improve integrated nutrient and water management (Tiffen *et al.*, 1994; Reij *et al.*, 1996; Reardon, 1997b; Clay *et al.*, 1998; Pender and Kerr, 1998; Reardon *et al.*, 1998; Savadogo *et al.*, 1998; Barrett and Reardon, 2000; Clay *et al.*, Chapter 8, this volume). Non-farm income can also serve as a substitute for collateral in facilitating access to credit, a feature that gains greater importance where property rights are not formalized and alienable. Some farmers can take advantage of contract-farming schemes, in which input packages are often extended on credit against a crop (e.g. French beans in Kenya, barley in Madagascar). But such opportunities are typically limited to or favour relatively large farmers proximate to metropolitan areas (Little and Watts, 1994).

If an improved NRM practice is labour-using, at least initially (e.g. in the construction of SWC structures) and local labour markets are incomplete, then household capacity to adopt may be constrained by family size and their capacity to mobilize reciprocal labour (Akinola and Young, 1985). This is especially a problem for female-headed households, who typically have less adult household labour available to them, poorer access to credit and, in patriarchal cultures, less capacity to mobilize reciprocal labour for NRM investments.[10] Labour availability often favours smaller farms, with lower land/labour ratios. Their lower unit-labour costs can help induce higher rates of adoption of improved NRM practices, as Byiringiro and Reardon (1996) found in the case of soil conservation structures in Rwanda.

Improved NRM adoption sometimes implies a labour allocation choice that many households are not willing or able to make. Where rudimentary agricultural technologies imply low agricultural labour productivity, investing further labour on farm is typically not very attractive, especially if viable off-farm income opportunities exist (e.g. in cities or mines). In such cases, labour-reducing practices, such as no-till cultivation, may appeal while more labour-intensive methods, such as mulching, may not. Where yield or price risk is great, smallholders tend to pursue diversified livelihoods through off-season employment, which often competes for the labour necessary to undertake land improvements outside the cropping season, especially when seasonal migration is a major source of off-farm income (Christensen, 1989; Reardon, 1997a). In risky environments, smallholders are not typically looking to concentrate their earnings in one activity, even if they can improve its productivity.

Land can also be a limiting factor to the adoption of land-extensive investments, such as traditional fallows, cover crops or green-manure banks, while fostering investments in land-intensive practices and structures, such as bunding, ridging, irrigation, terracing or hedgerows. The impact of land scarcity depends on the induced displacement of food-crop area relative to the

resulting change in crop yields, once the new NRM practice is established (Pagiola, 1996).

In the artificial, textbook world of complete, competitive markets, investment depends purely on preferences and economic incentives; capacity to invest is not an issue. Market imperfections therefore underlie all the problems of limited capacity to invest in improved NRM practices, and widespread evidence that liquidity, labour and land constraints impede adoption of improved NRM methods underscores the priority policy-makers must put on improving factor and product market performance. Much of this relates to improved institutional and physical infrastructure to reduce search and transactions costs and to facilitating the formation of self-help groups to mobilize resources locally, topics addressed by Barrett *et al.* in Chapter 22 (this volume). In some cases, this also requires competition policy where entry or exit barriers and strategic behaviours concentrate market power in the hands of one or a few market intermediaries (Barrett, 1997b).

Economic Incentives and the Importance of the Natural Resource Base

While many of the factors discussed to this point are specific to particular plots or farmers, many of the incentives to adopt improved NRM practices depend on broader incentives created by market and non-market institutions. Economic theory clearly shows that the incentive to invest in a new technology increases in the induced change in output (i.e. through higher yields and/or relative prices) and decreases in the relative cost of the investment. So relative prices and technologies matter.

Where rural infrastructure is thin, transaction costs tend to be high; therefore the shadow prices[11] on which smallholders base decisions vary markedly across households, depending on access to markets as well as on relative resource endowments. Poor transport or communications infrastructure reduces the net returns to products farmers sell, increases the cost of purchased inputs and increases the variability in those prices by impeding spatial and intertemporal arbitrage. Weak markets impede adoption of sustainable NRM practices and have helped fuel unsustainable intensification of SSA agriculture (Reardon *et al.*, 1999; Freudenberger and Freudenberger, Chapter 14, this volume). Those with poorest market access commonly appear less likely to adopt improved NRM practices (Akinola and Young, 1985; Lindner *et al.*, 1992; Spencer, 1996; Neill and Lee, 2001; Kristjanson *et al.*, Chapter 13, this volume). Market access is especially important to adoption of improved NRM methods that depend on marketing by-products, such as milk from ruminant livestock, fruit or other products from trees. Where households remain in semi-subsistence mode, adoption of techniques to stimulate production of greater surpluses holds little appeal when markets are remote, uncompetitive or otherwise inaccessible or unattractive. Sustainable agricultural intensification based on improved NRM practices depends on reliable, low-cost access to reasonably competitive markets.

Towards this end, institutions that support competitive markets are important. This includes regulatory bodies to oversee seed certification and fertilizer quality and to combat monopoly practices, as well as an effective and low-cost judicial system to enforce contracts so as to reduce expected losses due to fraud or malpractice. Many smallholder farmers in SSA lack confidence in the origin of seed sold by commercial distributors and the resulting reduced use of improved seed has a negative impact on demand for fertilizer and NRM investments.

Adoption also depends on the underlying productivity of local technologies. For example, even increasing yields by a half in cereals typically yielding 2 t ha^{-1} or less in SSA generates only an extra few hundred US dollars per hectare gross (i.e. not accounting for increased labour or other input costs), often not enough to make investments requiring any significant up-front or ongoing expenditures attractive.[12] Hence the appeal of higher-value improvements, such as trees yielding valuable fruit or medicine (Sanchez

et al., 2000). Access to complementary inputs can also pose a problem by limiting the productivity of improved practices. For example, while inorganic fertilizer and most biological methods of improved NRM are substitute means of providing nitrogen, few if any agroecological methods can remedy acidity or phosphorus-deficiency problems, so there is a certain complementarity between chemical and biological soil amendments. Fertilizer supply and distribution bottlenecks often make availability undependable in rural SSA and can thereby impede the productivity of improved NRM (Reardon *et al.*, 1999; Freeman and Coe, Chapter 11, and Kelly *et al.*, Chapter 15, this volume).

A household's broader livelihood strategy affects the stake it has in protecting the natural resource base, the pressure it puts on it and its capacity and willingness to invest in improved NRM practices (Reardon and Vosti, 1995; Vosti and Reardon, 1997). Households that depend more on agriculture for their livelihoods have stronger incentives to invest in improved NRM methods, while those that derive much of their income from non-farm sources tend to be less willing to divert cash, labour or both to NRM improvements (Scherr and Hazell, 1994; Boyd *et al.*, 2000). Even those who depend heavily on farming may be unwilling to invest in improved NRM methods if the costs of degradation under current practices accrue in relatively small measure to them or if the benefits of adoption accrue significantly to others (i.e. if externalities exist).

Externalities can be either cross-sectional – where one household's actions affect another's welfare today, as is the case with the adoption either of agroforestry systems that create wind-breaks or reduce erosion or of integrated pest-management strategies – or dynamic – where actions today affect the state of the underlying agro-ecosystem for tomorrow's users, which is typically the case for soil management. Cross-sectional externalities can be addressed either through voluntary collective action or government interventions. Dynamic externalities, however, rarely lend themselves to cooperative action since future generations are inevitably under-represented. Bequest motives can partly resolve the externality problem, as when a father invests in land improvements that will largely accrue to his sons. But even altruism rarely fully internalizes dynamic externalities, so asset markets and government play central roles.

Asset markets effectively internalize dynamic externalities in so far as the future productivity effects of NRM improvements or degradation get capitalized into the value of the land or livestock. This requires reasonably active markets with good, verifiable information available to both buyer and seller. Note that, while there may be considerable tenure security in traditional systems, in which land is allocated not by transactions between willing buyers and willing sellers but rather by the fiat of a local leader or council, such systems only occasionally remedy the dynamic externality problem.

Because contracting is costly and enforcement problematic in much of rural SSA, clear assignment of property rights and activation of asset markets cannot fully resolve externality problems. Rather, proper resolution demands voluntary collective action or policy interventions by local or national governments. What instruments are available to government? Taxes on practices that degrade the natural resource base, subsidies to conservation measures and regulatory controls through legal restrictions on use patterns or on production practices have at best a chequered history in rural SSA. The non-point, spatially distributed nature of soil degradation and limited bureaucratic capacity make it exceedingly difficult to monitor farmer behaviour and enforce regulatory standards or taxes, especially from the administrative centre (Boyd *et al.*, 2000; Shiferaw and Holden, 2000).

Where the resource is important at the more aggregate level, there may be further incentives for coordinated interventions across many farmers or villages if transactions costs (e.g. lorry-transport costs) or non-linear pricing (i.e. lower unit free on board (f.o.b.) cost for larger shipments) create external economies of scale.[13] This probably applies to activities such as liming to combat soil acidity, inorganic phosphorus applications and the adoption of integrated pest-management techniques. Such interventions can have 'crowding-in' effects by then increasing the expected productivity of individual-level adoption of improved NRM techniques – for example, nitrogen-fixing leguminous trees or cover crops in the wake of phosphorus applications. So public investment can induce increased private investment. Such larger-scale interventions could be financed by grants or longer-term 'resource restoration loans' from donors or central governments or by mobilizing local resources through cooperative financial institutions.

Scale also matters because even if the local NRM problem gets resolved – for example, if farmers on sloped lands are induced to adopt SWC structures – farmers may be inclined to reinvest their gains in clearing of additional forest lands or increasing stocking densities of livestock on already overgrazed common pastures, as Freudenberger and Freudenberger (Chapter 14, this volume) vividly describe in their chapter on Madagascar. In sum, the totality of households' livelihood systems and the broader agroecology need to be considered holistically.

The policy implications with respect to the economic importance of the resource base are several. First, government policies (e.g. agricultural marketing and pricing), along with public investments (e.g. in rural infrastructure), can have a profound effect on the profitability of agriculture and hence the economic importance of agricultural resources. Secondly, government must be more active where farmers' livelihood portfolios are more diversified out of agriculture or there are considerable negative (positive) externalities associated with inappropriate (improved) NRM practices. Thirdly, centralized regulatory controls and taxes are rarely an effective means of intervening in rural NRM problems in SSA agriculture. It is more effective for governments to foster the formation and maintenance of community groups to define and enforce resource-access rules appropriate to the local context and to facilitate community or large-scale financing of improvements exhibiting external economies of scale.

The Agroecological, Institutional and Policy Context

SSA's extraordinary biophysical variability limits the geographical scope over which any particular NRM practice proves effective. Even within a particular region, microvariability in hydrology, soils, climate, etc., can render techniques found effective on some farms ineffective on others. Similarly, local institutions, the efficacy of central government and the policies that apply to land and water management, much less to agriculture more broadly, vary considerably. Economists too often fail to take these non-microeconomic factors into adequate account when studying technology or NRM adoption patterns.

Innovations such as GMCC may not appeal to smallholders for ecological reasons unanticipated by agricultural researchers and extension agents. For example, the use of mucuna as a GMCC in West Africa has been rejected by some farmers out of concern that it provides an ideal habitat for poisonous snakes that will endanger those who must eventually manipulate the biomass (Galiba *et al.*, 1998; Sogbedji, 2000). Similarly, many leguminous cover crops fare poorly in phosphorus-poor soils or are low-growing and therefore shaded by taller, neighbouring cereals, thereby impeding their emergence and productivity when intercropped (Manson *et al.*, 1986; Kumwenda *et al.*, 1997). In drier areas, the scope for intercropping of nutrient providers with recipient crops, such as in an alley-farming system, is greatly reduced, due to moisture competition between the plants (Ong, 1994). In the Philippines, soil-fertility enhancement techniques had little effect because apparent infertility was primarily due to nematodes impeding plant growth (Fujisaka, 1994). Varietal choice and the biogeochemistry and topography of farmers' fields commonly differ from those in research stations but may significantly

affect the returns a given farmer might expect from adoption or may necessitate rotations rather than intercropping, in which case seasonal labour or financing constraints may bind.

The same principle applies to adoption of livestock to augment crop production. The productivity of livestock in harvesting nutrients through extensive grazing and in accelerating nutrient availability to soil macro- and microfauna and plants depends fundamentally on the biochemistry of local forage and cover-crop species, many of which contain anti-nutritional compounds toxic to animals (Pell, 1999). So the efficacy of crop–livestock integration and optimal species choice depends a great deal on local ecological factors.

Because men and women commonly control land of differing characteristics and quality, we suspect that a certain amount of the gender differences observed in the use of conservation practices really results from different agroecological conditions. Women commonly must work land that is less fertile than the land worked by men from the same household or community (Aboud *et al.*, 1996; Goldstein and Udry, 1999). But it appears that the key factor is not gender but the underlying agroecological conditions that affect the returns to alternative conservation practices.

The common denominator of these observations is that performance evaluation of NRM practices done on research stations inevitably cannot capture biophysical conditions faced on a large share of targeted smallholder farms. Furthermore, on-station experiments are unable to capture the effects of interactions between NRM practices and households' other farming and non-farm enterprises. Research on the performance of NRM practices must increasingly emphasize on-farm trials and acknowledge that heterogeneity in underlying conditions affects varietal productivity.

Sometimes the issues surrounding adoption are less related to information availability, the economic importance of the resource, farmers' willingness or capacity to invest for the long term or the economic incentives to do so, but rather revolve around weaknesses in the supporting institutional environment. Institutions – the rules that condition human behaviour (North, 1990) – largely derive from social norms at the micro level, local organizations (e.g. community credit unions, farmer cooperatives, resource-user groups) at the meso level and policies at macro level. Institutions can serve to internalize externalities. At the individual level, stewardship values can induce voluntary conservation efforts beyond what can be explained in material terms. At a more aggregate level, community organizations can establish effective access rules and monitoring and enforcement mechanisms for cooperative management of forests, land, livestock or water (Ostrom, 1990; Baland and Platteau, 1996). SSA enjoys a strong record of community-based initiatives, including cooperative construction, maintenance and operation of deep well systems among Borana pastoralists in southern Ethiopia (Coppock, 1994), formation of community tree nurseries (Kwesiga *et al.*, 1999), formation of community credit cooperatives that extend credit for land improvements (Zeller *et al.*, 1997), farmer marketing cooperatives, and water- and forest-user associations that regulate individual use patterns. Local organizations with strong management and clear rules have been shown to contribute to successful NRM, which can include stimulating adoption of improved methods for addressing problems on the commons (Veit *et al.*, 1995; Baland and Platteau, 1996; Reij *et al.*, 1996).

But strong, cooperative communities are by no means universal in SSA. Problems of institution formation and maintenance often relate to community heterogeneity and stability, and are often compounded by stress associated with sociopolitical turmoil or manipulation and severe food insecurity. Colonialism destroyed traditional community structures in many parts of Africa, neopatrimonial regimes since independence have done the same in others and, in some places, low population densities meant that households never had much reason to develop strong cooperative structures outside their extended-family network. Whatever the reason for the absence of a cohesive community, where communities cannot resolve externalities problems – including the provision of public goods, such as information, physical security, monitoring of local agroecosystem health or research on and evaluation of alternative NRM practices – there may be a role for others to play; in particular, government must be willing and able to fill in.

In some settings, governments or development projects have subsidized inputs (e.g. transport of material or food for work for constructing SWC structures), thereby inducing adoption where it is unattractive on straight market-pricing terms. Of course, subsidized adoption raises questions about both sustainability – will termination of subsidies induce disadoption, as has been observed in SSA (Reij *et al.*, 1996; Reardon *et al.*, 1999)? – and the potential for scaling up the effort where donor and government funding is limited. A mosaic of projects promoting improved NRM on different terms in different places is no substitute for a fiscally sustainable, coherent policy across broader landscapes.

Conclusion

The foregoing synthesis foreshadows several key insights that the 19 chapters of Parts I–IV develop in far greater detail: the importance of authentically participatory approaches to technique development and dissemination, the knowledge-intensive nature of NRM and the crucial role played by various economic incentives and constraints to undertake investment in natural capital. Several of these points have been identified previously in individual studies for particular sites. This collection of studies marks the first major attempt at synthesis, however, and thereby creates new opportunities to distil core lessons and to establish the policy and research priorities that must guide the next generation of work on NRM in African agriculture.

In Part V, we therefore return to draw out the implications of the findings and inferences of these preceding chapters, as well as those of the broader literature on which this volume builds. The penultimate chapter, by Place *et al.* (Chapter 21, this volume), focuses on lessons learned for the research establishment, particularly the technology development and dissemination cycle. The concluding chapter (Barrett *et al.*, Chapter 22, this volume) identifies policy priorities for improving NRM in African agriculture.

Improved NRM is a prerequisite for sustainable agricultural intensification, which is itself a necessary condition for economic growth, poverty alleviation and environmental conservation in Africa – indeed in much of the developing world (Lee and Barrett, 2000). Although the challenges of improving NRM and inducing sustainable intensification in African agriculture remain substantial, there has been much progress in recent years in identifying feasible and attractive options for smallholder producers and in understanding better what constraints impede adoption of sustainable practices. We therefore express cautious optimism that the progress evident in the following chapters and in fields throughout scattered villages around the continent foreshadow steady advances that will both improve the quality of life for the current generation and sustain the resource base on which future ones will depend.

Notes

[1]   We use the terms 'practices', 'techniques' and 'methods' interchangeably in this chapter and confine our discussion to what might be termed 'sustainable' means of NRM, i.e. ones that can be continued in perpetuity in the absence of exogenous disruption.

[2]   The term 'farmer' is used to refer to both farming households and to communities, since certain NRM problems and solutions apply at the community level and others at farm level.

[3]   This differs markedly from the relationship between agricultural productivity and ecosystem health being debated in East Asia, North America and Europe. In those places, concerns centre around high rates of chemical applications and the use of genetically modified organisms that may compromise the integrity of surrounding ecosystems.

[4]   Both the extraordinary heterogeneity of the human and natural landscapes of rural SSA and the dearth of detailed longitudinal data make it difficult to establish definitively cause-and-effect sequences in observational (not just experimental) data.

[5]   We note, however, that individuals may rationally reduce consumption in order to defend the asset base on which their future survival depends, and even poor households invest relatively heavily in the education and long-term health of their children whenever possible. Much remains to be learned about the investment behaviours of the poor.

[6]   Outside researchers can have a difficult time getting a clear sense of smallholders' true perceptions, as Peters (Chapter 3, this volume) vividly describes in the context of Malawian agriculture.

[7]   Because educational attainment is generally low, there is commonly insufficient variability within the data to be able to identify effects due distinctly to education with any precision. Formal education is also often negatively correlated with the age of husbands and wives and so it becomes difficult to disentangle these potentially opposing effects. Moreover, data on agricultural technology or NRM practice adoption invariably use sampling frames of farmers. Given that educational attainment affects livelihood choice, considerable self-selection may bias the estimated effects of education on adoption. Finally, there is also the issue of education of whom: Basu and Foster (1998) argue that the uneducated within a household make extensive use of educated co-residents, so the educational status of the household head may not really matter.

[8]   Place and Swallow (2000) emphasize three features of rights security that are especially important to adoption decisions: exclusivity, security and transferability.

[9]   Gladwin *et al.* (Chapter 9, this volume) also draw the important distinction between women in male-headed households, who are subject to the husband's authority, and women heads of households, who enjoy greater power and discretion.

[10]   Boyd *et al.* (2000) find this to be a factor in SWC adoption in Tanzania.

[11]   People make resource-allocation decisions based not on observable market prices but rather on 'shadow prices', the relative scarcity of goods once one accounts for transactions and search costs, risk premiums and rationing effects. Market and shadow prices are rarely identical.

[12]   Consider a 1 t ha^{-1} increase in rice fetching the equivalent of US\$0.20 kg^{-1}. The gross marginal yield is but US\$200 ha^{-1}, hardly a lot for

a household of six or seven persons cultivating only 1 ha in total (e.g. in Madagascar or West African rice systems). For crops returning still lower prices or for which a 1 t ha^{-1} increase is perhaps infeasible (e.g. cassava, millet, sorghum), the incentives would be even less. The basic logic also applies to non-marketed commodities, wherein modest gains in expected calories harvested do not justify a significant up-front investment of labour effort that consumes calories and risks injury or exposure to vectors carrying disease.

[13] External economies of scale exist when a larger scale of operation by others lowers my unit costs. This concept is distinct from the conventional notion of internal economies of scale, in which my unit costs are decreasing in the scale of my own operation.

2 Social Capital and Social Learning in the Process of Natural Resource Management

Jules Pretty[1] and Louise Buck[2]

[1]*Department of Biological Sciences, University of Essex, Wivenhoe Park, Colchester CO4 3SQ, UK;* [2]*Department of Natural Resources, Cornell University, 10B Fernow Hall, Ithaca, NY 14853, USA*

For as long as people have managed natural resources, they have engaged in forms of collective action. Farming households have collaborated on water management, labour sharing and marketing; pastoralists have co-managed grasslands; fishing families and their communities have jointly managed aquatic resources. Such collaboration has been institutionalized in many forms of local association, through clan or kin groups, traditional leadership, water users' groups, grazing-management societies, women's self-help groups, youth clubs, farmer experimentation groups, church groups, tree-growing associations and labour-exchange societies.

Although constructive resource management rules and norms have been embedded in many cultures, from collective water management in Egypt, Mesopotamia and Indonesia to herders of the Andes and dryland Africa, and from water harvesting in Roman North Africa and south-west North America to shifting agriculture systems, it has been rare, prior to the last decade, for the importance of such local groups and institutions to be recognized in agricultural and rural development. In both developing- and industrialized-country contexts, policy and practice in recent years have tended to be preoccupied with changing the behaviour of individuals rather than of groups or communities. Over this time, agriculture has had an increasingly destructive effect on the environment (Balfour, 1943; Huxley, 1960; Palmer, 1976; Jodha, 1990; Ostrom, 1990; Kothari *et al.*, 1998).

In some contexts, the loss of local institutions has led to natural resource degradation. In India, the loss of management systems for common-property resources has been a critical factor in the increased over-exploitation, poor upkeep and physical degradation observed over the past half-century. Jodha's (1990) study of 82 villages in seven states found that only 10% of villages still regulated grazing or provided watchmen compared with the 1950s; none levied grazing taxes or had penalties for violation of local regulations; and only 16% still obliged users to maintain and repair common resources.

At the same time as local institutions have disappeared, so the state has increasingly taken responsibility for natural resource management (NRM), largely because of a mistaken assumption that these resources are mismanaged by local people (Scoones, 1994; Pretty and Pimbert, 1995; Leach and Mearns, 1996; Ghimire and

Pimbert, 1997; Pretty and Shah, 1997). On the other hand, many studies of agricultural development have shown that, when people are well organized in groups and their knowledge is sought, incorporated and built upon during planning and implementation, then they are more likely to sustain activities after project completion (de los Reyes and Jopillo, 1986; Cernea, 1987, 1991; Uphoff, 1992; Pretty, 1995a, 1998; Pretty et al., 1995; Bunch and López, 1996; Singh and Ballabh, 1997; Röling and Wagemakers, 1998; Uphoff et al., 1998; Pretty and Hine, 2000).

One study of 25 completed World Bank agricultural projects found that continued success was associated clearly with local institution building (Cernea, 1987). Twelve of the projects achieved long-term sustainability, and it was in these that local institutions were strong. In the others, the rates of return had all declined markedly, contrary to expectations at the time of project completion. Projects were unsustainable where there had been no attention to institutional development and local participation.

There is a danger, of course, of appearing too optimistic about local groups and their capacity to deliver economic and environmental benefits. We are aware of the divisions and differences within and between communities and how conflicts can result in environmental damage. None the less, it is now clear that new thinking and practice are needed, particularly for developing and spreading forms of social organization that are structurally suited for NRM and natural resource protection at the local level. This usually means more than just reviving old institutions and traditions; it commonly means new forms of organization, association and platforms for common action. These are not the sole conditions for success, but are usually a necessary condition.

In this chapter we aim to generate insight into characteristics and effects of social capital and social learning in fostering more ecologically sustainable NRM. Our arguments are rooted in the logic that social acceptability is a critically important decision criterion which is often ignored, or at least underestimated, during the evaluation

and promotion of innovation in agriculture and NRM among smallholder farmers (Buck, 1995). While agroecological feasibility and economic viability are given predominance in designing intervention strategies and policy incentives, the extent to which a prospective new practice is consistent with social norms and social bonds can be a key determinant in anticipating and fostering adoption, particularly when commonly managed resources are at stake.

We begin our discussion by elaborating the concept of social capital and highlighting relations between capital assets and sustainability in agriculture, stressing linkages between social capital and human capital in bringing about the conditions for long-term improvements in the natural endowment. We argue that social learning is a key to unlocking the potential for social capital to be brought to bear on the challenges underlying more sustainable land-use practice.

We then offer some metaphors to improve insight into relations between social learning and improved knowledge and practice, and suggest that appropriately facilitated social learning fosters the co-evolution of practice, institutions and policies. Through these processes, ideas about what is socially acceptable will inevitably change, opening new possibilities for practice. Finally, we conclude on how to promote social learning at a larger scale than is currently practised.

Five Capital Assets for Sustainable Development

Economic and social systems at all levels – farms, livelihoods, communities and national economies – rely for their success on the value of services flowing from the total stock of assets they control. Five types of capital – natural, social, human, physical and financial – are now being addressed in the literature (Bourdieu, 1986; Coleman, 1988, 1990; Putnam et al., 1993; Putnam, 1995; Costanza et al., 1997, 1999; Carney, 1998; Flora, 1998; Grootaert, 1998; Ostrom,

1998; Pretty, 1998; Scoones, 1998; Uphoff, 1998; Pretty and Ward, 2001).

1. Natural capital produces nature's goods and services and comprises food (both farmed and harvested and caught from the wild), wood and fibre; water supply and regulation; treatment, assimilation and decomposition of wastes; nutrient cycling and fixation; soil formation; biological control of pests; climate regulation; wildlife habitats; storm protection and flood control; carbon sequestration; pollination; and recreation and leisure.

2. Social capital yields a flow of mutually beneficial collective action, contributing to the cohesiveness of people in their societies. The social assets comprising social capital include norms, values and attitudes that predispose people to cooperate; relations of trust, reciprocity and obligations; and common rules and sanctions mutually agreed or handed down. These are connected and structured in networks and groups.

3. Human capital is the total capability residing in individuals, based on their stock of knowledge skills, health and nutrition. It is enhanced by their access to services that provide these, such as schools, medical services and adult training. People's productivity is increased by their capacity to interact with productive technologies and with other people. Leadership and organizational skills are particularly important in making other resources more valuable.

4. Physical capital is the store of human-made material resources and comprises buildings (housing, factories), market infrastructure, irrigation works, roads and bridges, tools and tractors, communications and energy and transportation systems, which make labour more productive.

5. Financial capital is accumulated claims on goods and services, built up through financial systems that gather savings and issue credit, such as pensions, remittances, welfare payments, grants and subsidies.

These five assets are transformed by policies, processes and institutions to give desirable outcomes, such as food, jobs, welfare, economic growth, clean environment,

reduced crime, and better health and schools. Desirable outcomes, when achieved, feed back to help build up the assets base, while undesirable effects, such as pollution, deforestation, increased crime or social breakdown, reduce the asset base.

The basic premise is that sustainable systems, whether farms, firms, communities or economies, accumulate stocks of these five assets, thereby increasing the per capita endowments of all forms of capital over time. But unsustainable systems deplete or run down these various forms, spending assets as if they were income and so leaving less for future generations.

Social Capital

There has been a rapid growth in interest in the term 'social capital' in recent years. The term captures the idea that social bonds and social norms are important for sustainable livelihoods. Its value was identified by Jacobs (1961) and Bourdieu (1986), later given a clear theoretical framework by Coleman (1988, 1990) and brought to wide attention by Putnam *et al.* (1993) and Putnam (1995). Coleman describes it as 'the structure of relations between actors and among actors' that encourages productive activities. These aspects of social structure and organization act as resources for individuals to use to realize their personal interests. Local institutions are effective because 'they permit us to carry on our daily lives with a minimum of repetition and costly negotiation' (Bromley, 1993).

As it lowers the costs of working together, social capital facilitates cooperation. People have the confidence to invest in collective activities, knowing that others will also do so. They are also less likely to engage in unfettered private actions that result in negative impacts, such as resource degradation. Four central aspects have been identified (Pretty and Ward, 2001): (i) relations of trust; (ii) reciprocity and exchanges; (iii) common rules, norms and sanctions; and (iv) connectedness, networks and groups.

Relations of trust

Trust lubricates cooperation. It reduces the transaction costs between people and so liberates resources. Instead of having to invest in monitoring others, individuals are able to trust them to act as expected. This saves money and time. It can also create a social obligation – by trusting someone, this engenders reciprocal trust. There are two types of trust: the trust we have in individuals whom we know; and the trust we have in those we do not know, but which arises because of our confidence in a known social structure. Trust takes time to build, but is easily broken (Gambetta, 1988; Fukuyama, 1995) and, when a society is pervaded by distrust, cooperative arrangements are unlikely to emerge (Baland and Platteau, 1998).

Reciprocity and exchanges

Reciprocity and exchanges also increase trust. There are two types of reciprocity (Coleman, 1990; Putnam *et al.*, 1993): specific reciprocity refers to simultaneous exchanges of items of roughly equal value; and diffuse reciprocity refers to a continuing relationship of exchange that at any given time may be unrequited, but over time is repaid and balanced. Again, reciprocity contributes to the development of long-term obligations between people, which can be an important part of achieving positive environmental outcomes (Platteau, 1997).

Common rules, norms and sanctions

Common rules, norms and sanctions are the mutually agreed or handed-down norms of behaviour that place group interests above those of individuals. They give individuals the confidence to invest in collective or group activities, knowing that others will do so too. Individuals can take responsibility and ensure that their rights are not infringed. Mutually agreed sanctions ensure that those who break the rules know they will be punished.

These are sometimes called the rules of the game (Taylor, 1982), the internal morality of a social system (Coleman, 1990), the cement of society (Elster, 1989) or the basic values that shape beliefs (Collins and Chippendale, 1991). They reflect the degree to which individuals agree to mediate or control their own behaviour. Formal rules are those set out by authorities, such as laws and regulations, while informal ones are those that individuals use to shape their own everyday behaviour. Norms are, in contrast, preferences and indicate how individuals should act; rules are stipulations of behaviour with positive and/or negative sanctions. A high social capital implies high 'internal morality', with individuals balancing individual rights with collective responsibilities (Etzioni, 1995).

Connectedness, networks and groups

Connectedness, networks and groups and the nature of relationships are a vital aspect of social capital. There may be many different types of connection between groups (trading of goods, exchange of information, mutual help, provision of loans, common celebrations, such as prayer, marriages, funerals). They may be one-way or two-way and may be long-established (and so not responsive to current conditions) or subject to regular update.

Connectedness is manifested in different types of groups at the local level – from guilds and mutual aid societies, to sports clubs and credit groups, to forest, fishery or pest-management groups, and to literary societies and mother-and-toddler groups. It also implies connections to other groups in society, from the micro to the macro level (Uphoff, 1992; Flora, 1998; Grootaert, 1998; Ward, 1998; Woolcock, 1998; Rowley, 1999). High social capital implies a likelihood of multiple membership of organizations and links between groups. It is possible to imagine a context with large numbers of organizations, but each protecting its own interests with little cross-contact. Organizational density may be high, but intergroup connectedness low (Cernea, 1993). A better form of social capital implies high organizational density and cross-organizational links.

Connectedness, therefore, has five elements:

1. Local connections – strong connections between individuals and within local groups and communities.
2. Local–local connections – horizontal connections between groups within communities or between communities, which sometimes become platforms and new higher-level institutional structures.
3. Local–external connections – vertical connections between local groups and external agencies or organizations, being one-way (usually top-down) or two-way.
4. External–external connections – horizontal connections between external agencies, leading to integrated approaches for collaborative partnerships.
5. External connections – strong connections between individuals within external agencies.

Even though some agencies may recognize the value of social capital, it is common to find not all of these connections being emphasized. For example, a government may stress the importance of integrated approaches between different sectors and/or disciplines but fail to encourage two-way vertical connections with local groups. Another may emphasize the formation of local associations without building their linkages upwards to other external agencies. In general, two-way relationships are better than one-way, and linkages subject to regular update are generally better than historically embedded ones.

Social and Human Capital as Prerequisites for Natural Capital Improvements

To what extent, then, are social and human capital prerequisites for long-term improvements in natural capital? Natural capital can clearly be improved in the short term with no explicit attention to social and human capital. Regulations and economic incentives are commonly used to encourage

change in behaviour, and include establishment of strictly protected areas, regulations for erosion control or adoption of conservation farming, economic incentives for habitat protection and pesticide taxes (Pretty *et al.*, 2000). But, though these may change behaviour, there is rarely a positive effect on attitudes: farmers commonly revert to old practices when the incentives end or regulations are no longer enforced (Dobbs and Pretty, 2001).

The social and human capital necessary for sustainable and equitable solutions to NRM comprises a mix of existing endowments and that which is externally facilitated. External agencies or individuals can act on or work with individuals to increase their knowledge and skills, their leadership capacity and their motivations to act. They can act on or work with communities to create the conditions for the emergence of new local associations with appropriate rules and norms for resource management. If these then lead to the desired natural capital improvements, then this again has a positive feedback on both social and human capital.

Although there is now an emerging consensus that social capital and human capital manifested in groups can pay (Narayan and Pritchett, 1996; Rowley, 1999), there are surprisingly few studies that have been able to compare group with individual approaches in the same context (most have observed changes over time, with changing performance of groups being compared with earlier performance of individual approaches).

For farmers to invest in these approaches, they must be convinced that the benefits derived from group, joint or collective approaches will be greater that those from individual ones. External agencies, in contrast, must be convinced that the required investment of resources to help develop social and human capital, through participatory approaches or adult education, will produce sufficient benefits to exceed the costs (Grootaert, 1998; Dasgupta and Serageldin, 2000).

Ostrom (1998) puts it this way: 'participating in solving collective-action problems is a costly and time consuming process. Enhancing the capabilities of local, public entrepreneurs is an investment activity that needs to be carried out over a long-term period.' For initiatives to persist, the benefits must then exceed both these costs and those imposed by any free-riders in the group-based or collective systems.

Not all forms of social capital, however, are good for everyone in a community. A society may be well organized, have strong institutions and have embedded reciprocal mechanisms, but be based not on trust but on fear and power, such as feudal, hierarchical, racist and unjust societies (Knight, 1992). Formal rules and norms can also trap people within harmful social arrangements. Again a system may appear to have high social capital, with strong families and religious groups, but contain some individuals with severely depleted human capital through abuse or conditions of slavery or other exploitation. Some associations can also act as obstacles to the emergence of sustainable livelihoods. They may encourage conformity, perpetuate adversity and inequity and allow certain individuals to get others to act in ways that suit only themselves (Olson, 1965; Taylor, 1982). Social capital also has a 'dark side' (Portes and Landolt, 1996).

It is important for NRM to distinguish between social capital embodied in such groups as sports clubs, denominational churches, parent–school associations and even bowling leagues, and that in resource-oriented groups. It is also important to distinguish social capital in contexts with a large number of institutions (high density) but little cross-membership and high excludability from that in contexts with fewer institutions but multiple, overlapping membership of many individuals. In the face of growing uncertainty (e.g. economies, climates, political processes), the capacity of people both to innovate and to adapt technologies and practices to suit new conditions becomes vital. An important question is whether forms of social capital can be accumulated to enhance such innovation (Boyte, 1995; Hamilton, 1995).

Participation and Social Learning

The term participation is now part of the normal language of most development agencies. It is such a fashion that almost everyone says that it is part of their work. This has created many paradoxes. The term participation has been used to justify the extension of control of the state as well as to build local capacity and self-reliance; it has been used to justify external decisions as well as to devolve power and decision-making away from external agencies (Pretty, 1995b).

In conventional development, participation has commonly centred on encouraging local people to contribute their labour in return for food, cash or materials. Yet these material incentives distort perceptions, create dependencies and give the misleading impression that local people are supportive of externally driven initiatives. When little effort is made to build local skills, interests and capacity, then local people have no stake in maintaining structures or practices once the flow of incentives stops.

The dilemma for authorities is that they both need and fear people's participation. They need people's agreement and support, but they fear that such wider involvement is less controllable and less precise. But, if this fear permits only stage-managed forms of participation, distrust and greater alienation are the most likely outcomes. This makes it all the more crucial that judgements can be made on the type of participation in use.

'Participation' is one of those words that can be interpreted in many different ways: it can mean finding something out and proceeding as originally planned, or it can mean developing processes of collective learning that change the way that people think and act. The many ways that organizations interpret and use the term participation can be resolved into six distinct types. These range from passive participation, where people are told what is to happen and act out predetermined roles, to self-mobilization, where people take initiatives largely independently of external institutions (Pretty, 1995b).

Since the Brundtland Commission put 'sustainable development' on the map in the mid- to late 1980s, close to 100 definitions of 'sustainability' have been published. Most seek to put a value on the future stocks of natural resources, on the continued satisfaction of basic human needs for present and future generations, on processes to mediate among economic, social and biophysical needs and on development that protects both the natural environment and people. In detail, each emphasizes different values, priorities and practices. Clearly no reasonable person is opposed to the idea.

Does any of this help in the context of NRM? We all know that sustainability represents something good, but what exactly? In any discussion of sustainability, it is important to clarify what is being sustained, for how long, for whose benefit and at whose cost, over what area and measured by what criteria. Answering these questions is difficult, as it means assessing and trading off values and beliefs (Steer and Lutz, 1993; Viederman, 1994; Pretty, 1995a).

It is critical, therefore, that sustainable agriculture does not prescribe a concretely defined set of technologies, practices or policies. This would only serve to restrict the future options of farmers. As conditions change and as knowledge changes, so must farmers and communities be encouraged and allowed to change and adapt too. Sustainable agriculture is, therefore, not a simple model or package to be imposed. It should be seen more as a process for learning (Röling, 1988; Pretty, 1995b).

Modernist agricultural development begins with the notion that there are technologies that work and that it is just a matter of inducing or persuading farmers to adopt them. Yet few farmers are able to adopt whole packages of conservation technologies without considerable adjustments in their own practices and livelihood systems.

The problem is that the imposed models look good at first, and then fade away. Alley cropping – an agroforestry system comprising rows of nitrogen-fixing trees or bushes separated by rows of cereals – has long been the focus of research (Kang *et al.*, 1984; Lal,

1989). Many productive and sustainable systems, needing few or no external inputs, have been developed. They stop erosion, produce food and wood and can be cropped over long periods. But the problem is that very few farmers have adopted these alley-cropping systems as designed. Despite millions of dollars of research expenditure over many years, systems appear to have been produced that are suitable largely for research stations only (Carter, 1995). As we show below, the opposite occurs when farmers are allowed to choose and adapt technologies to their own conditions.

Enhancing Farmers' Capacity for Social Learning

It has become increasingly clear to practitioners that social learning is a necessary, though not the sole, part of the process of adjusting or improving NRM. The conventional model of understanding technology adoption as a simple matter of diffusion, as if by osmosis, no longer stands. But the alternative is neither simple nor mechanistic. It is to do with building the capacity of farmers and their communities to learn about the complex ecological and physical complexity in their fields and farms and then to act in different ways. The process of learning, if it is socially embedded and jointly engaged upon, provokes changes in behaviour (Argyris and Schön, 1978; Habermas, 1987; Kenmore, 1999) and can bring forth a new world (Maturana and Varela, 1982).

The metaphor we use here for this new sustainability science is that fields are full of megabytes of information and yet we collectively lack the operating system to understand and transform this information. This information is about pest–predator relationships, about moisture and plants, about soil health and about the chemical and physical relationships between plants and animals on farm. These are subject to manipulation – and farmers who understand some of this information and who are confident about experimentation have the components of an advanced operating system. Most of the time, though, this information remains locked up and unavailable.

However, the past decade has seen an increasing understanding of how to develop these operating systems through the transformation of both social and human capital. This is social learning – a process that fosters innovation and adaptation of technologies embedded in individual and social transformation. In the context of developing countries, most of this social learning is not embedded in hard information technology (such as computers or the Internet). Rather, it is associated, when it works well, with farmer participation, rapid exchange and transfer of information when trust is good, better understanding of key agroecological relationships in fields and farmers experimenting in groups. And large numbers of groups work in the same way as parallel processors – the most advanced forms of computation.

The empirical evidence tells us several important things. Social learning leads to greater innovation as well as an increased likelihood that social processes producing these technologies will persist. But it is also very difficult to promote, support and sustain.

Farmer field schools (FFS) have been one of the most significant models for social learning to emerge in the past decade and a half. Farmer field schools are 'schools without walls', in which a group of up to 25 farmers meets weekly during the rice season to engage in experiential learning for integrated pest management (IPM) (Eveleens *et al.*, 1996; van de Fliert, 1997; Kenmore, 1999). The FFS revolution began in South-East Asia, where research on rice systems demonstrated that pesticide use was correlated with pest outbreaks (Kenmore *et al.*, 1984). The loss of natural enemies and of the free services they provided for pest control was a cost that exceeded the benefits of pesticide use. The programme of FFS is supported by the Food and Agriculture Organization (FAO) and other bilateral development assistance agencies and has since spread to many countries in Asia and Africa. At the last estimate, some 1.8 million

farmers are thought to have made a transition to more sustainable rice farming as a result. FFS have given farmers the confidence to work together on more sustainable and low-cost technologies for rice cultivation (Pretty and Ward, 2001)

Elsewhere, the social capital and experimental capacity of farmers have been developed by the Comité de Investigación Agrículture Tropical (CIAT) in Latin America in the form of Comité de Investigación Agrícola Local (CIALs) (Braun, 2000). Some 250 groups have been established in six countries, and these develop their own individual pathways according to the motivations and needs of farmers. These groups decide upon research topics, conduct experiments and draw upon technical help from field technicians and agricultural scientists. Feedback is given to communities as a whole, and regional groupings of CIALs hold annual meetings to share their findings. Members talk about being 'awakened about their continuous learning process, and losing their fear of speaking out in public' (Braun, 2000).

There have been a wide range of benefits for those involved – more experiments by farmers, easier and quicker adoption of new ideas, plus improved food security. And not only do farmers benefit from their experimental findings, they also acquire increased status in the community at large. CIAT reports that maize yields in Colombia have increased from 800 to 1400 kg ha^{-1}, with the result that, during August–September, only 30% of CIAL households suffer food shortages, compared with 50–65% of non-member households.

It appears too that the process of learning is more likely to persist. Mangan and Mangan (1998) compared farmers in Sichuan, China, who had been trained either in FFS or by the economic threshold (ET) method (spray when a certain number of pest are present, or follow calendar spraying). There was good evidence to show that FFS farmers continued to learn in the years after training (continuous learning – vital for sustainability), whereas ET farmers showed no changes in knowledge. Incomes increased for FFS farmers by 23%, mainly because of large reductions in insecticide use, but also because of slightly increased yields compared with ET farmers.

One of the best examples of persistence of learning and its effect on innovation comes from studies of sustainable agriculture in Central America. The Guinope (1981–1989) and Cantarranas (1987–1991) programmes in Honduras and the San Martin Jilotepeque programme in Guatemala (1972–1979) were collaborative efforts between World Neighbours and other local agencies. Altogether, improvements were made in some 120 villages, using green manures, cover crops, contour grass strips, in-row tillage, rock bunds and animal manures. Staff of the Associaciòn de Consejeros una Agricultura Sostenible, Ecològica y Humana (COSECHA) returned to 12 villages and used participatory methods with local communities to evaluate changes after external support had been withdrawn (Bunch and López, 1996).

The first major finding was that crop yields and the adoption of conserving technologies had continued to grow since project termination. Surprisingly, though, many of the technologies known to be 'successful' during the project had been superseded by some 90 innovations. In one Honduran village, Pacayas, there had been 16 innovations, including four new crops, two new green manures, two new species of grass for contour barriers in vegetables, chicken pens made of king grass, marigolds for nematode control, use of legumes for cattle and chicken feed, nutrient recycling into fish-ponds, composting latrines, Napier grass to stabilize cliffs and home-made irrigation sprinklers.

Had the original technologies been poorly selected? It would appear not, as many that had been dropped by farmers are still successful elsewhere. The explanation appears to be that changing external and internal circumstances had reduced or eliminated their usefulness, such as markets, droughts, diseases, insect pests, land tenure, labour availability and political disruptions. Technologies had been developed, adopted, adapted and dropped. The study concluded that the half-life of a successful technology in these project areas is 6 years. Quite clearly

the technologies themselves are not sustainable. As Bunch and López (1996) put it, 'what needs to be made sustainable is the social process of innovation itself'.

Social Learning for Agroforestry in East Africa

Agroforestry is an area of practice in sustainable agriculture and NRM that offers a high potential for balancing often-competing interests in economic productivity and socio-ecological sustainability. It is an approach to land use that has been demonstrated to foster diversity, stability and resilience in agroecological systems (Buck *et al.*, 1999). Agroforestry practices can contribute importantly to the sustainable productivity of foods, fibres and medicines that households in many parts of the world depend upon for their livelihoods and well-being. To perform effectively, such that interactions among their multiple components are more synergistic than competitive, agroforestry practices need to be tailored to the local conditions under which they will be managed.

In Kenya, an initiative in agroforestry was pioneered in the early 1980s that fostered a participatory process of social and technical innovation that persists to the present. Framed as a 'learning process' approach to rural development (Korten, 1980, 1984), the CARE–Kenya Agroforestry Extension Project (AEP) invested in interactive training and knowledge generation (Buck, 2000). Appreciating that agroforestry technologies were not sufficiently developed to promote for adoption and that the scientific community was ill-prepared to generate experimentally an adequate range of technical options to suit the needs and diverse socio-ecologies of western Kenya, the project sought alternatives rooted in adult education (Freire, 1973; Roderick, 1986), communicative rationality (Habermas, 1987) and transformative learning (Hope and Timmel, 1984; Deshler and Selener, 1991; Mezirow, 1991).

Social learning networks were developed, comprising field workers recruited from local villages, field agents recruited from technical agencies (agriculture, forestry, energy), representatives of local women's groups and schools, household members representing diverse farming-system characteristics, the CARE agroforestry programme leadership and members of the formal scientific community. These diverse actors contributed to learning curricula (Buck, 1988a) that were organized around frameworks for diagnosing root causes of land-use and livelihood problems, designing potential agroforestry treatments and evaluating the proposed and adopted interventions by a variety of criteria, indicators and means of judgement and measurement (Buck, 1988b; Davis-Case, 1990).

The activity rapidly generated hundreds of observation trials, whose assumptions, objectives and design criteria were systematically documented both by local and professional field workers and by farm households, women's groups and schools. Information was pooled in facilitated group evaluation workshops to synthesize best-bet recommendations for managing the various types of emergent practices. Group-managed tree nurseries proliferated, with many varieties of indigenous and exotic species, selected for their known or potential suitability for a wide variety of agroclimatic conditions. While elsewhere debates raged on how to make tree nurseries self-financing, within the sphere of the AEP some 600 volunteer nurseries had already emerged without financial inducement.

The learning that supported this activity included classroom lectures from various specialists, field trips designed to foster critical thinking, group discovery and problem-solving around technical issues and focused interactions to role-play communication with local people. Farmer training focused on design principles and simple record keeping concerning plant performance, and cross-visits among farmers were facilitated by project staff. Schooled in the principle 'no technology before its time' (Raintree, 1987a,b), participants were guided in becoming attentive to 'pathways

of intensification' (Raintree, 1986) as a foundation for their land-use diagnosis and design activities. This included developing sensitivity to balancing subsistence and market-oriented production objectives. In practical terms, these concepts explained that landowners are ill-advised to move from an extensive, livestock-based, mixed agricultural system to intensive home-garden or alley-cropping management without some intermediate measures, such as protecting scattered trees in pastures, improving woodlot management or establishing boundary plantings.

From early evaluations, the project was acclaimed for numbers of trees planted, hectares treated, women's groups mobilized, farmers adopting and component technologies installed. An initial external review, conducted when the project was 2 years old, found that the third-year target for numbers of women's groups managing tree nurseries, for example, had already been exceeded by 100% (doubled) and demand for project assistance was outpacing the staff's ability to respond. These collective-action organizations were notably influential in their effect on innovation and the spread of knowledge (Feldstein *et al.*, 1989).

Scientific evaluators commented on the many adaptations to alley cropping and other technologies that the project has exhibited over the years (Scherr and Oduol, 1988). It has visibly and measurably contributed to land-use intensification processes that are agroecologically sound and improve rural welfare (Scherr and Alitsi, 1991). Also notable has been its longevity and ongoing contribution to the International Centre for Research in Agroforestry (ICRAF)'s agroforestry network in western Kenya. The experience has been drawn upon to illustrate theoretical and practical advances in several spheres of rural development, including gender and agriculture (Feldstein and Poats, 1989), non-governmental organization (NGO) and state relations (Wellard and Copestake, 1993) and planning for sustainable land use (Budd *et al.*, 1990).

Conclusions on Promoting Social-learning Operating Systems

We conclude that there are five core components of programmes that successfully promote social learning and sustainable NRM.

1. A conceptualization of sustainability as being an emergent property of systems high in social, human and natural capital.
2. The recognition that farmers can improve their agroecological understanding of the complexities of their farms and related ecosystems and that this better access to information and practices can lead to improved agricultural outcomes.
3. That the increased understanding is also an emergent property, derived in particular from farmers engaging in their own experimentation supported by external professionals.
4. That, if changes to individuals are embedded in social capital in the form of relations of trust, reciprocity and cooperation, then good ideas for improvements are more likely to spread from farmer to farmer and from group to group.
5. That social learning processes should become an important focus for all NRM programmes and that professionals should make every effort to appreciate both the complementarity of such social processes with sustainable technology development and spread and the subtlety and care required in their implementation.

What, then, can be done both to encourage the greater adoption of group-based programmes for environmental improvements and to identify the necessary support for groups to evolve to maturity and thence to spread and connect with others? Clearly, international agencies, governments, banks and NGOs must invest more in social and human capital creation. The danger is not going far enough and so being satisfied with any degree of partial progress. As Ostrom (1998) puts it: 'creating dependent citizens rather than entrepreneurial citizens reduces the capacity of citizens to produce capital'. The costs of development assistance will also inevitably increase – building human capital and establishing new organizations is not without cost.

But, although group-based approaches that help build social and human capital are necessary, alone they are not sufficient conditions for achieving sustainable NRM. Policy reform is an additional condition for shaping the wider context, so as to make it more favourable to the emergence and sustenance of local groups. This has worked well in India for the spread of joint forest management and in Sri Lanka with the national policy for water-user groups taking charge of irrigation systems.

One way to ensure the stability of social capital is for groups to work together by federating to influence district, regional or even national bodies. This can open up economies of scale to bring greater economic and ecological benefits. The emergence of such federated groups with strong leadership also makes it easier for government and NGOs to develop direct links with poor and excluded groups (though, if these groups are dominated by the wealthy, the opposite will be true). This could result in greater empowerment of poor households, as they draw better on public services. Such interconnectedness between groups is more likely to lead to improvements in natural resources than regulatory schemes alone (Röling and Wagemakers, 1998; Baland and Platteau, 1999).

But these policy issues raise further questions. What happens to state–community relations when social capital in the form of local associations and their federated bodies spreads to very large numbers of people? What are the wider outcomes of improved human capital, and will the state seek to colonize these new groups? And, finally, what new broad-based forms of democratic governance could emerge to support a transition to wider and greater positive outcomes for natural resources?

3 The Limits of Knowledge: Securing Rural Livelihoods in a Situation of Resource Scarcity

Pauline E. Peters

*John F. Kennedy School of Government, Harvard University,
79 John F. Kennedy Street, Cambridge, MA 02138, USA*

The mistakes and failures of many agricultural development programmes throughout Africa have been well documented. While the specific critiques vary, central explanations for the mistakes are the failure to take account of the sociocultural, political and institutional dimensions of production systems and the use of science-based knowledge that is inappropriate or misapplied.[1] Over the past two decades, some radically new approaches for agricultural research and action programmes have been developed that recognize the value of indigenous or local knowledge and of participation.[2] Not least among the contributions has been the demotion of so-called 'formal', 'scientific' or 'western' knowledge from its pedestal to being one among several types of knowledge. Yet the critiques of these new ideas and practices are also valuable, challenging the validity of the opposition between indigenous and western (scientific) knowledge (Agarwal, 1995) and pointing to the danger of separating knowledgeable practices from their political and economic contexts (Dove, 1996).

I want to argue here that both the knowledge of rural people and that of scientists and agricultural researchers have their limits. The problem for agricultural research and policy is the same for all proponents and practitioners of knowledge – to recognize that knowledge is never completely separated from routinization or bureaucratization (hence the seductive dangers of conventional wisdom and the stickiness of paradigms) and that it is entangled in power relations (hence the tendency for inappropriate models to disrupt not only theories but people's lives). The critiques of 'top-down' development and the call for more 'bottom-up' or participatory approaches should direct us not to oppose science/scientist to tradition/farmer but to help develop collaborative methods between rural producers and scientists/extension staff to identify, refine and circulate useful knowledge and 'best bets'. The aim is not to identify a single best solution for all times and places, but to recognize that multiple situations require multiple answers and that these necessarily change.

Using the example of Malawi, I wish to engage this set of issues by discussing the promise and limits of knowledge among farmers, on the one hand, and among agricultural researchers and extension officers, on the other. Hindsight tends to reveal more mistakes than successes, but certain current discussions and trends may indicate some hope for a more collaborative relationship between farmers and agricultural services than has been the case to date.

The Burden of the Past: Pushing Agricultural Modernization

Malawi is a small country of great beauty but limited wealth, hemmed in by the larger countries of Tanzania to the north, Zambia and Zimbabwe to the west and Mozambique around three sides (east, south and west). Nyasaland, as it was called by the British, was the poor third cousin in the briefly constituted Central African Federation of the two Rhodesias and Nyasaland, and a major source of labour for Southern Rhodesia and South Africa. Without any major source of minerals, Malawi has long depended on agriculture. Large estates were established in the southern region from the end of the 19th century and further north from the 1940s. They grew a range of crops for export, finally settling on tobacco and tea as the most profitable.

The first agricultural programmes developed by the British colonial administration focused in the 1930s on soil erosion, a concern that was deeply influenced by the debate about the 'dust-bowl' phenomenon in the USA (Beinart, 1984). One of the products of this focus was the effort to get farmers to plant on ridges rather than on their traditional mounds (*matutu*). But, beyond efforts to teach techniques of soil management to avoid soil erosion, to ban cultivation in sensitive areas and to promote certain crops, the administration made no efforts to fundamentally change the way farming was carried out. Part of this was due to the very small number of professional agriculturalists in the administration. Compulsion in promoting soil-conservation techniques became more marked by the 1940s and was one of the sources of resentment among rural people and fed into the political resistance of the 1950s (Beinart, 1984).

The first major attempts to bring fundamental changes to the systems of farming were initiated in the early 1950s and were driven by a postwar optimism regarding technological change, by an activist agriculture department with the explicit aim of achieving an agricultural revolution in Malawi through creating a type of 'yeoman farmer' and by a mounting concern about a decline in maize production and the spectre of food shortages as a result of ecological crisis.[3] During this period, too, many of the large European-owned estates were being broken up as the level of profits dropped and as political changes loomed. The colonial administration's decision to convert some of the estates into areas for 'resettlement' of the crowded population in the southern region, including former labourers from the estates, provided the chance for the more activist agricultural officers to design a model for transforming agriculture. This was modelled on the English 'mixed farming' method, which combined rotations of crops with intensively managed livestock. The farmer was assumed to be a man and, in a deliberate move to destroy the matrilineal mode of inheritance, which the influential Director of Agriculture, Richard Kettlewell, considered a disincentive to male farmers' investments, the farms were to be titled in the names of individual male farmers. In every aspect, this model was totally at variance with the social organization of farming, the existing organization of rights to land and the ecological and political-economic conditions in which small farmers sought to make a living. Resistance to the schemes was immediate, and deepened in the context of political moves to independence. A study conducted a decade later concluded that, by the mid-1960s, all the ambitious programmes to reorganize land and farming had 'collapsed', with settled families reverting quickly to their own preferred methods of working (McLoughlin, 1967).[4]

Today, Malawi is a densely populated country, heavily dependent on agriculture, producing maize and other food crops but whose foreign exchange comes from tobacco and tea, along with sugar, coffee, macadamia nuts and other minor products. Under President Banda, the 'dual' structure of agriculture was reinforced, whereby estates retained the privileged right to produce the valuable crops, while smallholders were required to sell their crops at lower prices to the marketing board. The result was a hidden subsidy to the estates, because the surplus built up by the board was funnelled through banks as soft loans to Malawian estate owners (Kydd and Christiansen, 1982). The supposed 'market miracle' of Malawi, as it was celebrated through the 1970s, was a smoke-and-mirrors method to privilege a small élite. Yet I would argue that the subsidy for maize channelled through the marketing board to the majority of smallholders who are in food deficit has also meant that hunger, though chronic, did not become famine. Only in the recent years of unprecedented drought and greater exposure to volatile price changes consequent on market liberalization, structural adjustment and political change has periodic acute food deficit been added to the situation of chronic deficit.

Shire Highlands of Southern Malawi: 1986–1997

Knowing something of the history so briefly sketched above is important because it informs the farming practices of today and the interpretations people of various social categories place on those practices. I learned this the hard way when I began to ask people about their farming practices. Over a period of many months spent trying to relate what people were telling me with what I was observing in the study villages, I realized that, at least in response to some of my enquiries, they were dividing their answers into two: the 'proper' way of farming and their own way of farming. I learned that they had become so used to their own ways being derogated that they sought to protect themselves from any further criticism. This was particularly clear in relation to intercropping, which I discuss below.

The research site and sample

The research on which I draw for this chapter has been conducted over a decade in the Shire Highlands area of the southern region of Malawi (Fig. 3.1). The research was begun in 1986/87: six sites were selected within an area of approximately 25 km east–west and 10 km north–south in the southern part of Zomba district. The area is bisected by a secondary dirt road, which was maintained quite well up to about 1994/95, and is served by two marketing-board centres, three major markets and several other smaller markets. The sites vary ecologically from hillier and wetter in the west to low-lying and drier in the east. The sample of some 200 households from six sites was purposive, selecting households with at least one child under the age of 6, with one-third to be tobacco growers (because a major interest in the initial research was to compare non-food cash crops with food crops in relation to nutritional status). Although another criterion was to sample the full range of farm sizes, the selection of one-third tobacco growers resulted in higher average landholdings and income for the sample households than the averages for the district or region. Thus, the mean sample household landholding is 1.49 ha (ranging from 0.2 to 9.5 ha), compared with 0.8 ha for the neighbouring Zomba Project area.

Farming and diversified income

Farming provides the bulk of their livelihood for the sample households: full year-long surveys in 1990 (and 1986) showed that the value of home-produced and consumed maize made up 31% (32%) of total household income; 16% (17%) of income came from food-crop sales, 11% (5%) from tobacco sales and 8% (8%) from non-crop agriculture (about half of this from sales of poultry, rabbits, goats, pigs or sheep or their products[5]). The rest of household income came from 'off-farm' sources, which included 14% (15%) from gifts and remittances,[6] 6% (8%) from self-employment,[7] 6% (8%) from agricultural wages,[8] 5% (4%) from non-agricultural wages and the remainder of 4% in both years from pensions, sale of household assets and loans/repaid loans. Households in the bottom income quartile earn more from casual labour, while those in the top quartile earn more from agriculture, especially burley tobacco.[9] Both this pattern and the skewed distribution of income have intensified over the decade from 1987 to 1997.

While agriculture, including home-produced foods, provides most of the

Fig. 3.1. Research site in Malawi. Map drawn by Nancy Leeper, a graduate student in the Department of Geography, University of Oregon, and kindly made available by her adviser, Dr Peter Walker.

income, the sample families are similar to many being documented for Africa in being increasingly dependent on off-farm income (mainly from casual labour and small-scale businesses), especially for cash. While off-farm income is proportionally higher for the families with less land and lower for those with more land (who have more agricultural income), in no cases is it unimportant. The result is that family members are constantly juggling a range of tasks in farming and in non-farming activities, which, of course, greatly complicates an already complex set of farming practices.

Two aspects of smallholder cultivation are most significant: the highly labour-intensive character of farming practices and the diversity of crops grown. The aims of cultivation are to provide as much of a family's food as possible and products that can be sold. The central strategy of rural households, both in terms of crops grown and in terms of income sources, is to diversify, seeking in this way to manage the variability in climatic and ecological conditions as well as in political and economic circumstances. My main focus here is on one significant part of villagers' intensive agriculture – intercropping – which displays both the positive sides of farmers' 'knowledge' and its limitations. In discussing cropping patterns, I shall also briefly consider two other dimensions of cultivation: land preparation and soil management, and seed selection.

Although Malawi is a small country, it is far from homogeneous; the practices I discuss here are those found in the Shire Highlands, which has one of the highest people/land densities in the country but which also contains several large urban centres that are major markets for food supplies and other rural products, with significant effects on villagers' production and income strategies. Farming systems along the lake shore in the Southern Region and in much of the Central Region tend not to have such complex intercropping practices and have more animals. There is little research from the Northern Region, but there, too, fragile soils, lower population densities, more live-stock and greater distances from major urban

centres result in very different farming systems. Finally, the Lower Shire has a quite distinct ecology and associated farming systems. Nevertheless, as later consideration of recent reports on farming by national teams suggests, the practices described here are typical of a substantial part of the rural population of Malawi, and certain dimensions are relevant to broader categories of rural producers.

Land preparation and soil management

Cultivation practices in southern Malawi have changed over the centuries (Mandala, 1990), but the most rapid changes have occurred during the 20th century. In soil management, a critical change has been from mound cultivation to almost universal cultivation on ridges. Bunding and box ridging used to control the flow of water is much more sporadic in the sample area. The crop ridges are remade every planting season by turning over the residues in the furrow to make new ridges, a practice that adds a degree of compost and thereby helps maintain soil structure and fertility.[10] Some people also burn part of the residues and the ash is incorporated into the soil.[11]

Research on soils in the country shows shortages of all nutrients (Matabwa and Wendt, 1993; MPTF, 1998), and the consensus among researchers as well as among the farmers I know is that the situation has worsened in living memory. Most of the land in the research area is densely used and intercropped, so, during the growing season, when rains may be heavy, soils are generally covered by plants and are therefore somewhat protected from the worst runoff and erosion. Weeding is done in January and February and the weeds are generally left to rot in the furrows, while some are incorporated into the mini-ridges that some farmers make to plant peas in February. Banking or mounding up the ridges around plants also adds to the management of soils. The practice of tying goats to sections of fields after harvest also adds some nutrients from their manure, although the goats' eating some of the crop residues also takes away some of the beneficial effects of leaving them to rot in the fields.

While the population density on the land does not allow for regular fallowing, most farmers move their crops around their fields from year to year. Since most cultivate the majority of their land with maize, there is an obvious limit to their ability to change the main crop on their land but the intercrops do change and the small plots making up the fields tend to hold different crops from year to year. Farmers try out different crops or varieties, decide to try a different spot in their fields because of the past year's pests or other vicissitudes and generally change the composition of crops in the fields. Obviously, families with the smallest amounts of land are the most constrained. The larger producers, who grow burley tobacco as well as maize and many intercrops, tend to rotate their fields, growing burley on the past year's maize fields and vice versa. Most are consciously taking advantage of the residual fertilizer from tobacco for the new year's maize but also have had the dangers of not rotating burley and other tobaccos dunned into them by messages from the agricultural department and the Tobacco Association. Nevertheless, some, especially those who grow small amounts of tobacco, do not bother or are unable to rotate their crops, and thereby are courting the greater likelihood of nematodes infesting their fields. I shall return to problems requiring attention later.

Patterns of intercropping

When I first began asking about intercropping – how did people decide to intercrop, what crops to plant in what patterns or in what parts of the fields – some farmers assured me that they intercropped because they had too little land but that they knew it was much better to plant the main crops, particularly maize, in 'single stand', that is, as sole crop. It took months for the villagers to get used to me (and to begin to trust that I was not going to race off and inform the authorities about the failure of farmers to follow agricultural-extension messages) before they, or at least some of

them, were prepared to acknowledge that they felt their own methods were better than those promulgated by the extension staff.

Some of these farmers were also conscious experimenters, prepared to try something new, whether a new crop or crop variety or a new technique, and interested in comparing the outcomes with their other practices. Some, I think, had become so accustomed to being seen as backward or wrong-headed that, at least at some level, they accepted the extension messages about the proper way to farm, but nevertheless retained their own ways, trying to explain them away to me with such rationalizations as not having enough land or enough family members to work in the fields.

Most fields in 1986/87 were divided into small plots, with an average of 1.5 plots per field (just under 900 fields cultivated by 220 households).[12] The single most important crop is maize – the staple food but also a cash crop for many – grown on approximately 80% of the cultivated land. Seventy per cent of all plots were intercropped, with a median of two to three crops per plot, though 55% contained between two and five crops and 17% contained even more. In succeeding years (up to 1997), this basic pattern has been retained, although there have been shifts in the relative importance of particular crops over the decade. The local maize varieties are mostly intercropped (81%), compared with approximately half (53%) of the hybrid maize. By 1997, both the numbers of growers and the acreage of hybrid maize had declined (due to the disappearance of the government-subsidized sources that had driven the earlier increase), and there had been an increase in the amount of burley tobacco grown (because smallholders were formally allowed to grow burley from 1990 and able to sell on the auction floor, at world prices, from 1992), but the pattern of intercropping remained very similar.

There are probably as many crop mixes as there are plots or growers, but some patterns are much more common than others and some mixes dominate. The intercrops to maize are most often pigeon peas, beans, pumpkin, cowpeas and groundnuts, in that order. A common method of planting is for several seeds (maize and intercrop) to be planted in holes several feet away from each other along a ridge; another intercrop, such as groundnuts, is planted some time later[13] in the interval spaces. The maize–cowpea–pumpkin mix is one of the oldest, with a history that appears to date back to at least the 19th century and probably before, both in southern Malawi and in neighbouring areas of Mozambique (from where the ancestors of many of the families came in the first quarter of the 20th century). There are two main varieties of cowpeas, several varieties of pigeon peas, the most popular being the one called *nandolo wa research*, which is resistant to wilt and whose name indicates its provenance, and many varieties of beans.

The typical intercropping pattern for the area is that maize, pigeon peas, with either cowpeas, beans or groundnuts, and pumpkin are planted on ridges; vines, such as climbing beans and cucurbits, are allowed to climb up maize and sorghum, sweet reed and other shrubs. The more vigorous vines, such as the popular hyacinth bean (*mkhungudzu*) and the velvet bean (*kalongonda*), are usually isolated in discrete areas such as old termite mounds where they will not strangle other crops. A variation is a main-crop mix of maize + beans + pumpkin, with cowpeas and pigeon peas scatter-planted across the field. Sorghum, an important secondary grain crop to maize, is grown either along the edges or in a cross-hatched pattern laid like a lattice across the entire field. Farmers explain that sorghum and maize do not get on well together – sorghum deprives maize of 'food' (nutrients), so they separate them by the widely dispersed form of planting. The few people who grow millet also plant it along the edges of fields. Similarly, the several varieties of cassava were planted along the edges of maize fields by those who did not have enough land to set aside entire plots or fields to cassava.

A detailed analysis by Shaxon (1990) of the cropping data collected in our survey during 1986/87 suggests that, while crop diversity increased with both landholding size and available household labour, the relation was concave. That is, diversity was less for the smallest holdings, increased as more land (labour) was available but then dropped off to a maximum. In reverse, crop density had a convex relation to land size and labour, being higher among the smallest farms, then decreasing and then rising slightly. I would interpret these findings as indicating that people with very little land do not have the space or labour or the seeds/cuttings for a wide range of crops but try to cram in as many as possible of those they are able to plant.[14] As constraints of land and labour (and of seeds) are relaxed, so the range of crops grown increases and the density drops somewhat. Those with the larger holdings (who are those more able to obtain the labour and other inputs they need) have sufficient land to grow a wide range of crops but are able to pick and choose.

In addition to intercropping, farmers use the method of crop sequencing or relay cropping. Thus, in February, when the maize is maturing and pumpkin and some bean varieties are ripening, farmers strip off some of the maize leaves to create more space for air and sunlight, clear small areas and plant green peas halfway down the ridges. Green peas have become a major cash crop, which sell for high prices in the local and town markets early in the season. As crops mature and are harvested bit by bit, parts of the fields and sometimes new ridges created at the edge of fields are planted with sweet potato.[15] In March and April, when maize stores have been depleted but before the maize harvest is ready, common meals in the area are boiled fresh groundnuts and sweet potato or dishes of boiled pumpkin. Other crops planted as other crops ripen and small parts of fields are cleared include gram, chickpeas[16] and soybeans, all of which are used mainly as cash crops sold to traders from the towns. For most people, these are grown in very small quantities.

The main cash crops that are not local food crops are different types of tobacco, mostly burley with some dark-fired tobacco, as well as chilli peppers and sunflowers. The latter two are grown in a single stand in small plots. Tobacco is supposed to be grown alone, according to the extension agents, but, by 1990, about 30% of the tobacco grown by the research families was interplanted with food crops. These included green leafy vegetables, such as rape, sweet potatoes and some tomatoes. One man, who was constantly trying out any new interesting idea he had heard or a new variety or technique he learned about, was experimenting with different types of intercrops with his burley crop. Because the agricultural extension staff told farmers that burley should not be intercropped, this man deliberately kept one plot of burley 'single stand' to which he took the agricultural assistants. But his other burley plots were intercropped.

When farmers responded to my questions about intercropping or showed me around their fields, they described how they select different parts of their land on which to grow certain crops depending on the types of soil they identified (sandy, loam, clay, rocky), the relative dampness or dryness of the soil, the slope of the land, the balance of sun and shade at different points in the growing season, and so forth. They also judged the characteristics of the crop in relation to the former set of factors, and also compared different varieties of the same crop. Thus, certain farmers considered that particular varieties of maize or beans did better in particular sites. Most farmers seem to develop a particular set of cropping practices they follow year by year, with changes according to what seeds they were able to obtain, to changes in climate, to pest infestations and disease or to changes in fields (which are sometimes exchanged among family members or loaned). Also important are shifts in market conditions. Crops grown exclusively or mostly as cash crops are those most affected; sunflower, for example, has gone up and down in farmers' fields over the past decade in response to changing markets, and there has been an overall increase in burley growing, though with declines after years of poor prices and sharp rises in cost of fertilizer.

Chilli peppers provide another example of crop shifts. Only a few families grew chilli in 1986/87 but by 1990/91 it had been taken up by more, including some who switched from growing small quantities of dark-fired tobacco. The reason given was that traders

had started coming into the villages looking for chillies to buy and farmers found they could make more cash from chillies, especially since they were finding it more and more difficult to find the firewood needed to dry the tobacco. The chilli traders were on-selling to companies in the towns of Limbe and Blantyre; some of these companies were exporting to Europe and others were processing for domestic use and export. However, after the first flush of increasing prices offered by traders and then by the parastatal marketing board, the response of farmers was so great that the markets apparently became saturated. Representatives of several companies and the parastatal told me that they found themselves at the end of the 1989/90 season with unsold surpluses of chillies rotting in their warehouses. The farmers had been pleased with a rapid rise in price over the previous 4 years (from about 15t to K2 kg^{-1}),[17] but they told me in 1990/91 that the dropping prices led them to cut back on production. In fact, during 1992/93, prices again surged to K5 kg^{-1} in a year when sunflower seed was being bought at 50t and pigeon peas around 80t kg^{-1}. Prices were also good in 1995, but once again fell so that, by 1997, very few were growing chillies and most former growers had switched to burley and other crops. Another factor was the means of selling. In one of our sites, farmers had been approached by a small non-governmental organization (NGO) to establish an association of chilli sellers but, after giving deposits and after several meetings, the farmers said the NGO failed to come through with the arrangements, leaving them disappointed and with no desire to continue producing chilli.

Explaining intercropping

The prejudiced view so common in colonial times that intercropping was 'hit and miss planting in mixtures' which reflected the irrational or lazy habits of farmers is now fading, though not entirely gone.[18] Although the advantages of intercropping have been given a great deal of attention in recent years by agronomists and other researchers, there remain lingering notions about the backward character of the practices. While many researchers and policy-makers in Malawi now recognize the importance of intercropping as a strategy, this view is by no means universal among the local extension staff. Moreover, most of the crop research has been done in research fields and usually with only one intercrop (such as maize and pigeon peas). Although, in recent years, research in farmers' fields has increased, there is very little, if any, research on multiple intercrops *in situ*.

The overriding concern about Malawi's agricultural production (apart from the main foreign-exchange earning crops of tobacco, tea and sugar) has been to increase maize production, with the chosen strategy being to promote high-yielding varieties and fertilizer. Initiated in the last years of the colonial era and greatly intensified under Banda in the 1970s, this strategy remains central to agricultural research and extension. The spread of hybrid varieties was slow and limited to the top 25% of smallholders (in terms of landholding, capacity to purchase inputs and membership of the credit clubs). An earlier assessment (Kydd, 1989) attributed much of this to a failure of the maize research programme, in being underfunded, understaffed and impervious to farmers' preferences for flint-type maize (as opposed to the dent types in hybrids). A more recent opinion is more charitable, pointing to researchers' efforts in the 1950s and again in the 1970s to develop flint or semi-flint varieties (Smale and Heisey, 1994). The big push to develop more flinty hybrids, however, did not take off till 1987. It paid off by the early 1990s, when several semi-flint varieties were released and quickly embraced by farmers (Smale and Heisey, 1994).

In my study, only 2% of the households were growing hybrid varieties in 1986/87, the year that proved to be the bottom of the national trough in the smallholder use of hybrids (nationally, only 5% of the total maize area was in hybrids). The particular reasons for the decline that year, according to our study farmers, were that they had become discouraged not only by the poor

producer price offered by the marketing board but by the failure of the board's centres to buy their maize the previous year. This was because of the marketing board's increasingly serious cash-flow problems. This particular problem was set against longer-standing problems referred to by Kydd (1989) and other researchers: there were perfectly good reasons why farmers were not switching wholesale from 'local' varieties to the hybrids.

The main reasons are food security and cost. First, having to purchase new seed every year is a problem for a seriously poor population, and even more of an obstacle is the cost of obtaining fertilizers to apply to hybrids (as is highly recommended by extension and researchers). The second critical difficulty for farmers is that the dent hybrids are much more vulnerable to damage by weevils in storage than the flinty local varieties. A further problem was that the hybrids, again because they were soft types, did not pound well and, according to some people, did not taste as good either. The reasons for the first remarkable increase in hybrid production in the late 1980s were the increase in government efforts and subsidized credit. Thus, by 1990/91, 52% of the sample households were growing hybrid maize, a pattern that was found in other parts of the country (Smale *et al.*, 1991). This upward trend was again reversed when the credit system collapsed as part of the upheaval with the change in government in 1994 and with the removal of subsidies on fertilizer. Only in years when there has been free distribution of seeds (plus or minus fertilizer) and food has this trend reversed.[19]

More instructive than the aggregate figures is the analysis of why different categories of farmers take up different varieties of maize. The extension programme up to the end of the 1980s was directed to farmers as sellers of maize; the entire system of high-input production based on credit assumed that only the better-off or 'commercial' farmers would grow and sell maize. Hence, the poorer storage quality of hybrid varieties was of less importance. In parallel, it was assumed that hybrid varieties were not of interest to the poor because they were not surplus growers. The recent years, in which free distribution of seeds has enabled more of the poor to grow hybrids, have revealed that, on the contrary, poorer, severely food-deficit families are very interested in obtaining hybrids. This is because many of the varieties are early-maturing and thus can be used several weeks earlier than the local varieties. Our surveys showed that most of the hybrid harvested by poorer families was eaten immediately, whereas the richer families ate some but sold most.

Another assumption made was that the better-off farmers would switch completely from local to hybrid varieties because of the obvious yield gains. Why, then, do most farmers – those with more land and more ability to obtain inputs – still grow local varieties of maize as well as hybrids? The better-storing local maize is used partly for food (as for all families) but, in addition, it constitutes an income and employment strategy. The stored local maize is used by the larger farmers to employ labourers in the peak growing months (November to February). Here is one of the examples where behaviour that is apparently irrational to the outsider proves to have its own set of rationales. Whereas extension officers, researchers and others professed not to understand why the larger-scale farmers would bother to grow local varieties when they could specialize in the higher-yielding hybrids, their failure to link the planting practices to the entire round of farming, including acquisition of labour, and to the central link between production and consumption (food security) obscured the rationale of bigger farmers using local maize to attract poorer people as labourers at the time of both peak labour needs and the worst food deficits.

The introduction in the mid-1990s of the semi-flint varieties, which are more resistant to weevils, has made hybrids more popular than before. Yet, because they are still not as resistant as local varieties, most farmers still prefer to grow the latter as well as hybrids, when they are able to obtain them. Nevertheless, my research revealed a trend among the small group of the wealthiest farmers (those with sufficient land and sources of income to cover the costs of new hybrid seed and fertilizer) to increase hybrid and decrease local maize production. One or two had stopped growing local varieties altogether. Such trends may reverse themselves, of course, in that they are influenced by the relative cost and availability of hybrid seed and fertilizer, market conditions and climate. Another new trend observed by 1997 was that a handful of the larger-scale burley farmers in our sample have reduced their production of surplus maize (i.e. surplus to their own food and other needs) in favour of increased burley production. They said the higher incomes from tobacco justified the shift. The recent downturns reported for burley prices in Malawi may well reverse this again.[20]

The breakthrough in breeding semi-flint hybrids is important, but the escalation in the cost of fertilizers and other inputs has put a considerable damper on the ability of most farmers to grow more hybrid than local maize. Some evidence suggests that, even without fertilizer, hybrids can yield more than local maize, though other research is less optimistic.[21] Farmers in my sample (as in other studies) found that the drop in yields from reusing hybrid seed is outweighed by the advantage of earlier harvests, at least for the first year of reuse.

In light of these difficulties, researchers have looked to other, complementary methods of increasing maize productivity. Through the 1990s, mounting worry about declining soil fertility and about the domination of maize has led to greater interest in crop diversity and in the benefits of crop mixes, including intercropping. The recent reports and discussion papers produced by the Maize Productivity Task Force (MPTF) are important examples of these newer trends. While the main conclusion of a recent report is that 'high yielding variety [HYV] maize seed and fertilizer technology is essential to the survival of most Malawians in the foreseeable future', the report also recommends support for 'organic strategies', particularly by 'increasing the grain legume component in maize based cropping systems through rotations or pigeon pea intercropping' (MPTF, 1998).

One main impetus behind the new focus on 'organic strategies' is surely the realization that Malawi's poverty is a serious obstacle to sole reliance on the expensive HYV maize and fertilizer technology, rather than an appreciation of farmers' knowledge and experience. Nevertheless, official (government and research/extension) recognition of the benefits of such common strategies as intercropping is a positive step. It could provide a new basis on which to construct a more equitable working relationship between farmers and research/extension and to address many farmers' problems that have been largely neglected to date (such as identifying intercrop mixes that work best for different sites and purposes, developing more effective pest and disease controls, improving selected characteristics of crops, etc.). Most obviously, it comes closer than any past approach by government to recognizing the reality of most farmers' practices and expressing a wish to build on these – at least as part of the overall strategy – rather than sweeping them away in the name of modernization or progress. Rather obvious, too, from a reading of the reports is that researchers have more to learn from and about farmers. I shall indicate some examples first and then, later, suggest where farmers may be able to learn from researchers. In both cases, the main change has to be in the way in which researcher/extensionist and farmer interact.

The MPTF report places more hope in rotations of grain legumes with maize than in intercrops, citing the considerable gain in fertility and yields with such rotations (of groundnuts and soybeans with maize), though there are constraints in Malawi, with low seed multiplication, poor market prices and labour demands. Here one needs to stress that the overwhelming preference for intercropping as compared with rotations among the vast majority of Malawi's farmers is because of their small landholdings and their consequent need to ensure annual production of a range of food crops and crops for sale (often the same). Rotations, then, are likely to be followed only by the tiny minority of farmers with larger than average landholdings.[22] In contrast, any support for farmers' existing practices of intercropping maize with grain legumes will be very welcome to most.

The intercrop that receives the most enthusiastic recommendation by the MPTF is pigeon pea. The reasons given are its considerable biomass production, deep tap root and non-competitiveness with maize because it matures after maize: in short, it has 'excellent temporal and spatial complementarity' with maize (MPTF, 1998).[23] The pigeon pea is one of the most important crops, after maize, among my research sample and in much of the Shire Highlands because it is a significant food crop and, increasingly, a cash crop.[24] This explains why farmers in these areas would not be interested in replacing pigeon pea with an intercrop that provided the soil-fertility benefits but was not a food crop (such as some of the crops promoted by agroforestry programmes). While the researchers' concern with maintaining and improving soil fertility is justifiable and is one echoed by farmers in my conversations with them, that goal must be seen in combination with the farmers' strategies for achieving food security and income.

A similar point may be made in respect of the report's judgement that 'the soil fertility benefits of other common intercrops', such as cowpeas, 'are limited'. The major reason for this, at least in respect of cowpeas, is that their yields are often low because of pests and disease (MPTF, 1998). Farmers have told me as much. But a critical fact about cowpeas is that their leaves are (again, in my sample) the single most important source of 'relish' (*ndiwo*) eaten with the staple maize porridge (*nsima*), along with pumpkin leaves and bean leaves (and a range of 'wild' greens). In addition, cowpeas are an essential item in several key rituals. Here, again, researchers have to recognize the multiple uses of crops in order to be able to work most productively with farmers to improve overall productivity.

A final instance of where collaboration between researchers and farmers has promise concerns the use of undersown green manures as another means of improving overall soil fertility. The report mentions *tephrosia* (fish bean), *crotalaria* (sunn hemp) and *mucuna* (velvet bean)[25] but, on the basis of research conducted so far, favours *tephrosia* (MPTF, 1998). Again, this judgement is made on the basis of improving soil fertility. Farmers are more likely to favour the velvet bean (*kalongonda*) and hyacinth bean (*mkhungudzu*, *Lablab purpurea*), which they grow as reserve foods, a category of crops that is particularly important in years of poor or failed rainfall. Moreover, the report's comment that *mucuna* has the disadvantage of being overcompetitive 'unless carefully managed' seems to be made without knowing that farmers recognize this danger and carefully watch such vigorous vines, often separating them from the maize.

In summary, the recent recognition by researchers and others in Malawi of the considerable benefits of intercropping may augur a new beginning for appropriate research and extension. In addition to current concern with reversing the decline in soil fertility, however, researchers and extension agents must also take account of the multiple purposes for which farmers plant crops, which, necessarily, influence the selected crop mixes. Among these purposes, food security and food diversity are signally important.

Such a new beginning also brings Malawi researchers in line with a growing appreciation of intercropping in other parts of Africa and the world. The documented advantages of intercropping, which has been referred to as 'one of the great glories of African science' (Richards, 1983), and of sequential and relay cropping include the following (see Innis, 1997, for non-African cases). First, intercropping of grains and legumes helps preserve the soil fertility of land that is in permanent use. Secondly, the diversity of crops grown reduces risk in the face of variability in the natural phenomena of weather, pests and crop disease. Farmers exploit different micro-niches on their land for different plants and different crop mixes. A third advantage is the adaptability built into the farming system. The temporal dimension of intercropping and relaying enables 'sequential decision-making' as a response to changes in growing conditions,

markets (a rise or collapse of prices), other constraints (such as unexpected changes in policies affecting crop cultivation or selling) and the farmers' needs.[26] Staggered planting, for example, enables staggered harvesting, which is useful for spreading tasks where labour supply is tight, as well as for management of family food needs and of income from crop sales.

Fourthly, intercropping, relay and sequential cropping are labour-intensive and land-saving in ways that suit a land-short population. Finally and importantly, agronomic research suggests that intercropped systems have higher productivity than monocropped ones, because they facilitate a more efficient use of the 'growth resources' of light, water and nutrients and encourage complementary or 'facilitative relationships' among plantings.[27] Other advantages identified by agronomists are the positive effects of intercropping on weed control, in stemming soil erosion (in heavy rains or harsh winds) and on pest control (Innis, 1997).[28]

All of these findings, as well as those highlighted in the MPTF report on soil fertility, apply to Malawi. Moreover, in the present context of a persistent decline in the use of fertilizers since the mid-1990s because of the removal of subsidies, the rise in prices and the collapse of most credit sources, these benefits of intercropping become even more significant.

Limits on farmers' knowledge

Many problems have derived from the application of scientific knowledge in agricultural programmes without requisite attention to the lived realities and knowledge of farmers. One of the most common errors has been to assume that 'one model fits all'. This lesson is being learned, albeit slowly. The MPTF report discussed above concludes that 'crop management research' is too quickly translated into 'a highly distilled format in the form of farmer recommendations', which 'frequently fail to take adequately into account the diversity of farmer circumstances' (MPTF, 1998). The new turn to including such 'organic strategies' as intercropping in routine agricultural research and extension is surely produced out of this realization.

It would be wrong, however, to assume that farmers' knowledge (so-called 'indigenous' or 'local' knowledge) does not suffer from limitations. Several anthropologists have pointed out that much of what might be elicited as 'indigenous knowledge' by outsiders may be oversystematized interpretations by the researchers. Much of what farmers do is improvisational, adapting their resources and knowledge to season and changeable conditions (Fairhead, 1993; Richards, 1993). Many intercropping and crop-relay patterns are the product less of planned design than of sequential adaptations. It seems likely that practices that are many years (or even generations) old are more sustainable than those improvised within a shorter time frame. This may be the way to consider some examples from my own research of where farmers' developing practices appear not to be appropriate for sustainable use, but where the existing research and extension services do not encourage the kind of collaboration needed to develop better practices.

The first example concerns intercrops with burley tobacco. Since 1990, when the government removed the prohibition on smallholders' growing and selling burley tobacco, the numbers of growers have accelerated. As the cultivation of burley has increased, so some growers have interplanted the tobacco with food crops. This is in direct contravention of the recommendations of extension assistants, as well as the Tobacco Association of Malawi (TAMA). Farmers explained to me that they intercrop because they want to increase the crops they can harvest from their fields, and some say that the burley does not suffer in any way. In the early part of this period, the intercrops tended to be rape or similar leafy vegetables. In the most recent years, some farmers have been growing aubergine and tomatoes in their tobacco fields, crops which, according to agronomists, should not be planted in proximity to tobacco, since they encourage

the nematodes that are destructive of the tobacco and other crops.

My sense here is that, because the agricultural extension service has taken a blanket approach by seeking to ban all intercrops in tobacco fields, no room has been created for discussion of which intercrops might be indifferent for the burley's health, which might be beneficial and which negative. Despite the shift from the autocratic, one-party regime of the late Dr Banda to the 'multi-party' democracy of today, the extension service continues to take a highly autocratic stance towards farmers, telling them what to do rather than recasting themselves as advisers who work with farmers to develop the best practices for the circumstances in which farmers of different categories find themselves. Even when researchers in the national centres are informed and supportive of farmers' techniques, this is frequently not the case among the local extension staff (as I had many occasions to observe throughout the fieldwork).

A second example of farmers' limited knowledge and the missed opportunity for collaborative research and practice concerns the stream-bed gardens (dimba), which are extremely valuable and, in my sample area, are used mainly for production of vegetables for sale, although out-of-season maize and other family food crops are also grown. As burley tobacco cultivation has increased, dimba have become even more desired, since they are used as nurseries for the tobacco seedlings in September–October before the rains arrive. Yet, at the same time, vegetable selling has burgeoned with the rapidly expanding urban and peri-urban areas (an expansion driven by the sudden lifting of controls consequent on the political changes). An indication of the increasing value of dimba and the mounting competition over them is that the money exchanged between people who lend and borrow has increased steadily over the past decade and is more often referred to as 'renting' than before. One consequence of this increasing use has been an increase in the use of pesticides. Although some vegetable growers have begun to put pesticides on their *dimba* crops, virtually all makers of burley nurseries do so.

Together, these trends result in more pesticides being used in *dimba* than ever before. Since *dimba* are made right on the banks of rivers and in the seasonally dry river-beds, there is runoff directly into the stream. I have never heard anyone raise questions about the use of pesticides so close to rivers, which are, in part, sources of water for consumption. Yet this is surely a huge problem in the making, at least equivalent to the problems of nematodes mentioned earlier. Much is now known about the detrimental effects of pesticide use, especially in areas where people are poor and struggling to increase the returns to their efforts whether in the form of food crops or products to sell and where information about pesticides is woefully limited.[29] Here again, then, is an example where limitations on knowledge are clear and where an effort to inform and work with stream-bed cultivators to avert any health hazards is needed.[30]

These few examples suggest the limitations of farmers' knowledge, as well as that of the agricultural extension and research services, and the continuing obstacles in the way of developing collaborative exchange on agricultural problems, rather than the routinized top-down imposition of 'improved methods'.

Conclusion

In this exploratory chapter, I have said that recognizing the many mistakes made by agricultural research and development programmes is important and that we should welcome the turn towards taking more seriously 'indigenous' or 'local' knowledge. But I have warned, too, that all knowledge has its limits. In trying to come to grips with the challenges facing smallholder farmers in Malawi, which I think are not unique to that country or that population, I have argued that careful attention to what small farmers actually do and describe suggests, first, that their methods of intensifying production and improving productivity do not follow the model assumed in

most agricultural scientists' definitions of intensification. Modern, scientific agriculture places at the heart of agricultural intensification a process of specialization and simplification, with the aim of increasing yields and thereby total output. In contrast, smallholder farmers in Malawi and in many other parts of Africa have developed highly intensive methods that are premised on diversity and diversification rather than specialization and on complexity rather than simplification and that are directed to a range of aims that, while not excluding yields, privilege flexibility, multiple use and adaptability to climatic and environmental shocks (Guyer, 1997). In the case of Malawi, the management of shifts in crop mixes within crop regimes (Dommen, 1988) that are adapted to relatively poor, fragile soils of low moisture-holding capacity, and to unpredictable rainfall (Guyer, 1997) is central. In addition, however, the farming practices in the research area of southern Malawi do display considerable investment of labour and expertise in 'techniques for managing the soil'. This, I propose, is because the shortage of land results in all cultivable land being permanently under crops, with virtually no fallowing, so that farmers have sought ways of stemming the decline of soil fertility. In turn, however, research has shown worrying declines in soil fertility that put a severe limitation on what can be achieved through even these labour-intensive methods of farming.

In describing some of the intensive cropping patterns used by farmers in a land-pressured part of Malawi, I have pointed to the successes they have achieved in the face of enormous resource constraints and of seriously mistaken attempts to change their practices. But I have also indicated points at which their practices appear to reveal serious limitations in their knowledge, too. The challenge is how to provide a better analysis of such complex systems and more appropriate modes for researchers and extension staff to work with farmers to overcome their difficult circumstances. The very recent turn among part of the Malawi agricultural researchers towards including organic strategies, such as intercropping, in the preferred means to improve maize productivity is a good sign that one step has been taken in this direction. Let us hope that the more difficult next steps will be taken: of researchers and extension workers treating farmers as prospective, knowledgeable clients with differentiated problems to solve, rather than as a homogeneous, backward group needing to be hauled, willy-nilly, into modern agriculture.

Notes

[1] A tiny sample from this large body of literature includes Chambers and Moris, 1973; Horowitz, 1979; Korten, 1980; Robertson, 1984; Cernea, 1985; Grillo and Rew, 1985; Long and Long, 1992; Crush, 1995.

[2] Some examples are Brokensha *et al.*, 1980; Cohen and Uphoff, 1980; Chambers, 1983; Richards, 1985; Pottier, 1993; Brush and Stabinsky, 1996.

[3] See Peters (1997) for sources on this period, and Vaughan (1987).

[4] One of the research villages in my study was located in one of the major resettlement schemes of the early 1950s and I, like McLoughlin, found that virtually all the planned changes (permanent field boundaries, single-stand crops in annual rotations, male ownership of land, etc.) had been redirected to local notions of proper practice.

[5] Very few people own cattle.

[6] Remittances are moneys sent by a husband working away from home (many in 1986 in South Africa, most in 1990 in Malawi); gifts are moneys sent by family members working away – almost all of these being adult children.

[7] This includes a wide range of activities: retailing crops, processing food/beer, basketry and mat-making, pot-making, brick-making, carpentry, shoe and bicycle repair, and tailoring. The first three are agriculture-based.

[8] Most of this work is casual or temporary labour. The Chinyanja term *ganyu* is often used as a synonym for casual agricultural labour in papers written in English, but I do not use it because, in village use, the term refers to any temporary employment (digging wells, roofing, pounding grain, etc.), even though most references are to agricultural labour.

[9] In 1990, the proportion of total household income from agricultural wages was 10.7% for the bottom income quartile and 1.6% for the top quartile, and the proportion from crop sales was 14%

for the bottom-quartile households and 45.5% for the top quartile (29% was from tobacco).

[10] A recent report points out that most residues are of 'poor quality' because much of the woody part is removed for fuel and some is eaten by goats, thus ideally requiring the addition of green manures or residues richer in nitrogen (MPTF, 1998: 29).

[11] Many villagers are quite defensive about burning crop residues, due, it seems, to the efforts of colonial agricultural officers and more recent extension staff to prevent it. Ironically, recent research shows that 'burning of maize stover gave a significant increase in maize yield over removal, incorporation, or [as] mulch' (MPTF, 1998: 29).

[12] The mean number of fields was 2.8 for households in the smallest land class (under 0.7 ha), 4.1 in the middle class (0.7–1.5 ha) and 5.6 in the top class (over 1.5 ha). Obviously, fields are generally small and plots are 'micro' areas.

[13] This is a day or so for small areas but a week or more for someone with large fields or who has other tasks to attend to.

[14] One of the difficulties faced by poor families throughout the decade of research was obtaining seeds for planting.

[15] There was a palpable increase in sweet-potato planting on all scraps of land by 1997, indicating, I believe, a rising demand from urban and peri-urban people as well as a food-security measure by rural families.

[16] Gram (*mphodza*) is *Vigna radiata*; chickpea (*nchana, tchana*) is *Cicer arietum* (Williamson, 1975).

[17] 1 kwacha = 100 tambala.

[18] This was said by Alvord, an American missionary who became Chief Agriculturalist for the Instruction of Natives in Southern Rhodesia in 1926 and later Head of the Department of Native Areas and Reserves until his retirement in 1950. His view was that Africans practised a 'primitive agriculture that wastes and destroys' (cited in Page and Page, 1992). As Page and Page (1992) show, nothing could have been further from the truth. More waste and destruction were caused by inappropriate agricultural policies and, of course, appropriation of land, than by indigenous methods themselves.

[19] The figures for my sample are: 1986/87 2%; 1990/91 52%; 1992/93 68%; 1993/94 42%; 1994/95 75%; 1995/96 47%; 1996/97 39%; 1997/98 61%.

[20] This shift, while small in our area, raises a concern expressed by other commentators on the effect of burley and other higher-value crops (such as legumes) squeezing out surplus maize

and the implications for national production (Carr, 1996).

[21]   See Jones and Heisey (1993) and Smale and Heisey (1994) on the optimistic side and Carr (1996) on the pessimistic side of this argument.

[22]   At present, at least in my sample, only estates use rotations of single-stand crops with any regularity. As noted earlier, the larger-scale burley growers among smallholders rotate their fields (maize plus intercrops rotated against burley with or without intercrops) but never grow legumes single-stand.

[23]   Also see Soko (1998), who refers to pigeon pea as 'the most versatile grain legume . . . in Malawi', and who recommends learning more about the 'wide range of traditional landraces' grown by farmers. Another insightful example of the failure of crop scientists to investigate farmer selection of bean landraces is described in several publications by Anne Ferguson (e.g. 1991, 1994, and one co-authored with Bill Derman in 2000).

[24]   Since market liberalization in the late 1980s, pigeon peas have become one of the major cash crops of the region, sought after by traders and destined to both domestic and export markets.

[25]   This is named *Stizolobium aterrimum* by the botanist Jessie Williamson, who did the most extensive research on plants and their uses (1975 [1955]: 226).

[26]   Shaxon, 1990. This thesis was based on an analysis of the farm data from my 1986/87 survey.

[27]   See specific references in Shaxon, 1990; cf. Dommen, 1988. Shaxon also says that most of the agronomic research on intercropping has been 'conducted under research station conditions' with 'relatively little reporting of on-farm trials' (p. 18). As I note above, this is beginning to change in Malawi.

[28]   The Science Section of the *New York Times* (22 August, 2000) reports on the 'exciting' research conducted in China, which shows that by including two different varieties of rice in fields instead of the usual monocrop, pest infestation dropped precipitously and yields rose. The article goes on to refer to a long-standing argument along these lines by ecologists, which was now being more seriously considered in light of these spectacular experiments. I assume the former say 'About time'!

[29]   See, for example, Chapin and Wasserstrom (1981) on the links between pesticide use and the resurgence of malaria in several parts of the world. Within Malawi, the tea estates have been escalating the levels of the pesticides applied to tea, which, because they are aerially sprayed, have a wide range of coverage. Similarly, many of the estates growing tobacco and other crops use large amounts of pesticides. Among smallholders, pesticides are used most on cotton and, increasingly, on vegetables.

[30]   The situation is even more complex, since stream beds are also the focus of competitive uses and competitive categories of users – water for consumption, washing, watering crops, brickmaking, etc., and gravel and sand for construction by villagers and by urban builders, who come into the rural areas looking for sources. A very recent interest in small-scale irrigation by the government is likely to add to this competition.

4   Farmers' Use and Adaptation of Alley Farming in Nigeria[1,2]

Akinwumi A. Adesina[1] and Jonas Chianu[2]

[1]The Rockefeller Foundation, Southern Africa Office, 7th Floor Kopje Plaza, PO Box MP 172, Harare, Zimbabwe; [2]International Institute of Tropical Agriculture (IITA), Ibadan, Nigeria, c/o IITA, L.W. Lambourn & Co., Carolyn House, 26 Dingwall Road, Croydon CR9 3EE, UK

Introduction

Agricultural production systems in many parts of sub-Saharan Africa are characterized by the slash-and-burn system, wherein farmers use bush-fallow to restore soil fertility. But rapid growth in population and land-use pressure have led to a reduction of fallow duration below the minimum threshold required for the system's sustainability (FAO, 1985; Conway, 1997). Facing declining land productivity, farmers have adjusted by expanding cultivation into marginal lands and bringing new forest areas under slash-and-burn, with significant negative environmental effects. The seriousness of the problem led to the formation of Alternatives to Slash-and-Burn – a global consortium aimed at developing and diffusing to farmers more sustainable land-use alternatives to the slash-and-burn system (ICRAF, 1996; ASB, 1997).

One of the alternatives to the slash-and-burn agricultural system is alley farming (Kang et al., 1984). Based on the principle of nutrient recycling, the technology involves continuous cultivation of annual crops within hedgerows formed by leguminous trees and shrubs. These species are periodically pruned and their biomass is applied as mulch to the crops in order to maintain or improve soil fertility. In alley-farming systems, the most commonly used species are the woody legumes, e.g. Leucaena, Calliandra, Acacia and Grilicidia. These legumes help to fix nitrogen, enhance nutrient cycling, because of their deep roots, and provide biomass for use as mulch and fodder for livestock (Atta-Krah and Francis, 1987; Kang et al., 1990). In addition, the technology has been shown to reduce soil erosion (Ehui et al., 1990; Kang et al., 1995), improve soil organic matter and nutrient status (Kang et al., 1990) and sustain crop yields under continuous cropping (Kang et al., 1990, 1995).

In Nigeria, economic analyses of alley farming have shown that the system is financially profitable (Ngambeki, 1985; Ehui et al., 1990). However, because early adoption constraints were many, these led several studies to assess constraints to the adoption of the technology (Atta-Krah and Francis, 1987; Whittome et al., 1995; Dvorak, 1996). Constraints identified include non-conducive property rights over land and trees (Francis, 1987; Lawry et al., 1995; Fabiyi et al., 1991; Carter, 1995; Whittome *et al.*, 1995), high labour requirements (Atta-Krah and Francis, 1987; Dvorak, 1996), long periods between establishment of hedge-rows and accrual of benefits, above- and below-ground competition between trees and crops for light, water and nutrients (Carter, 1995) and non-adaptability of some of the leguminous trees and shrubs (Atta-Krah and Francis, 1987; Whittome *et al.*, 1995; Dvorak, 1996).

However, it has now been several years since the technology has been extended to farmers. It is therefore important to investigate the adoption of the technology by farmers. In an extensive review of the literature on the socio-economic research on agroforestry technologies, Mercer and Miller (1998) found that one of the reasons that agroforestry projects failed was the lack of attention to socio-economic issues in the development of the systems as well as in the extension of the technologies. Their analysis of all the literature published from 1982 to 1996 in *Agroforestry Systems* showed that only 9% of studies of agroforestry systems were published on alley cropping. They noted that the most important research gap identified in the literature was 'understanding factors affecting adoption behaviour'. This issue was also noted by Sanchez (1995), who called for more papers that develop models to predict farmer adoption behaviour and its determinants as a way of guiding technology development and targeting. This chapter contributes to filling this gap in the socio-economic literature. The objective of this chapter is to use an econometric analysis to determine the socio-economic factors that influence farmers' adoption and modification of the alley-farming technology in Nigeria.

Survey

A farm-level study on the adoption of alley farming was conducted in Nigeria between July and September 1996. Two hundred and twenty-three farmers were surveyed in 14 villages, 142 in the south-west and 81 in the south-east. The two zones have distinct

agroecological characteristics. While the south-west is in the forest–savannah transition zone, with an average annual rainfall of about 1252 mm, the south-east is in the humid forest zone, with an average rainfall of about 1800 mm. The survey was done in two stages. In the first stage, focused group discussions were used to obtain background information on adaptations as well as the adoption of alley farming. This information was used to design a structured questionnaire administered to respondents during the second stage of the survey.

Selection of survey villages was accomplished through a stratified random-sampling procedure. A complete list of villages where alley farming had previously been introduced was available. Sample villages were selected based on the number of years of alley-farming interventions, the number of farmers exposed to alley-farming technology and an informed assessment by key informants on the extent of adoption of alley farming in each village. From each selected village, lists were developed: (i) of all farmers who had been exposed to alley farming; and (ii) of those without such knowledge. A random sample of farmers was taken from each of the two groups of farmers.

Adoption Processes and Dynamics of Alley-farming Use in the Sample Villages

Across all the sample villages, alley farming was introduced to them starting in 1980, with about 1.4% of the farmers, mainly in the south-east region of the country (Table 4.1). The establishment of alley farms peaked at 31% of farmers in the national sample, but the use pattern was different between the south-east and the south-west regions. Alley-farming establishment increased to 23% of the sample farmers in the south-east in 1985, peaking at 37% of the sample farms in 1986. In the south-west, the establishment of alley farms was much later, starting in 1984, but with 18% of the farmers in the sample, compared with only 5% of the farmers in the south-east sample when its establishment started. Establishment of alley farms then grew slowly, peaking at 28% of the farmers in 1986. While in the south-east sample the establishment cycle ended in 1987, farmers in the south-west sample continued alley-farm establishment through 1990.

Researchers have had a strong role to play in the establishment of many of these alley farms. For the entire sample of farmers, researchers accounted for helping 64% of the farmers to establish their alley farms.

However, about 15% of the farmers established their alley farms themselves, while 21% were established through researcher and farmer partnerships.

The source of information and the patterns of use of the technology are given in Table 4.2. It is interesting that the majority of the farmers (93%) had heard of alley farming by the time of the survey, with the figure higher in the south-west region (97%). This may reflect the higher level of intensity of interaction by researchers in this zone, due partly to the location of the International Institute of Tropical Agriculture (IITA) in the region. By the time of the survey, 64% of the sample farmers indicated that they had adopted alley farming, while 33% had not, with 2.4% still experimenting with the technology (see Table 4.2). The majority of those that established alley farms did so on their personal fields (83%), with 14% establishing them on family fields and 3.6% establishing them on rented parcels.

Of the 139 farmers who initially established alley farms, only 53% continued to use the technology during the survey year (Table 4.3). The high level of adoption discontinuity raised serious questions about the technology and its appropriateness for a

Table 4.1. Historical dynamics of farmers' establishment of alley-farming fields in surveyed villages in Nigeria (from survey data, 1996).

	Nigeria (n = 139)	South-east (n = 43)	South-west (n = 96)
1980	2 (1.4%)	2 (4.7%)	–
1981	1 (0.7%)	1 (2.3%)	–
1984	21 (15.1%)	3 (7.0%)	18 (18.8%)
1985	19 (13.7%)	10 (23.3%)	9 (9.4%)
1986	43 (30.9%)	16 (37.2%)	27 (28.1%)
1987	19 (13.7%)	11 (25.6%)	8 (8.3%)
1988	14 (10.1%)	–	14 (14.6%)
1989	5 (3.6%)	–	5 (5.2%)
1990	15 (10.8%)	–	15 (15.6%)

Table 4.2. Characteristics of alley-farming information and field establishment in Nigeria (from survey data, 1996).

	Nigeria	South-east	South-west
Heard of alley cropping	(n = 223)	(n = 81)	(n = 142)
Yes	208 (93.3%)	71 (87.7%)	137 (96.5%)
No	15 (6.7%)	10 (12.3%)	5 (3.5%)
Source of information	(n = 208)	(n = 71)	(n = 137)
Researchers	164 (78.8%)	47 (66.2%)	117 (85.5%)
Extension	3 (1.4%)	1 (1.4%)	2 (1.5%)
Other farmers	41 (19.7%)	23 (32.4%)	18 (13.1%)
Immediate use status	(n = 208)	(n = 71)	(n = 137)
Experimented	5 (2.4%)	1 (1.4%)	4 (2.9%)
Adopted	134 (64.4%)	42 (59.2%)	92 (67.2%)
Not planted	69 (33.2%)	28 (39.4%)	41 (29.9%)
Field where established	(n = 139)	(n = 43)	(n = 96)
Personal	115 (82.7%)	30 (69.8%)	85 (88.5%)
Family	19 (13.7%)	13 (30.2%)	6 (6.3%)
Borrowed	5 (3.6%)	–	5 (5.2%)

significant proportion of the farmers. For those that continued to use alley farming, their reasons for adoption included: soil-fertility improvement (82%), production of staking materials and poles (66%), fuel wood (51%), reduction of fallow length (45%), feed for animals (26%), and erosion control (20%) (Table 4.4).

Even for the adopters, they continued to face some important difficulties. The major problems continued to be the high labour demand of the technology (60%), tree competition with crops (49%), root obstruction (42%) and too many volunteer seedlings from the trees, which leads to bush (31%). For the farmers that had abandoned the technology, their reasons included: too many volunteer seedlings (45%), high labour demand (40%), trees not being well adapted (34%), lack of knowledge of how to manage alley fields, and loss of tenure of the land (14%).

For those farmers that never used alley farming, their major reasons were insufficient information about the technology (64%), lack of enough land (32%) and lack of any perceptible advantage over the traditional bush-fallow rotation system (25%). Farmers that continued using alley farming were closely monitored to see if they had made any modifications to the conventional system introduced to them by farmers. Due to space limitations, the detailed result of this analysis is beyond the scope of this chapter (for details see Chianu et al., 2000).

Farmers were found to have made significant modifications to the important configurations of the technology to suit their different situations. The components mostly affected are the recommended land-use intensity (83% of farmers), the height at which the hedgerow trees were cut back (43%), pruning intensity (33%) and intra-row spacing between the trees (25%). Others

Table 4.3. Characteristics of alley-farming plot establishment and adoption patterns in surveyed villages, Nigeria (from survey data, 1996).

	Nigeria (n = 139)	South-east (n = 43)	South-west (n = 96)
Establisher			
Researcher	89 (64.0%)	28 (65.1%)	61 (63.5%)
Farmer	21 (15.1%)	3 (7.0%)	18 (18.8%)
Researcher/farmer	29 (20.9%)	12 (27.9%)	17 (17.7%)
Still using alley cropping?			
Yes	74 (53.2%)	25 (58.1%)	49 (51.0%)
No (abandoned)	65 (46.8%)	18 (41.9%)	47 (49.0%)

Table 4.4. Distribution of reasons for alley-cropping adoption in Nigeria* (from survey data, 1996).

Reason for adopting	Nigeria (n = 74)	South-east (n = 25)	South-west (n = 49)
Improvement of soil fertility	61 (82.4%)	21 (84.0%)	40 (81.6%)
Fuel-wood production	38 (51.4%)	20 (80.0%)	18 (36.7%)
Erosion control	15 (20.3%)	10 (40.0%)	5 (10.2%)
Production of staking materials and poles	49 (66.2%)	18 (72.0%)	31 (63.3%)
Reduction of fallow length	33 (44.6%)	10 (36.0%)	23 (46.9%)
Materials for construction	13 (17.6%)	2 (8.0%)	11 (22.4%)
Feed for animals	19 (25.7%)	18 (72.0%)	1 (2.0%)
Mulching	2 (2.7%)	1 (4.0%)	1 (2.0%)
Future usefulness	1 (1.4%)	–	1 (2.0%)
Trees smother weeds	1 (1.4%)	–	1 (2.0%)

*More than one reason was given by most farmers.

are inter-row spacing between the hedge-rows (6.7%), pattern of planting in rows (5%) and the use of foliage from the plots (4%). The factors that affect farmers' decisions on these modifications will be discussed later in the chapter.

The next section of the chapter now turns to the examination of the factors that determine farmers' adoption of the alley-farming technology. This is followed by econometric analysis of factors determining farmers' decisions to modify the technology into the variants that were observed above.

Analytical Model

Farmers' welfare- or utility-maximization framework has been used in a number of studies to model farmers' adoption decisions, including contingent-valuation models (Lohr and Park, 1995), the Tobit model (Norris and Batie, 1987; Adesina and Zinnah, 1993; Adesina and Baidu-Forson, 1995) and discrete-choice models (Gould *et al.*, 1989; Adesina, 1996; Adesina *et al.*, 2000). The choice of any of these models depends on the issues of interest. Where the interest is in examining the role of farm and operator characteristics affecting adoption decisions, studies have used discrete-choice models, such as Tobit (Norris and Batie, 1987; Adesina and Zinnah, 1993; Adesina and Baidu-Forson, 1995) or logit (Adesina, 1996; Adesina *et al.*, 2000). Most studies have used the utility-difference model as the basis for selecting the discrete-decision models that they used. Using the utility-difference model, Lohr and Park (1994) modelled farmers' participation in acreage filter-strip programmes in the USA. The factors in their model included economic factors (i.e. net gain from the programme and the ability to share in the cost), attitude of the farmer to environmental issues and contact with environmental agencies. Lohr and Park's (1995) model included payments offered to farmers to participate in the programme, as well as the opportunity cost of the land enrolled under the programme, and the cost of revenue lost for participation in the programme.

Conceptual Model

Following Rahm and Huffman (1984) and Adesina and Zinnah (1993), let farmers' adoption be based on an assumed underlying utility function. Since the farmer has an option to adopt alley farming or any other natural resource management technology, let technology choice be represented by j, where $j = 1$ for alley-farming and $j = 2$ for non-alley-farming options. The latter may include the use of the conventional bush-fallow rotation system instead of alley-farming. The non-observable utility function that ranks the ith farmer's preference is given by $U(M_{ji}, C_{ji}, A_{ji})$, where M_{ji} represents a vector of farmer-specific characteristics, C_{ji} represents a vector of economic factors, and A_{ji} represents a vector of the farmer's village-specific or locational variables. The underlying utility function for the farmer can then be represented as:

$$U_{ji} = \alpha_j F_i (M_{ji}, C_{ji}, A_{ji}) + e_{ji},$$
$$\text{where } j = 1, 2; i = 1, 2, \ldots n \qquad (1)$$

Equation (1) in no way implies that the underlying utility function is linear. Its form will depend on the assumed distribution of the error term e_{ji} and it can be non-linear. Since utilities are random, the farmer will adopt alley farming if the preference comparison is such that $U1_i > U2_i$ or if the non-observable (latent) random variable $y^* = U1_i - U2_i > 0$. The probability of adoption of alley farming can then be represented as Equation (2):

$$Pi = Pr (Y_i = 1) = Pr (U1_i > U2_i)$$
$$= Pr \{(\alpha1)F_i(M1_i, C1_i, A1_i) + e1_i >$$
$$(\alpha2) F_i(M2_i, C2_i, A2_i) + e2_i\}$$
$$= Pr \{(e1_i - e2_i) > F_i (M_i, C_i, A_i)(\alpha2 - \alpha1)\}$$
$$= Pr \{u_i > - F_i (M_i, A_i, C_i, \beta\}$$
$$= F_i(X_iB) \text{ or } Y_i(X_iB) \qquad (2)$$

where $X_i = n \times k$ matrix of explanatory variables and $B = k \times 1$ vector of parameters to be estimated; u_i = random error term; $Y_i(X_iB)$ is the cumulative distribution function for u_i estimated at X_iB. The probability that a farmer will adopt alley farming is thus a function of the explanatory variables and the unknown error term. To estimate Y requires that one specifies the nature of the distribution of the error term. If it is assumed that the error term follows a logistic distribution, then Y can be estimated using a logit model, which assumes a logistic distribution. Following Equation (2) and using a logistic distribution, the logit model to capture the above underlying utility maximization can be given as:

$$Y_{ik} = F(I_{ik}) = e^{Z_{ik}}/(1 + e^{Z_{ik}}) \text{ for }$$
$$Z_{ik} = X_{ik\beta ik} \text{ and } -\infty < Z_{ik} < +\infty$$

where Y_{ik} is the dependent variable, which takes the value of 1 for the ith farmer that has adopted alley cropping in zone k and 0 if no adoption occurred. X_{ik} is a matrix of explanatory variables related to the adoption of alley cropping by the ith farmer in zone k, and β_{ik} is the vector of parameters to be estimated. I_{ik} is an implicit variable that indexes adoption. The logit model was estimated by a maximum-likelihood method using LIMDEP 6.0©.

In the logit model we use in the empirical model, we assume that farmers' adoption is influenced by M, C and A vectors. However, due to lack of data on C (economic variables, such as prices, taxes, subsidies on inputs, etc.), it was impossible to consider these in the analysis. However, we assume the village-specific variables (e.g. land-use pressure, market access, etc.) can be used as proxies for underlying economic factors, since they also reflect underlying differential economic and institutional incentives that farmers in various villages or locations face. While economic factors such as income and 'inducements' to use technologies are important in influencing decisions of farmers, such variables are very difficult to capture in cross-sectional surveys such as ours. It is possible that farmers receive 'free' seedlings or extension advice on the technology, but imputing the price for such inputs is problematic. Incomes of farmers in Africa are also very difficult to determine. Given the bias in using wrongly reported incomes and 'prices' for free seedlings they might have received, we left these out in this study. The details on these variables and the variables in the vectors M and A are specified and justified in the next section. We also examine, later in the chapter, the factors that affect farmers' modification of alley-farming technology, using a similar logit-model approach.

Empirical Model

Variables selection from evidence from prior studies

Numerous adoption studies of soil-conservation technologies have shown that the characteristics of the farmer are often the most significant in influencing the adoption of technologies. For studies outside Africa, see Ervin and Ervin (1982); Saliba and Bromley (1984); Norris and Batie (1987); Lynne et al. (1988); and Lohr and Park (1995). For evidence in Africa, see Atta-Krah and Francis (1987); Tonye et al. (1993); Franzel (1999); and Adesina et al. (2000). Some of these factors include operator's age, family size, informational variables such as contact with agencies that can educate about and administer conservation programmes (Norris and Batie, 1987; Lohr and Park, 1995) and attitudes towards conservation (Lynne et al., 1988). Land tenure or ownership of land has been found in several of the studies to influence soil-conservation decisions. Lynne et al. (1988) found that renters displayed less conservation effort than owners did. In Africa, several studies cite the importance of land ownership rights. However, there is no consensus on whether security of rights significantly influences investments (see Francis, 1987; Fabiyi et al., 1991; Place and Hazell, 1993; Tonye et al., 1993; and Place et al., 1994). Based on the convincing evidence that farmers' socio-economic factors influence adoption behaviour, we tested the influence of these variables in the empirical model.

Variables in the empirical model

The list of variables used in the main empirical model is given in Table 4.5. The dependent variable is indexed if the farmer has adopted alley farming. The variable takes the value of 1 if the farmer currently uses alley farming and 0 otherwise. The explanatory variables are discussed below.

SEX refers to the sex of the farmer and takes the value of 1 if the farmer is a male and 0 if a female. It has been reported that women in south-west Nigeria face constraints in using alley-farming technology (Fabiyi *et al.*, 1991). In many areas, women are not allowed to own land or plant trees. Furthermore, there also exists gender bias in the selection of farmers for on-farm tests of alley farming. Women are often excluded and, when included (Versteeg and Koudokpon, 1993; Vabi *et al.*, 1995), they are usually old widows (Fabiyi *et al.*, 1991). It is hypothesized that SEX is negatively related to the adoption of alley farming.

FHHSIZE is the size of the farm household. Labour constraints are critical in farmers' use of agroforestry technologies (Dvorak, 1996; Franzel, 1999). Alley farming is a labour-intensive technology and high labour requirements can discourage farmers from using the technology (Carter, 1995; Dvorak, 1993). It is expected that the larger the family size, the greater will be the availability of labour for alley farming.

EDUC measures the level of education of the farmer. Alley farming is a knowledge- and management-intensive technology, requiring ability to manage the hedgerows properly to achieve optimal results. Lack of proper understanding of the technology leads to poor tree performance and abandonment of alley plots (Atta-Krah and Francis, 1987; Carter, 1995; Koudokpon *et al.*, 1992).

FAS measures if the farmer is a member of a farmer association and takes the value of 1 if the farmer belongs to a farmer association or group and 0 otherwise. Agroforestry

Table 4.5. Descriptive statistics for variables in the empirical econometric models.

Variables	Mean	Standard deviation	Min.	Max.
SEX (1 = male; 0 = female)	0.83	0.38	0	1
FHHSIZE (family size)	10.40	9.28	1	50
EDUC (educational level of farmer)	0.43	0.50	0	1
FAS (membership of farmers' associations)	0.85	0.35	0	1
CONT (contact with agroforestry research, extension or NGO)	0.48	0.50	0	1
FORIGIN (if farmer is native of the village)	0.83	0.38	0	1
VLANDP (village land-pressure index)	1.58	0.78	1	3
VEROS (village erosion index)	2.47	0.77	1	3
VFUELW (village fuel-wood scarcity index)	1.40	0.64	1	3
VLST (importance of livestock in village)	2.41	0.75	1	3
VDISTOWN (distance of village to town)	2.50	0.81	2	4
VFOD (village fodder situation)	2.93	0.25	2	3
HDIV (divided-inheritance field type)	0.65	0.48	0	1
TRENT (if crop field was rented)	0.18	0.39	0	1
VDISTNGR (interaction between village distance to nearest town and contact with researchers)	1.47	1.38	0	4
VDISTCONT (interaction between village distance to nearest town and contact with researchers, extension or NGOs)	1.24	1.43	0	4
VDISTNGE (interaction between village distance to nearest town and contact with extension)	0.22	0.63	0	2
REGFRGIN (interaction between region of location of village × FORIGIN)	1.15	0.68	0	2
AFEXP (number of years of experience in alley farming)	9.66	1.99	6	16
RLANDACX (whether or not the farmer has sufficient land)	0.80	0.40	0	1

extension and development agencies have higher success rates on adoption when working with farmer groups (Atta-Krah and Francis, 1987; Versteeg and Koudokpon, 1993). It is expected that FAS will be positively related to the adoption of alley farming. CONT is a dummy variable that measures the contact of farmers with research and development agencies or extension agencies that work on agroforestry. Because contact with extension allows farmers to be able to get information on the technology and possibly see or participate in demonstration tests (Atta-Krah and Francis, 1987; Carter, 1995; Whittome et al., 1995; Dvorak, 1996), it is hypothesized that this will positively influence adoption decisions.

FORIGIN is a dummy variable that indexes whether or not the farmer is a native of the village or a migrant. Village studies in Nigeria show that migrants tend to be more active in agriculture and often tend to be more aggressive in the use of new technologies (Polson and Spencer, 1991). This suggests that migrants are more risk-taking than natives are and have a higher likelihood of adopting alley farming. However, in areas where migrants face a highly inelastic land-supply situation, this may not necessarily be the case, as they may be able to get access to land only on a temporary basis. TRENT is a dummy variable that indexes whether the crop field is a rented plot or not. It takes the value of 1 if it is and a value of 0 if otherwise. It is expected that TRENT will be negatively related to adoption, as renting a food-crop field may be an indication of land scarcity or limited access to land.

VLANDP measures the population density in the village. Some studies have suggested that alley farming should be targeted to areas with a high population density (Ehui et al., 1990; Carter, 1995; Reynolds and Jabbar, 1995). It is expected that the higher the population density, the higher the likelihood of adoption of alley farming. VEROS measures the extent of erosion in the village where the farmer is located. It takes on the value of 1 if erosion is a major problem for farmers in the village, 2 if it is a minor problem, and 3 if it is not a problem. Since alley farming has been shown to reduce

runoff (Ehui et al., 1990; Kang et al., 1995), it is expected that farmers in villages that have more erosion problems will be more likely to adopt the technology.

VFUEL measures the extent of fuel-wood scarcity in the village. It takes the value of 1 if fuel wood is easily available, 2 if it is scarce, and 3 if it is very scarce. Trees used in alley farming can supply significant quantities of fuel wood (Kang et al., 1984). It has been suggested that alley farming should be encouraged in areas where fuel wood is scarce (Whittome et al., 1995). VLST measures the importance of livestock as a source of income for villages where the farmers are located. It takes the value of 1 if livestock income is not important, 2 if livestock income is important, and 3 if it is very important. It is expected that farmers located in villages where livestock is very important may have a greater likelihood of using alley farming. VDISTOWN measures the distance of the village where the farmer is located to the nearest town. Several of the research and development efforts on agroforestry technologies tend to focus their attention on promoting farmer demonstrations in villages with relatively better road access. This is to allow them to reduce technology-testing or demonstration costs and improve chances of farmer adoption. Farmers located in villages close to town are also better able to capture economic benefits from the use of the technology, due to better market access, which encourages a 'market-driven' intensification process. It is hypothesized that the greater the distance of the village from the town, the less the likelihood of farmers in the village adopting alley farming.

VFOD measures the extent of abundance of fodder in the village where the farmer is located. The variable takes the value of 1 if fodder is very scarce, 2 if fodder is abundant, and 3 if fodder is very abundant. It is hypothesized that, as fodder supply gets scarcer, farmers will have greater incentives to adopt alley farming. The leguminous trees and shrubs used can be good sources of fodder (Sumberg et al., 1987; Jabbar et al., 1992). HDIV is a dummy variable that measures whether or not the farmer's plot is obtained from a divided inheritance or otherwise. Generally, lands from a divided inheritance are more likely to be used for planting trees.

A number of interaction terms were included in the model to account for possible interaction effects between some of the village-level variables and the contact variables for research and extension effort. REGFRGIN is an interaction term between region and farmer's origin in the village or not. VDISTNGR, VDISTNGE and VDISTCONT are interaction terms between the distance of the village to town and contact with research, with extension, and with extension, researchers and/or non-governmental organization (NGO) groups in the 5 years prior to the survey. AFEXP is the number of years the farmer has been practising agroforestry and it is expected that this will have a positive effect on the adoption of alley-farming technology. Exposure to agroforestry technologies before may influence uptake of tree-based technologies. RLANDACX is a dummy variable that indexes whether or not the farmer has sufficient land. It takes the value of 1 if there is sufficient land and 0 if the response is otherwise. It is expected that farmers who consider themselves as having sufficient land may face less pressure to change from the conventional bush-fallow system and this may negatively affect adoption of alley farming.

Results and Discussion

The results of the empirical model are given in Table 4.6. The model gave 81% correct predictions of adopters and non-adopters. The model also has very strong explanatory power. Eleven explanatory variables were significant in explaining adoption decisions of farmers on alley farming.

SEX is significant at the 10% level and is positively related to adoption of alley farming. This suggests that men are more likely to adopt alley farming than women. This result may reflect the traditional bias against women either inheriting lands or having secure land or tree rights. Although alley farming is targeted to farmers for soil-fertility management, this result may suggest that its adoption is not gender-neutral and that technical change is biased towards men. This result is similar to the finding of Adesina (1996), who found that the probability of adoption of chemical fertilizers is higher for male farmers than for female farmers.

CONT is significant at the 5% level and positively related to adoption. This suggests that farmers with contact with research–development or extension agencies have a greater likelihood of adopting alley farming. This result is corroborated by findings on alley farming and other improved-fallow technologies in Cameroon (Adesina *et al.*, 1997). FORIGIN is significant at 5% and negatively related to adoption. This result suggests that migrants are more likely to adopt alley farming. This unexpected negative sign is, however, corroborated by evidence from other adoption studies in Nigeria. In a study of the adoption of improved cassava technologies in south-west Nigeria, Polson and Spencer (1991) found that adoption was higher among migrants than among indigenous farmers of the study villages. Migrants move into areas in search of new lands for productive uses and are motivated for productive investments. In addition, because migrants often have poorer lands, they may have greater incentives to use natural resource management technologies for soil-fertility improvement in order to increase production and incomes. Dvorak (1993) found that, in areas of south-east Nigeria where land pressure was low, migrants had no difficulty in getting land on similar terms to those for indigenous farmers. Our results here may indicate that land pressure is not so high as to represent disincentives for migrants to use alley farming.

VLANDP is positive and significant at 5%. This suggests that, as the land-use pressures increase in the village, the likelihood that the farmers in that village will adopt alley farming increases. VEROS is significant at 1% and negatively related to the adoption of alley farming. This suggests that the greater the erosion problem in the village, the greater the likelihood that farmers in that village will adopt alley farming. This is borne out by the survey result, which

Table 4.6. Logit-model result of factors affecting farmers' adoption of alley farming and its variants in Nigeria.

Variables	Parameter estimate	Standard error	Pr > chi-square
Intercept	4.4018	10.8710	0.6855
SEX	1.2501	0.7482	0.0948*
FHHSIZE	−0.0154	0.0291	0.5976
EDUC	0.4686	0.4652	0.3137
FAS	0.3817	0.6466	0.5549
CONT	5.0395	2.0950	0.0161**
FORIGIN	−3.7530	1.5202	0.0136**
VLANDP	1.6223	0.6483	0.0123**
VEROS	−1.3645	0.4322	0.0016***
VFUEL	−1.4452	0.8123	0.0752*
VLST	−0.8559	0.5084	0.0923*
VDISTOWN	0.1712	0.9825	0.0303**
VFOD	−2.5443	2.3770	0.2844
HDIV	−0.0176	0.6378	0.9780
TRENT	−0.3800	0.7851	0.6284
REGFRGIN	2.7102	1.1980	0.0237
VDISTNGR	1.9671	0.5851	0.0008***
VDISTNGE	0.3946	0.4000	0.9732
VDISTCON	−2.0518	0.8822	0.0200**
AFEXP	0.4420	0.1979	0.0255**
RLANDACX	−0.9735	0.5960	0.1024

***, Significant at 1%; **, significant at 5%; *, significant at 10%.

Association of predicted probabilities and observed responses

Concordant	=	81.3%
Discordant	=	18.6%
Tied	=	0.2%
(4672 pairs)		

showed that 20% of the adopters adopted alley farming for the control of erosion. VFUEL is significant at 10% and negatively related to the adoption of alley farming. It suggests that, as fuel wood becomes scarce in the villages, the likelihood of adoption of alley farming decreases.

While adoption of alley farming for fuel wood is important for farmers, it is possible that, in villages where fuel wood scarcity is a major problem, farmers have developed alternative supply sources for fuel-wood provisioning. This may include increased reliance on specialized wood lots. Alley farming may not have a comparative advantage for fuel-wood supply in such situations. VLST is negative and significant at the 10% level. This suggests that, as livestock becomes more important as a source of income in the village, the likelihood of adopting alley farming reduces. Where livestock is a very major activity, it is likely that farmers may prefer to invest directly in the use of fodder banks rather than in alley farming, as there may exist greater competition between cropping and livestock.

VDISTOWN is significant at 5% and positively related to the adoption of alley farming. This implies that the further the village is from an urban centre, the greater is the probability of farmers' adopting alley farming. This may suggest that farmers in closer proximity to urban centres may face greater land pressure, which causes their production practices to be more intensive. Given the high wages in such areas, labour-demanding technologies such as alley farming are less likely to be adopted by farmers. In areas that are further away from towns, two factors may play complementary roles in influencing the use of alley farming. The low rural wages favour labour-intensive systems like alley farming, while the distance from the urban centres significantly raises the costs of chemical fertilizers. Due to poor infrastructure and the consequent high transaction costs, farmers in areas far from urban areas are less likely to be able to afford the high fertilizer prices. These factors interplay to create economic incentives for them to invest in labour-using and less cash-demanding technologies, such as alley farming.

The positive and significant signs on VDISTNGR and VDISTCONT suggest that increased interaction between researchers, extension and NGO groups with more distant villages will increase adoption of the technology in such more marginal areas. It is often in these more remote villages, with limited road and market infrastructure, that farmers face even more serious soil-fertility problems. Research, extension and NGO groups should ensure that their technology dissemination or research/experimentation activities have a clear equity perspective at the early stages. If not, the outcome of such research on technologies will certainly have a distributional or equity bias. AFEXP is positive and significant at 5%. This suggests that the number of years that the farmer has been practising agroforestry positively influences adoption decisions on alley farming. Farmers who have been practising agroforestry are more likely to be aware of different types of agroforestry technologies, possibly due to better contacts with agroforestry extension projects or through learning from other farmers.

Determinants of farmers' modification decisions on alley farming

While studies have identified that farmers are making modifications to the conventional technology, no study has attempted to econometrically determine the factors that influence such decisions. It is hypothesized that a farmer's decision to adapt the technology is a function of human-capital and farmer-household characteristics. Rahm and Huffman (1984) showed that human-capital variables significantly affected farmers' ability to use reduced-tillage technology in the USA. Operator education has also been found to significantly influence decisions of farmers in Iowa to adapt fertilizer decisions. Following this literature, we assume that a farmer's decision to adapt is a function of human-capital, economic and farmer-specific attributes. The modifications evaluated were: (i) changes in the pruning frequency from that recommended by researchers; and (ii) the introduction of fallow periods into the conventional alley-farming system.

For the factors affecting changes in the pruning regime, the following variables were included in a logit model. The dependent variable was whether or not the farmer modified the pruning regimes recommended by researchers. This variable takes on the value of 1 if they changed the pruning regime, and 0 otherwise. The explanatory variables used in the model are farmer's age (AGE), farmer's educational status (EDUC), number of years of experience in alley farming (AFEXP), contact with agroforestry research, extension or NGOs (CONT), ownership of non-agricultural income-generating assets (NFIC), gender of the farmer (SEX) and whether or not the farmer has sufficient land (RLANDACX).

The results of the logit model on the factors affecting farmers' modification of the pruning regime under alley-farming system are shown in Table 4.7. Two variables significantly affect farmers' decisions to make this modification. Farmer education (EDUC) and family size (FHHSIZE) both positively and significantly affect farmers' decisions. Other factors that positively influence this decision include the farmer's age, number of years of experience with alley farming and contact with agroforestry research and/or researchers. These results suggest that better-educated farmers are more likely to modify the pruning regime from that recommended by researchers. This is most

probably due to their better ability to understand the technology and its demands. The
larger the family size, the greater the likelihood of modifying the pruning regime.
Farmers with larger family sizes have more
labour to prune the trees more often than
recommended by researchers. Pruning is a
labour-intensive activity.

To model the factors affecting the
decision of farmers to make modifications to
the conventional alley-farming technology
by introduction of fallow periods in the system, a logit model was used. The dependent
variable takes on the value of 1 if the farmer
made modifications to the technology by
introducing fallow periods and 0 otherwise.
The explanatory variables in the model were

age of the farmer (AGE), gender (SEX),
educational status (EDUC), contact with
extension (CONT), family size (FHHSIZE),
membership of farmers' associations (FAS),
village land pressure (VLANDP), tenure status of the field (TRENT), origin of the farmer
(FORIGIN), village erosion index (VEROS),
importance of livestock in the village (VLST)
and whether or not the farmer has access to
sufficient land (RLANDACX).

The model results for factors affecting
farmers' introduction of fallow periods into
the alley-farming system are shown in Table
4.8. Four factors were significant in the logit
model. These are the farmer's age (AGE),
farmer education (EDUC), whether or not
the farmer has access to sufficient land

Table 4.7. Logit-model result of factors affecting farmers' adaptation of alley-farming pruning frequency in Nigeria.

Variables	Parameter estimate	Standard error	Pr > chi-square
Intercept	1.997	2.1202	0.5715
AGE	0.0005	0.0253	0.8169
AFEXP	−0.0824	0.1583	0.6026
EDUC	1.4697	0.7371	0.0462*
CONT	0.9243	0.6984	0.1857
NFINC	0.6823	0.6603	0.3015
SEX	−1.5628	0.9899	0.1144
REGG	−1.0671	0.9605	0.2666
RLANDACX	−0.9766	0.7712	0.2054
FHHSIZE	0.1206	0.0587	0.0398*

*, Significant at 10%.

Table 4.8. Logit-model result of factors affecting farmers' introduction of fallow-period adaptation into the conventional alley-farming system in Nigeria.

Variables	Parameter estimate	Standard error	Pr > chi-square
Intercept	−7.7176	5.2645	0.1427
AGE	0.0916	0.0465	0.0488**
SEX	−0.2941	1.5566	0.8501
EDUC	2.3433	1.1799	0.0470**
CONT	1.8227	1.1532	0.1140
RLANDACX	−2.1764	1.2280	0.0763*
FHHSIZE	0.0803	0.0738	0.2776
FAS	−1.0364	1.7606	0.5561
VLANDP	1.5699	1.0415	0.1317
TRENT	2.1762	2.8014	0.4373
FORIGIN	2.3943	2.5089	0.3399
VEROS	−1.5778	0.7662	0.0395**
VLST	0.6786	0.7313	0.3535

*, Significant at 10%; **, significant at 5%.

(RLANDACX) and the village erosion index (VEROS). The positive sign on EDUC shows that farmer education positively influences the probability of introduction of fallow into the alley-farming system. This may be because better-educated farmers are able to understand the nutrient-cycling processes underlying the technology. They modify the technology to incorporate the benefits of fallow, which they traditionally do in their bush–fallow rotation system. The positive sign on AGE suggests that older farmers have a higher probability of introducing fallow into their alley farms. This may be due to several reasons. Older farmers may have accumulated more knowledge of the benefits of fallow, from their years of experience. Secondly, older farmers may find the management of the conventional alley-farming system too labour-intensive. Introduction of the fallow phase lowers the labour demands of management of the fields, except the additional cost of clearing the bush and cutting the trees after the fallow phase is over.

The negative sign on RLANDACX suggests that farmers who have enough land are less likely to introduce a fallow into their alley plots. If land is not a constraint, such farmers are more likely to try out the conventional alley-farming system. The negative sign on the village erosion index (VEROS) indicates that the higher the erosion intensity in the village, the higher the likelihood that the farmer will introduce a fallow into the alley-farming system. Villages that experience significant erosion are likely to face significant topsoil loss, which negatively affects productivity. One of the ways that farmers have traditionally controlled erosion is by the use of fallow periods to allow for vegetation regrowth on their fields. Tree and vegetation cover helps to reduce rainfall intensity and soil loss. This result, when taken together with the aforementioned finding on VEROS in the earlier section, strongly suggests that erosion is a very important issue in the use of alley farming and its farmer-modified variants. This result interestingly concurs with the observation from the modelling results of Ehui *et al.* (1990), who found in south-west Nigeria that alley farming led to significant economic

benefits from reduction of erosion and conservation of topsoil. Our results show that, by modifying the conventional alley-farming system with incorporation of fallow periods, farmers are able to achieve greater soil-erosion control.

Conclusions

Despite initial slow interest in alley farming, the technology is being adopted by some farmers in Nigeria. Importantly, farmers are experimenting with different configurations of the technology (Chianu *et al.*, 2000). There is a need for continued efforts to adapt the technology variants to better fit the needs of farmers. In particular, support for farmer participatory development of variants of the alley farming will further encourage wider adoption. However, this requires careful targeting of the technology. Many of the earlier efforts to target the technology were based on the biophysical characteristics of agricultural systems. The non-consideration of socio-economic factors has led to inappropriate targeting of the technology into areas with a lower likelihood of adoption in much of West Africa (Whittome *et al.*, 1995). Results from this chapter suggest a number of such factors that should be used for better targeting of alley farming and its variants in Nigeria.

Some conclusions can be drawn from the analysis in this chapter. First, our results show that the adoption process of the alley-farming technology is not gender-neutral. The probability of adoption was higher for men than for women farmers. This may be the result of the lack of consideration of gender-equity issues in the design and introduction of the technology to farmers. In many parts of West Africa, women do not have secure land and tree tenure, due to the largely patrilineal inheritance systems. Only older women, widows and female-headed households are often able to have such access to more secure land rights. It is important to address this inequity by introducing to women farmers other technologies that do not require secure long-term land and tree rights. This could include improved fallow technologies with leguminous shrubs or mixed intercropping with leguminous-shrub species.

Secondly, it appears that village-level proxies for economic and institutional factors play a very significant role in influencing incentives for technology adoption. It is therefore important for studies to use spatial analysis to better understand where to target these agroforestry technologies, based on the relative incentive structures across villages, as determined by market and non-market factors. It is important to focus more attention on extending this technology and its variants to farmers in areas that need them most. These would be those found in the more remote and distant villages, whose economy and environment are probably more fragile.

Thirdly, efforts to promote agroforestry technologies should focus on not just these locational issues, but also on their interaction with policy-amenable variables. By improving the access of such distant villages to public goods from research, through increased interaction with extension, research and NGO groups, the likelihood of technology adoption will rise significantly. This implies that there is a need to restructure the process of engagement with rural communities. There is need for researchers and development agencies to move away from the cosy confines of peri-urban areas and reach the poor and excluded farmers in more distant and remote locations. Regardless of the quicker impacts that are achieved in areas closer to towns, the majority of the poor will be unreached by such a focus of agroforestry projects in such areas.

Finally, it is important for researchers to carefully monitor and assess how farmers are using technologies. The result shows that farmers were already making modifications and adaptations to the conventional alley-farming technology that was introduced

to them. Human-capital variables were particularly important in influencing such decisions. Farmers' adoption of such technologies must be seen as a continuum that considers the fact that farmers may be making modifications to the technology. Part of the adoption process is the ability of farmers to use their knowledge to modify and adapt technologies. When trying to understand adoption decisions, researchers should make sure that they spend enough time evaluating the entire sequence of adoption processes from initial adoption to technology modification/adaptation.

In the case of alley farming in Nigeria, researchers were unable to see or appreciate the modifications farmers were making, as their focus was on finding the 'perfect technology' they gave farmers. The presence of 'protective researchers' earlier on may have contributed to the stifling of farmers' creativity and innovation on this technology. The lesson for future studies on technology development and evaluation is that researchers should become learners and not controllers. Researchers should try to facilitate the process of farmer learning and technology adaptation. Regardless of how protective researchers are of their technologies, what will remain at the end of the day is what farmers consider appropriate to their social, economic and institutional endowments.

Note

1 An expanded version of this chapter was presented at the international workshop on 'Understanding Adoption Processes of Natural Resource Management Practices for Sustainable Agricultural Production in Sub-Saharan Africa', 3–5 July 2000, at the International Centre for Research in Agroforestry, Kenya.

2 A version of this chapter is expected to appear in *Agroforestry Systems*.

5 Farmers as Co-developers and Adopters of Green-manure Cover Crops in West and Central Africa

G. Tarawali,[1] B. Douthwaite,[1] N.C. de Haan[1] and S.A. Tarawali[1,2]

[1]International Institute of Tropical Agriculture (IITA), Ibadan, Nigeria, c/o IITA, L.W. Lambourn & Co., Carolyn House, 26 Dingwall Road, Croydon CR9 3EE, UK; [2]International Livestock Research Institute (ILRI), Oyo Road, PMB 5320, Ibadan, Nigeria

International research institutions, together with national and international partners, have, over the years, developed promising technologies, such as improved crop varieties or soil- and pest-management practices, that are ripe for dissemination to end-users (IITA, 1999). Such interventions address the challenge of enhancing the food security, income and welfare of resource-poor farmers and their families in sub-Saharan Africa. Despite the high potential of these technologies to improve the sustainability of African agriculture, their transfer to the poorest farmers has, to date, been limited (CGIAR, 1997). Factors often cited for this lack of adoption include the following.

- Appropriate inputs (materials and/or labour, etc.) are needed.
- Timing of proposed operations does not fit with farmers' calendar of operations.
- Farmers do not benefit immediately from the introduced technology.
- Farmers' economic circumstances prevent them from adopting the technologies.
- Interventions are not targeted to areas where there is real need.

These factors often arise because the intended beneficiaries were not involved in the technology-development phase. This would have given them the opportunity to adapt these interventions to their socio-economic circumstances, thereby making them co-developers of the innovations. In the past, development agencies applied the transfer-of-technology approach (Chambers and Jiggins, 1986), where research is seen as separate from extension and the job of the researcher is seen as the innovator and the job of the extension as 'spreading the message' (Ruthenberg, 1985). With this approach, farmers are passive recipients of ready-made interventions, a strategy that is unrealistic, as technologies need to be adapted to the farmers' complex farming systems and circumstances before the final product is adopted by smallholders. A successful technology represents a synthesis of the knowledge sets of different researchers and beneficiaries (Douthwaite *et al.*, 2001), which can only be derived through farmer–researcher partnership using participatory approaches. While there needs to be a starting-point for any intervention (scientists have a role to play in identifying appropriate technologies and developing them to a certain level), working in partnership with farmers to develop and refine interventions is essential if appropriate, robust and adoptable technologies are to result.

This chapter refers to four technologies involving cover crops and integrated crop–livestock interventions developed under varying social, ecological and production systems, in which farmers and researchers, working in partnership to combine indigenous knowledge and circumstances with research interventions, have contributed to the development of the final innovation. For each innovation, a brief account is given of the technical-development history and the lessons that have been learned through the interaction of researchers and farmers, together, in some cases, with extension services. Experiences from these examples are pooled in the discussion to highlight the importance of partnerships in the development of agricultural interventions suitable for small-scale farmers in sub-Saharan Africa.

Improved *Mucuna* Fallows

In sub-Saharan Africa, 389 million ha are classified as lowland moist savannah (Fig. 5.1), with a growing period of 150 to 270 days, especially suited to annual crops (Jagtap, 1995). Soils are relatively infertile, fertilizer use is low and soil is easily degraded under intensified agriculture. Shifting cultivation, which allows fallow to restore the land and has formed the basis of the traditional agricultural system, can no longer be sustained because of the rapidly growing population. Of the alternative soil-management strategies that have evolved, one of the most promising is the use of *Mucuna* (*M. pruriens*) as a weed-smothering and soil-improving cover crop.

The *Mucuna* technology was primarily introduced in the bimodal rainfall zone in southern Benin, and two different management systems have developed for integration of the legume into the cropping systems. *Mucuna* can be planted either in pure stands on an uncultivated piece of degraded land or in association as a relay with an annual crop, such as maize. For intercropping, the sole-crop seeding rate of 10–15 kg ha^{-1} is reduced to minimize competition with the companion crop. *Mucuna* plants achieve nearly 100% ground cover in 2 months, with vines up to 6 m in length if soil fertility is adequate (Carsky *et al.*, 1998).

The bimodal rainfall distribution in southern Benin allows two cropping seasons a year. Planting *Mucuna* during one growing season therefore leads to the loss of

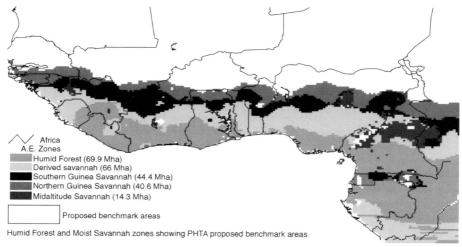

Africa
A.E. Zones
Humid Forest (69.9 Mha)
Derived savannah (66 Mha)
Southern Guinea Savannah (44.4 Mha)
Northern Guinea Savannah (40.6 Mha)
Midaltitude Savannah (14.3 Mha)
Proposed benchmark areas

Humid Forest and Moist Savannah zones showing PHTA proposed benchmark areas.

Fig. 5.1. Agroecological zones in West Africa (from Jagtap, 1995).

production of edible crops in the same cropping season (Versteeg and Koudokpon, 1991). However, mulch left behind on the soil surface quickly decomposes and supplies soil nutrients to the subsequent crops, which results in higher yields in the second season of cropping. In addition to increasing crop yields (Versteeg and Koudokpon, 1990) through an improvement in soil physical and chemical properties (Hulugalle *et al.*, 1986), *Mucuna* also has the ability to suppress weeds, in particular, spear-grass (*Imperata cylindrica*) (Galiba *et al.*, 1998), provide livestock feed (Yai, 1998) and provide an income for adopters through the sale of seed (Honlonkou *et al.*, 1999).

When researchers and extensionists started to promote the dissemination of *Mucuna*, they initially focused on restoration of soil fertility in Mono Province of Benin. However, after 2 years of demonstrations, participants became more involved in the evaluation. Out of the 20 recruits, 14 successfully established a dense stand of *Mucuna*, which they discovered was very effective in suppressing *Imperata* (spear-grass), one of the most noxious weeds in the derived savannah. The *Mucuna* intervention enabled these adopters to cultivate land after 2–3 years, rather than abandoning it for many more. Further implications of the intervention include the reduced requirement for weeding, an arduous task usually carried out by the women in the household. Freeing more time for women has other social implications, which can best be understood by working in close partnership with farm households. Based on this farmer discovery of controlling spear-grass and subsequent farmer-to-farmer exchange of experience, 103 adopters planted *Mucuna* the following year (Versteeg and Koudokpon, 1990). The government extension service (Institut National des Recherches Agricoles du Benin (INRAB)) began testing *Mucuna*'s weed-suppression abilities with other participants and the number involved grew to 500.

The early success in identifying a promising technology and then targeting it to farmers with *Imperata* problems can be attributed to very close researcher–extensionist–farmer interactions. If farmers had not participated in redefining the emphasis for *Mucuna* use, the uptake might not have been so successful. Involvement of development non-governmental organizations (NGOs) was also important and, in 1992, the international NGO Sasakawa Global 2000 (SG2000) became involved and purchased 4 tons of *Mucuna* seed from adopters to give to 128 farmers in other provinces where *Imperata* infestation and soil depletion were problems (Vissoh *et al.*, 1998). SG2000 also produced a technical bulletin on *Mucuna* establishment and management to guide extension workers and worked through the existing government extension services. SG2000 continued with the strategy of planting demonstration plots in villages and buying *Mucuna* seed from collaborators to give to new farmers, increasing the amount of seed it purchased each year. In this artificial market system, most participants benefited from the sale of *Mucuna* seed, which became an incentive for adoption (Vissoh *et al.*, 1998). This was reflected in the fact that, by 1996, about 10,000 adopters were growing *Mucuna* (Galiba *et al.*, 1998).

The adoption dynamics of *Mucuna* in southern Benin showed that there was a rapid adoption rate until 1996, and Houndékon *et al.* (1998) found that three-quarters of adopters were using the technology for *Imperata* control. In 1997, the adoption rate, measured in terms of area planted to *Mucuna*, fell by more than a quarter. The study also showed that, even though there was a big rise in the adoption rate in 1995, there was also an increase in the rate at which farmers stopped using the technology. If interpreted at face value, such figures might be interpreted to mean the intervention had failed. However, the understanding, gained by working closely with the adopters, that the *Mucuna* could eliminate *Imperata* after 2–3 years (Galiba *et al.*, 1998) suggests that the decrease in adoption is, in fact, a measure of the success with which the weed had been controlled. Once it has done the job, farmers will plant food crops instead. Another factor for the observed decline in adoption was that, in 1996,

SG2000 stopped buying large amounts of seed and this led to a collapse in the market. This meant that some initial adopters stopped growing *Mucuna* (Honlonkou *et al.*, 1999), emphasizing the influence that market conditions can have on technology adoption.

Best-bet Options for Crop–Livestock Production in Dry Savannahs

The dry savannahs can be defined as regions where the rainfall is between 400 and 1200 mm per annum (Grandi, 1996). They constitute more than 50% of the total land area of sub-Saharan Africa, with a significant proportion located in West and Central Africa. With more than 40% of the rapidly expanding human population (FAO, 2000) and some 50% of the total ruminant population of West/Central Africa (Winrock, 1992), the demand for agricultural production and the resultant pressure on the resource base are high. The long, harsh dry season, poor, sandy soils and a lack of readily available and cost-effective agricultural inputs mean, however, that there is a considerable challenge for all those involved in agriculture in the region, from the small-scale farmers, who are by far in the majority, through to researchers and extension workers. With sorghum and millet dominating, cropping is generally cereal-based, with intercropping on at least 90% of the fields, usually with grain legumes, cowpea or groundnut being the most common. Ruminant livestock form an integral and essential part of the farming system, feeding on crop residues and providing meat, milk, manure, transport and a source of cash.

The farmers' agricultural practices, involving a mixture of crops, as well as livestock, including soil management, have prevailed over many decades, but the farmers are now under increasing pressure to produce more, without detriment to the natural resource base. This presents both an opportunity and a challenge to research seeking to respond to the pressure to intensify production. Mixed crop–livestock systems are known to be one of the best ways of addressing increases in productivity without jeopardizing the natural resource base (de Haan *et al.*, 1997). The challenge is to bring together the right mix of expertise to contribute to the solution. In the late 1990s, a group of scientists in the dry savannah region recognized that working in isolation on components of the system led to approaches vastly different from what was actually happening on farmers' fields. They began to take a more integrated approach to addressing agricultural problems. The result was a project that has two basic tenets: to work in close partnership with farmers and to combine appropriate complementary interventions from a range of institutions into 'best-bet options'. This combination includes plant breeding (especially cowpea), socio-economic and natural resource management from the International Institute of Tropical Agriculture (IITA), livestock production and management from the International Livestock Research Institute (ILRI) (formerly the International Livestock Centre for Africa (ILCA)), crop and cropping systems improvement from the International Crops Research Institute for the Semi-Arid Tropics (ICRISAT), integrated nutrient management from the International Fertilizer Development Centre (IFDC), village-level resource management from the Centre for Overseas Research and Development (CORD), University of Durham, and component research and links with development and extension from the national agricultural research and extension stations (NARES). Further details have been presented elsewhere (Tarawali *et al.*, 2000a,b).

The trial began at Bichi, a village near Kano in northern Nigeria in 1998, where 11 farmers participated in the evaluation of two best-bet options, identical except for the presence of inputs (minimum fertilizer to the sorghum and pesticide to the cowpea) in one (BB+) and not in the other (BB). These were compared with the farmers' traditionally managed fields of sorghum–cowpea intercrop (L). The best-bet options consisted of a combination of the best sorghum variety (ICSV 400), the best dual-purpose cowpea variety (IT90K-277-2), planted in a 2 : 4

sorghum : cowpea row arrangement with close spacing inter-row (75 cm) and intra-row (25 cm). This contrasts with the farmers' practice, where sorghum and cowpea are planted in a 1 : 1 row arrangement, with spaces up to 1.5 m between plants. In BB+, the inputs were applied selectively to the sorghum or the cowpea, as appropriate, and the row geometry used enabled this to be done with maximum efficiency. Double-cropping of cowpea was also recommended (planting a second crop of the same cowpea variety on the same four rows after harvest-ing the grain and fodder from the first). This practice was known to give more fodder and some grain from the second crop (Singh and Tarawali, 1997). The farmers managed the plots themselves, with minimum technical guidance from research and extension staff, but close interaction with these co-developers throughout the year ensured that their opinions and advice were continually contributing towards the interventions. Grain and fodder were sampled at the time the adopters harvested in order to estimate yields. The fodder was stored and, based on the amounts available, fed to an appropriate number of the participants' own small ruminants during the dry season. Manure and feed refusals collected during the feed-ing period were returned to the same fields, where the same crop combinations were planted the following season. Nutrients and economic components were also monitored during the trials, to gain a complete picture of the implications at every level.

In 1999, a further 13 adopters joined in at Bichi and 23 adopters began the trial for the first time at Unguwan Zangi, a village near Zaria in northern Nigeria. During the 2-year period, estimated grain and fodder yields, especially of the cowpea, in the best-bet options were more than double those from the traditional fields. However, the adopters have contributed considerably to developing the best-bet options. At Bichi, in 1998, the farmers preferred the option that included inputs of fertilizer and pesticide, even though these were not provided free, but were only reimbursed in grain equiva-lent at harvest time. This option gave the highest yields, although the option without

inputs still gave better yields than the local fields. In response to the adopters' prefer-ence for an option with inputs, the BB option was refined for 1999. This was done in relation to another first-year observation; the local sorghum in the traditional plots seemed to have yields of grain and fodder very similar to those of the improved sorghum in the BB plots. Hence, in 1999 BB+ and BB both included inputs, but BB had local rather than improved sorghum. From a researcher's point of view, it would have been preferable to keep a 'treatment' without inputs, to answer those frequent questions about areas where inputs are not available or out of the reach of participants' financial means. The close partnership with the co-developers meant that the farmers' views were important, as well as noting that these were not researcher trial plots of a few square metres, but substantial parts of farmers' fields (in some cases up to 0.5 ha) and therefore essential for the livelihood of their families.

Feeding livestock and collecting manure in the other phase of this study pre-sented some logistical challenges, and it was implemented only as a result of the farmers playing a significant role as co-developers of this aspect. The initial intention was to tether animals on the treatment plots early in the dry season so that they would benefit from any remaining stubble and weeds. However, the adopters pointed out that there would be no way to prevent other animals also grazing the same plots, as livestock roam freely once the crops are harvested. It was therefore decided to follow the farmers' usual practice and allow the animals to graze freely until all *in situ* residues were used up and only then to begin feeding them with the harvested residues within the compound. This meant that the feeding period was from about the middle to the end of the dry sea-son. The farmers again played the role of co-developers and provided suggestions as to how it would be possible to keep separate the animals, feed refusals and manure from each of the three treatments. They readily came up with the idea of providing separate compartments within their compounds. Furthermore, in an effort to ensure that the refusals and manure from each treatment were kept separate, the researchers' plan had been to collect these materials monthly and store them. Again, the adopters knew better and pointed out that the manure quality would be considerably reduced if this happened, because the urine and effects of trampling on the material would be lost. Powell and Williams (1993) reported that crop yields were up to 52% greater when urine as well as manure was captured. Hence, all the refusals and manure were retained in the respective compartments until it was time to take them to the fields at the beginning of the next growing season. Almost all the participants provided sheep and not goats for the crop-residue feeding. When asked the reason, they replied that the manure was better from sheep. In this example, there are again many more implications for the farm households than the productivity of the crops and livestock, and research is ongoing to better understand these. For example, in many households, the small ruminants are managed by women and more fodder has implications for the way these animals are managed. Similarly, a greater yield of cowpea grain is likely to mean that more income will be derived, either from the sale of grain directly or from small-scale snack processing, the latter being in the hands of the womenfolk.

Stylosanthes as a Feed and Fallow Crop

The ILRI Subhumid Programme, based in Kaduna, Nigeria, adopted a farming-systems research approach and identified poor nutrition in the dry season as the main constraint to cattle production in the sub-humid zone of Nigeria. During this period, the natural pasture is not only scarce but also poor in quality, with a crude protein (CP) content of about 3%, below the critical 7% CP level required in the ruminant diet (Crowder and Chheda, 1982). Cattle may lose up to 15–20% of their body weight during this season (Otchere, 1986). Also,

milk yields are low, calf mortality is high and many cows are unable to conceive because of nutritional anoestrus.

A contributory factor to the low quality and productivity of the herbage in the sub-humid zone is the poor nature of savannah soils. Any attempt to promote livestock production in the subhumid zone should, therefore, include aspects of soil-fertility maintenance in addition to improving the nutritional value of the pasture. For instance, the poor carbon and nitrogen content of these soils could be improved by the incorporation of forage legumes into the cropping and fallow system. The use of agroindustrial by-products, such as cottonseed cake, groundnut meal, urea and molasses, can improve the productivity of lactating and pregnant cows. However, supplies of these feeds are not readily available and prices are escalating. In view of these ecological and financial constraints, researchers considered a sustainable enterprise such as planted forage legumes (fodder banks) to be a more appropriate long-term option.

Based on the strategies outlined above, the initial research concentrated on the introduction of *Stylosanthes* into cereals so as to increase the nutritive value of the succeeding crops through undersowing (Mohamed-Saleem, 1985). This on-farm research exercise involved planting *Stylosanthes* into crops owned and already established by agropastoralists. For one of the participants, *Stylosanthes* was introduced into the cereal too early and the results were not very encouraging, as there was excessive competition between the cereal and the legume. However, this unlucky but enterprising collaborator, instead of getting angry with the research team, fenced the area affected on his farm and allowed his animals access to the nutritious fodder during the peak of the dry season. This worked out well. Although the agropastoralist did not call it a fodder bank, the research and development team adapted this farmer's innovation into the fodder-bank concept, which was later developed and disseminated.

The fodder-bank concept

A fodder bank is a concentrated unit of forage legumes established and managed by pastoralists near their homesteads for dry-season supplementation of selected animals (Mohamed-Saleem and Suleiman, 1986; Otsyina *et al.*, 1987). These legumes can fix soil nitrogen, and the protein content can stay above 8% for a greater part of the dry season. To date, *Stylosanthes guianensis* cv. Cook and *Stylosanthes hamata* cv. Verano have been the two main species recommended, although a wider range of species is now being considered.

The guidelines for the establishment, management and utilization of fodder banks are as follows:

1. Depending on the availability of land and number of animals, select an area (normally about 4 ha) close to the homestead (size varies according to animals and land available).
2. Prepare the seed-bed by confining a herd overnight in the area for several weeks.
3. Broadcast scarified seeds mixed with 150 kg ha^{-1} of single superphosphate fertilizer.
4. Control fast-growing grasses by early-season grazing.
5. Allow the forage to bulk up by stopping grazing until the dry season.
6. Ration the fodder bank by selecting the appropriate type and number of animals and limiting grazing to 2.5 h day^{-1}.
7. Ensure sufficient seed drop and adequate stubble for regeneration of the *Stylosanthes* in the following season.

Animals with access to fodder banks in the dry season produced more milk, lost less weight and had shorter calving intervals and better calf survival (Tarawali *et al.*, 1999). In experiments with legume–legume mixtures, including combinations of *Aeschynomene histrix*, *Stylosanthes*, *Centrosema pascuorum*, *Centrosema pubescens* and *Centrosema macrocarpum*, the differences were dramatic between heifers grazing the legume mixtures and those on unimproved pasture. For instance, in the dry season,

heifers gained an average of 140 g per animal day^{-1} on a mixture of *C. pascuorum*, *A. histrix* and *C. pubescens*, and lost 58 g per animal day^{-1} on the unimproved pasture (Tarawali *et al.*, 1999).

Tarawali (1991) reported that maize yields after *Stylosanthes* were greater than yields on natural pasture at three levels of applied N in trials conducted at four locations in central Nigeria. Without fertilizer N addition, the average grain yields were 1700 kg ha^{-1} in the leguminous area and 800 kg ha^{-1} in the natural pasture.

The benefits of fodder banks have been acknowledged by adopters. An impact-assessment study conducted by ILRI using data from 15 West African countries identified nearly 27,000 adopters (Fig. 5.2) growing forage legumes involving mainly *Stylosanthes* innovations on some 19,000 ha (Elbasha *et al.*, 1999). The baseline analysis indicated that, for an expenditure of research resources of just over US$7 million, the total benefits to society that accrued up to 1997 amounted to US$16.5 million, with an internal rate of return of some 38% (Elbasha *et al.*, 1999). A contributory factor to the wide adoption of the fodder-bank technology, especially in Nigeria, was the promotional effort provided by extension systems, such as the National Livestock Projects Department, a World Bank-sponsored project, in addition to the robustness of the technology.

Among the factors known to influence farmers' interest in adopting the fodder banks, fencing and land tenure often came to the fore, but the farmer–researcher partnerships also made possible some interesting approaches to solving these aspects, initially perceived as constraints.

The fencing factor

Fencing continued to be a necessity in the fodder-bank system, as adopters who did not fence their *Stylosanthes* usually suffered losses in terms of both herbage productivity and subsequent crop yields, as a result of intrusion from stray animals

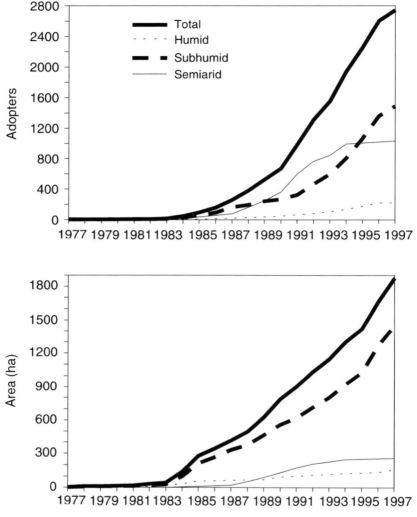

Fig. 5.2.   Adoption of *Stylosanthes* in the subhumid zone of Nigeria (from Elbasha *et al.*, 1999).

and/or bush fires. Fencing a *Stylosanthes* fodder bank constituted 40–70% of the cost of establishing a 4 ha leguminous pasture. This is exorbitant, especially when the imported metal posts and barbed wire initially recommended are used. To overcome this constraint, adopters used their ingenuity to find cheap alternative fencing materials, which varied from the use of old metal sheets to live fences with species such as *Newbouldia*, *Ficus*, *Gmelina*, *Euphorbia*, citrus and cashew. Except for the additional labour required, the use of live fencing

seems to be appropriate; this practice is already part of the culture of the West African farmers.

Another classic example of how farmers have played a co-developer role in devising strategies for overcoming the fencing problem was in Mali, where villagers are penalized by local law and custom if their animals damage crops. In this context, a farmer in southern Mali planted 2 ha of *Stylosanthes* in the middle of his cotton field and this prevented animals from entering the area. At the end of the growing season, the pasture was available for his traction animals.

Land tenure and the mini-fodder bank

Where land rights are insecure, farmers are reluctant to make long-term commitments to land development. In many places in Nigeria, where the fodder-bank concept was originally developed, and in other countries in West Africa, cattle owners, who are the primary beneficiaries of fodder banks, have settled only recently, usually on marginal fields, and they do not have land rights. The land belongs to crop farmers, who have no interest in cattle production (although they keep small ruminants) and are sometimes unwilling to give their unused fallow land to pastoralists for pasture development. This land-tenure constraint initially affected the adoption of fodder banks. This limitation was partly resolved by addressing issues important to these crop farmers and demonstrating the beneficial effects of fodder banks for both livestock and crop production.

Despite years of contact with fodder banks (in the fields of their cattle-owning neighbours), demonstration trials enabled the farmers to see for the first time that legume-based pastures could benefit their crops as well as livestock. As a result, farmers asked for *Stylosanthes* pastures to improve soil fertility in their fallows and, since they do not own cattle, requested that the legume be used for providing feed for their small ruminants, with an average herd size of three to five animals. These flocks are usually owned and managed by the women and children in the household, and there may be social implications of the technology that need to be understood through working closely with the participants. Furthermore, through interaction with the farm families, it became apparent that for small ruminants the feed constraint was in the wet season. This is a period when the traditional method of tethering goats in fallow areas and in houses to prevent them from damaging crops creates feed stress and restricts movement, leading to undernutrition and weight losses

in breeding females, with consequent low reproductive performance (ILCA, 1991). Farmers and researchers worked together in order to determine that, for the typical flock size, the minimum area of *Stylosanthes* pastures would be 0.1–0.3 ha (mini-fodder bank), which could adequately supplement the feed of three to five small ruminants for both the wet and the dry season. A comparison of the wet-season live-weight changes of West African Dwarf goats showed reduced weight loss ($P < 0.05$) and improved kid survival ($P < 0.05$) through legume supplementation (Ikwuegbu and Ofodile, 1992).

There are several features of the mini-fodder-bank concept that are attractive to adopters and have been derived as a result of the joint development of the idea by farmers and researchers. The plots are small and compatible with average flock sizes. The package is simple and can also be adopted by women, providing them with greater security of land use. Smallholder crop farmers are familiar with the use of local materials for fencing and readily applied this to their *Stylosanthes* pastures. Finally, positive effects of the pastures on the subsequent cereal yields are also accrued. In the two case-study areas in Nigeria, where the mini-fodder bank originated, almost 100 participants joined in during the first 3 years; currently, the concept is expanding through partnership with farmers and NGOs in central Nigeria.

Green-manure Cover-crop Systems for Smallholder Farmers in Igalaland, Nigeria

Igalaland is situated on the eastern side of the Rivers Niger and Benue in central Nigeria. The region is within the Guinea savannah but is better described as a transition from rain forest to savannah with oil-palms as the predominant trees. Rainfall varies between 1400 and 1600 per annum and follows a bimodal pattern, with an early growing season (April to July) and a late season (July to September). The two seasons are divided by a few days of no rain.

Agriculture supports 95% of the estimated 1.2 million people. There is limited
industry in the area and with Nigeria's
rapidly growing population, agriculture
makes a major contribution to livelihoods
(McNamara and Morse, 1992). Farmers in
the region keep livestock, such as goats,
sheep, chickens and ducks, as a secondary
activity to cropping. The cropping systems
are based on intercropping and include
cassava, cowpea, melon, yam, maize and
rice, together with a bush-fallow system that
varies from zero to 10 years or more. Such
indigenous systems are now under pressure
to increase productivity. However, cropping
activities have expanded to all appropriate
land and this has resulted in shortened
fallow periods and a consequent decline
in soil fertility and therefore in crop yields.
The majority of farmers cannot compensate
for this shortcoming by using inorganic
fertilizers, as these are not affordable. In this
region, the two main problems highlighted
by farmers were weeds, especially spear-
grass (*I. cylindrica*), and poor soil fertility.

The Diocesan Development Services
(DDS), a religious NGO, has been working
directly with over 50,000 farmers for the
past 30 years. During this period, the main
activities have involved farmer evaluation of
improved crop varieties (mainly developed
by IITA), with the best being multiplied by
the NGO to distribute materials to Igala farmers. This extension programme has proved to
be successful, as many farmers have adopted
high-yielding crop varieties, such as cowpea, cassava, yam and maize. However, with
the scarcity of fertilizers and high levels of
weed infestation, the adopters recognized
that the potential benefits of these interventions are not maximized but need to
be complemented by sustainable soil- and
weed-management practices. In response to
this need, the use of green-manure cover
crops has been considered recently in order
to improve soil fertility, control weeds and
erosion, improve livestock feed, etc. In line
with this objective, DDS initiated a cover-
crops project in 1997 which included:

- introduction of six cover crops on 17
  farmers' fields;

- simultaneous screening of 29 cover-
  crop species or combinations at the
  DDS seed-multiplication farm.

Results from both the on-station and
on-farm trials (Tarawali and McNamara,
1998) showed that the highest performance
in Igalaland in terms of biomass were species
of *Stylosanthes*, *Aeschynomene*, *Mucuna*
and *Chaemacrista*, with biomass yields
varying between 3863 and 5838 kg ha^{-1}
(on station) and 2234 and 2467 kg ha^{-1} (on
farm). Among the lowest yielders were the
Centrosema spp., *Vigna* and *Lablab*, with
biomass ranging between 1842 and
2373 kg ha^{-1} on station and 630 and
823 kg ha^{-1} on farm.

In evaluating the on-farm trial (Fig. 5.3)
it was shown that some species yielded
more biomass than material less adapted to
the area. Participants who received these
species had higher yields. For instance, the
best cover crop in 1997 was *S. guianensis*,
with biomass yields of 3196–3353 kg ha^{-1},
grown by Paul Drisu and Drisu Sule (I).
Other cover crops with good yields were
Centrosema rotundifolia (2528–2783 kg
ha^{-1}) and *A. histrix*, yielding 2368 kg ha^{-1},
grown by Drisu Sule (II), Sam Adah
and Achimi Shaibu. The lowest producers,
Lablab purpureus and the *Centrosema* spp.
(242–948 kg ha^{-1}), were grown by DDS Idah,
Terimu Usman, Isaac Obaka, Friday Ekele,
Daniel Jekeli, Joseph Drisu and John
Ochimana.

These results and the ensuing discussions with these participants showed how
the material provided from a research or
extension institution can influence the outcome and revealed farmers' impressions of a
technology. It would be more appropriate for
farmers and researchers to work together
with the available material to make joint
selections and subsequently to develop the
intervention best suited to the farmers.

Despite the limitations of the approach
used in this instance, the trend in the uptake
of cover crops, although low in numbers, is
still very promising; 17 farmers started in
1997, and this number had risen to about 60
in 1999. This accelerated dissemination was
due to a combination of factors. One is that

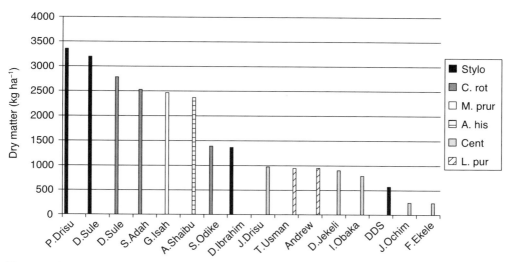

Fig. 5.3. Performance of cover crops in farmers' fields. Stylo, *Stylosanthes* spp.; C. rot, *Centrosema rotundifolia*; M. prur, *Mucuna pruriens*; A. his, *Aeschynomene histrix*; Cent, *Centrosema* spp.; L. pur, *Lablab purpureus*. (From Tarawali and McNamara, 1998.)

the cover-crops technology was targeted to Igalaland, where farmers perceive a real need for such an intervention. Also of great importance was the presence of a genuine grass-roots NGO, supported by qualified consultants, which had gained the confidence of and had a close partnership with the collaborating farmers, developed over a 30-year period.

Discussion

The cases illustrated above provide clear examples of how partnerships involving farmers, researchers and extension agents can later enhance the dissemination and performance of researcher-developed technologies. Many lessons were learned during researcher–extensionist–farmer interactions.

- Most adopters need to realize some short-term benefits. In the examples considered here, these included weed control and seed (or grain) to sell. Policies that enable the provision of low-interest or short-term credit (as in the crop–livestock intervention) can also enhance technology adoption.

- New technologies cannot be developed or introduced in isolation from the wealth of participants' existing knowledge. Options that recognize many related aspects of farming and wider household activities are well appreciated and farmers are ready to come out with suggestions if given the opportunity. Joining traditional wisdom with new research findings may contribute substantially to the progress of African agriculture.

- Early adopter participation in testing interventions enables the adopters to contribute to realistic solutions for innovations which they believe have value, even if constraints are initially perceived. The several options presented by adopters in seeking cheap and affordable fencing materials underscore this effect.

- Farmers and researchers need to work together to identify baskets of suitable options that may be applicable to a range of social, economic and cultural circumstances. Providing a single technology, as was done in the on-farm evaluation of selected legumes, can sometimes lead to embarking on the wrong course and has the danger of killing farmers' interest.

- Technologies should be targeted where the demand is high in relation to farmers' perceived constraints. This became clear from the enthusiasm shown by adopters from Igalaland and Benin, who, over the years, have been deprived of good crop yields due to poor soils and weeds.

- Partnership with farmers can ensure that both the redefined intervention and the emphasis for dissemination are appropriate for farmer-to-farmer promulgation. Advantages of this include the clients considering themselves to be among the innovators and hence keen to convince their colleagues. Also, successful farmer-to-farmer diffusion is a good indicator that the technology is robust and sustainable and has a potential impact.

- Intervention promotion involved national and international research centres, NARES (such as INRAB) and both local (DDS, Idah) and international (SG2000) NGOs. These partnerships are pivotal for the effective dissemination of improved technologies to farmers. Inclusion of all stakeholders as partners in the development process ensures that the research-to-development continuum is maintained and can also promote continuation of new technologies by national counterparts when project cycles finish.

- One of the best strategies in the collaborative approach with farmers is to use the 'follow the technology approach', in which the technology is developed with a few self-selected farmers (say 20 in the first instance) to produce the final product. It is hoped that this will then propagate throughout the community.

- Cover crops are attracting considerable attention from the West African smallholders because of the various benefits highlighted in the examples above. Nevertheless, their adoption as a profitable agricultural practice faces many constraints. These include land scarcity, insecure land-tenure systems, toxicity of *Mucuna* grain for human and animal consumption due to substantial quantities of 3-(3,4-dihydroxyphenyl) alanine (L-DOPA), destruction of cover crops by fire, especially in the dry season, and unavailability of locally produced seed. Other constraints include a lack of credit facilities for the purchase of inputs, distorted pricing policies that do not guarantee economic returns to farmers, and the poor condition of the infrastructure, which disrupts both delivery of inputs and removal of outputs. On farm, there can be severe competition between the aggressive legumes and some of the dwarf crops (cowpea) and/or with long-duration food crops, fear of disease incidence and acute shortage of labour and agricultural mechanization, making it difficult to include additional enterprises, such as cover crops, in subsistence farming (Vissoh *et al.*, 1998). Despite such limitations, these examples show that promising cover-crop interventions are best developed in partnerships between farmers, researchers and extension agents.

6 Sustainable Management of Private and Communal Lands in Northern Ethiopia

Berhanu Gebremedhin[1] and Scott M. Swinton[2]

[1]International Livestock Research Institute, PO Box 5689, Addis Ababa, Ethiopia;
[2]Department of Agricultural Economics, Michigan State University, East Lansing,
MI 48824-1039, USA

Land degradation is one of the fundamental problems confronting sub-Saharan Africa in its efforts to increase agricultural production, reduce poverty and alleviate food insecurity. With the land frontier shrinking, future increases in agricultural production will have to come from yield increases rather than area expansion. Yet the production potential of the land resource is declining, due to soil erosion, nutrient depletion, soil-moisture stress, deforestation and overgrazing. The continent confronts the challenge of how to increase current agricultural production while maintaining the future productive capacity of the natural resource base.

Land degradation is especially severe in the East African highlands. In Ethiopia, it stands out as one of the major contributors to the slow growth rate of agricultural production. Land degradation has been particularly damaging in the highlands – those areas over 1500 m above sea level – which account for more than 90% of the cultivated land, 75% of the livestock and more than 80% of Ethiopia's farming population.

Public intervention to halt land degradation in Ethiopia started in the early 1970s (Campbell, 1991). However, a top-down approach, the inadequate scientific and technical base of the recommended practices and a lack of involvement of local people rendered the efforts ineffective. These experiences emphasize the importance of understanding how and why individual farms adopt soil-conservation measures if they are to be diffused successfully.

Apart from private, household-level conservation measures, some natural resource conservation is most usefully done at the community level. Hence, for communal hillsides, grazing lands and wood lots, the community-level motives and impediments to resource conservation are important.

In this chapter, we synthesize results of recent research conducted in the northern Ethiopian region of Tigray, which has experienced severe land degradation. We examine the technological and institutional factors determining the adoption of natural resource conservation at both the household and the community levels. Using 1995/96 data from 250 Tigray farm-household interviews, we first examine private land management, focusing on the following questions:

1. What factors determine farmer perceptions of the severity and yield impact of soil erosion?
2. Is soil conservation profitable? Under what conditions?
3. What determines farmers' willingness to invest in soil conservation?

Using 1998/99 data from a survey of 100 Tigray villages, we next examine the management of communal lands (grazing lands and wood lots), focusing on two additional questions:

4. What makes communities engage in collective natural resource management (NRM) activities?
5. What determines the effectiveness of collective NRM?

The Setting

The study area, Tigray, is the northernmost region of Ethiopia, located in the semi-arid Sudano-Sahelian zone (Warren and Khogali, 1992). It covers an approximate area of 80,000 km², with a population of more than 3.3 million and an estimated annual population growth rate of 3%.

The region lies on a mountainous plateau with a tropical semi-arid climate, characterized by erratic and unreliable rainfall. The average annual rainfall in the region is 600 mm. Most of the precipitation falls intensively within the 3 months of June–August, contributing to soil erosion, and is characterized by high spatial and temporal variability. Soils are shallow and infertile and frequent outbreaks of crop pests and diseases are a major problem of agricultural production.

Agriculture is the mainstay of the economy of Tigray. More than 85% of the regional population depends on mixed crop–livestock subsistence agriculture, with oxen power supplying the only draught power for ploughing. Most of the region either produces just enough for subsistence during good-rainfall years or faces chronic food deficit.

As in many semi-arid settings, livestock are a key element of farming systems.

According to the 1998 livestock census, Tigray has about 3.04 million cattle, 0.94 million sheep, 1.47 million goats, 0.41 million equines and 1300 camels (Bureau of Agriculture and Natural Resource Development (BoANRD), 1999). Communal grazing lands of about 3.2 million ha have been important sources of livestock forage in Tigray. Recently, however, the free and unrestricted access has resulted in severe degradation of the grazing lands.

Deforestation is very severe in Tigray. Cutting trees for fuel, timber, and agricultural implements and clearing forests to expand agricultural lands have exhausted the forest cover of the area. Currently, only about 1.6% of the region is covered with forests or wood lots (BoANRD, 1995).

Since 1991, the Tigray region has embarked on a regional development strategy for natural resource conservation based upon popular participation. The strategy focuses on soil and water conservation, the development of irrigation and environmental rehabilitation through area enclosures, reforestation and development of community wood lots, through public, communal and private efforts.

Adoption of Soil-conservation Practices on Private Lands

The existing literature on technology adoption identifies adoption determinants associated with expected profitability, farm characteristics, household characteristics and technology characteristics (Feder *et al.*, 1985; Feder and Umali, 1993), as well as awareness and perception of the soil-erosion problem and the practices that can treat it (Ervin and Ervin, 1982).

Our research examined the determinants of erosion perceptions and the adoption of soil-conservation practices on 250 farms in the rural Tigray region in Ethiopia during 1995/96. Purposive selection of villages based on topography, followed by random sampling of households, ensured representation of the diverse agroecological conditions. In the following subsections, we present results from analyses of determinants of perceived soil erosion, profitability of investments in stone terraces and determinants of soil-conservation adoption apart from perceptions.

The perception of soil erosion and its yield impact

Soil erosion is an insidious and slow process. Yet farmers need to perceive the severity of soil erosion and the associated yield loss before they can consider investing in its prevention. In Tigray, where soil erosion is generally severe, understanding of the level and determinants of farmer perceptions of soil erosion and its impact is important for policy purposes. Prior research in the USA (Ervin and Ervin, 1982; Bultena and Hoiberg, 1983) and Ethiopia (Shiferaw and Holden, 1998) has highlighted the importance of perceptions for enhancing the adoption of soil conservation technologies.

Farmer perceptions of the severity of soil erosion on each plot were solicited in four subjectively assessed categories (1 = severe, 2 = moderate, 3 = slight, 4 = none). Farmers were also asked to estimate the likely yield impact of erosion that would occur on their fields in a normal year without any soil-conservation measures, using five possible levels (1 = no yield reduction, 2 = 20% reduction, 3 = 25% reduction, 4 = 33% reduction and 5 = 50% reduction). Following Ervin and Ervin (1982), we specified explanatory variables in three categories: physical factors, socio-institutional factors and demographic characteristics. Physical factors include those natural physical elements that make soil erosion more likely, such as rainfall, soil texture and topography (Yoder and Lown, 1995). The socio-institutional variables include land tenure and the existence of related conservation projects (for demonstration or substitution effect). The demographic characteristics include human capital as well as other conditioning factors, such as age and gender.

At least moderate erosion was perceived on 58% of the 565 plots surveyed (Gebremedhin, 1998: 168–169). Statistical analyses of the determinants of these perceptions

used as dependent variables both the four levels of erosion (ordered probit) and a binary variable distinguishing between some erosion perceived and none perceived (probit). A separate ordered-probit model examined determinants of the yield-loss estimates due to erosion. Plot-level physical characteristics that aggravate erosion are important determinants of farmer perceptions of soil loss and its yield impact (Table 6.1). Younger farmers tended to recognize erosion better, perhaps due to better education or a longer planning horizon. Experience with prior public campaigns that constructed bunds or terraces on private lands detracted from perceived erosion. Plots operated longer and those close to the homestead were perceived to have worse erosion, suggesting that more frequent observation and more cultivation activity add to awareness. Farmers with more extension-service contacts tended to perceive less erosion and yield loss.

Profitability of soil conservation: the case of stone terraces

Given that soil erosion and its yield impact are recognized as problems among most Tigrayan farmers interviewed, the next question is whether investment in conservation practices is likely to be profitable. Prior research in Ethiopia and elsewhere has found that profitability is central to the farm-level adoption and maintenance of soil-conservation practices. Failure to adopt or maintain conservation practices occurs because: (i) socially desirable projects are not privately profitable (Lutz and Pagiola, 1994); or (ii) privately profitable projects fail to offer immediate benefits or generate a positive cash flow (Gebremichael, 1992).

In order to evaluate the return on soil-conservation investments in Tigray, we conducted a capital budgeting analysis of an investment in stone-terrace construction. The results were driven by the changes in wheat and fava-bean grain and hay yields as observed in on-farm research plots. The plots were divided equally between wheat and fava bean on 70 terraced plots, as

compared with 70 unterraced plots planted
to the same two crops. In order to capture
accurately the effect of terracing, each terraced plot included one 8 m² quadrat just
above the terrace (in the soil-accumulation
zone) and one just below the terrace (in the
soil-loss zone). Likewise, each unterraced
plot had one quadrat (designated the control
treatment). Crop yields were measured in all
quadrats and converted to quintals per
hectare (q ha⁻¹). Raw yields were also
adjusted for planted area lost to terracing,
assuming 5 and 15% levels of loss to planted
area. Based on the 5% planting-area loss
scenario, inter-treatment yield differences

were regressed on two farm-management
variables (tillage frequency and weeding
frequency) in order to correct for management differences.

The corrected yield gains between treatments were incorporated into budgets based
on constant 1995/96 farm-gate crop prices at
harvest, input costs (including family labour
at rural daily wage of 6 Birr (= US$1)) and
terrace investment and maintenance costs.
Finally, these partial budget data were incorporated into capital budgets to calculate the
net present value (NPV) of investments in
stone terraces (Gebremedhin *et al.*, 1999).
Given that the terraces were observed *in situ*

Table 6.1. Statistically significant* determinants of farmer perceptions of soil erosion and its yield impact (signs in parentheses) (from Gebremedhin, 1998: 168–169, 172–173).

Variable	Erosion severity		Yield impact (ordered probit)
	Ordered probit	Probit	
Village physical factors	Location in rainier upper highland[†] (−) Hilly topography (+) Dung used as major fuel source (+)	Location in rainier upper highland (−) Hilly topography (+) Dung used as major fuel source (+)	Dung used as major fuel source (+)
Plot physical factors	Loam soil[‡] (−) Distance from homestead (−) Plot slope degree (+) Convex slope[§] (+) Concave slope (+) Age (+) Area (+)	Loam soil (−) Distance from homestead (−) Plot slope degree (+) Convex slope (+) Concave slope (+) Area (+)	Distance from homestead (−) Plot slope degree (+) Convex slope (+)
Socio-institutional factors	Extension contact (−)	Extension contact (−)	Extension contact (−) Beneficiary of public campaign for conservation (−)
Demographic characteristics	Age of HH head (−)	Age of HH head (−)	Age of HH head (−)
Chi-square	128.3	89. 8	72.4
Prob. > chi-square	0.000	0.000	0.000
Pseudo R-square	0.084	0.135	0.047
Predicted probability at mean	n/a	0.583	n/a
N	565	565	487

*Significant at least at 10% level.
[†]Upper highland is defined as a location at or above 2500 m above sea level.
[‡]Soil dummies were compared against clay soil.
[§]Slope dummies were compared against rectilinear slope.
HH, household; n/a, not applicable.

and already stabilized, the capital budgets assumed (conservatively) that terraced fields would not show a yield advantage until the fourth year after terracing, at which time they would obtain the full yield advantage from terracing. Due to differences between government agricultural-loan interest rates of 15% versus prevailing informal interest rates around 50% (Shiferaw and Holden, 2000), both rates were applied in separate NPV scenarios.

The results of the on-farm experiments reveal dramatic differences between yields in the soil-accumulation zone and both the soil-loss zone and the control plots, as shown in Table 6.2. The yield advantage of the soil-accumulation zone is consistent across both wheat and fava-bean crops and also across both grain and straw yields. Moreover, the coefficient of variation shows that yield from the soil-accumulation zone is more stable than that from the unterraced and control zones.

The capital budgeting analysis showed that returns to investments in stone terraces are highly sensitive to the discount rate applied. As illustrated in Fig. 6.1, the payback period at a 15% discount rate was 5 years, versus 14 years at a 50% discount

rate. Over the 30-year time horizon projected, the NPV was 3907 Birr (US$650) at a 15% discount rate versus 12 Birr (US$2) at a 50% discount rate, indicating that investment in stone terraces results in an internal rate of return (IRR) of 50%.

Determinants of investment in soil conservation: the value of secure land tenure

Our investment-profitability analysis was predicated upon the assumption of secure land tenure. Yet the 5–14-year range of payback periods highlights the minimum period of land tenure over which land must be held to make terracing investments financially worthwhile. Having established the potential profitability of investments in terraces via a capital budgeting analysis, it was fitting to analyse determinants of soil-conservation adoption in Tigray in the broader context of farm resources and their physical and institutional setting.

The conceptual model underlying the soil-conservation investment analysis focuses on six classes of investment determinants that have proved influential in rural

Table 6.2. Mean wheat and fava-bean grain and straw yield in soil-accumulation zone, soil-loss zone and control zone, 100 kg ha^{-1} units (standard deviations in parentheses)* (reprinted from Gebremedhin et al., 1999: 570–571).

Output	Treatment	Unadjusted	Adjusted for 5% area loss to terraces
Wheat			
Grain	Accum. zone	16.1 (6.09)[a]	15.3 (5.79)[a]
	Loss zone	8.5 (3.35)[b]	8.1 (3.18)[b]
	Control	6.6 (4.08)[b]	6.6 (4.08)[b]
Straw	Accum. zone	27.9 (9.84)[a]	26.5 (9.35)[a]
	Loss zone	14.5 (5.42)[b]	13.8 (5.15)[b]
	Control	12.0 (6.05)[b]	12.0 (6.05)[b]
Fava bean			
Grain	Accum. zone	8.0 (3.13)[a]	7.6 (2.97)[a]
	Loss zone	5.5 (2.37)[b]	5.2 (2.25)[b]
	Control	5.4 (4.19)[b]	5.4 (4.19)[b]
Straw	Accum. zone	11.8 (4.07)[a]	11.2 (3.87)[a]
	Loss zone	7.5 (3.22)[b]	7.1 (3.05)[b]
	Control	6.4 (5.09)[b]	6.4 (5.09)[b]

*Figures followed by different letters were significantly different within each crop and product at the 5% level using the Bonferroni multiple range test (Watson et al., 1990).

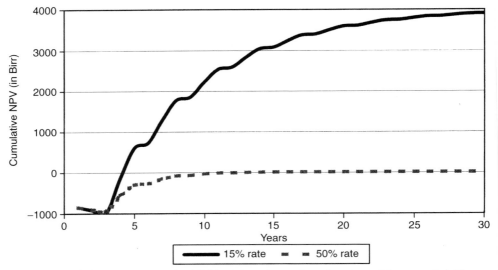

Fig. 6.1. Cumulative net present value (NPV) of stone-terrace investment over 30-year period, Tigray, Ethiopia (in Ethiopian Birr as of 1995/96 (US$1 = 6 Birr)): discounted at 15% and 50 rates (from Gebremedhin *et al.*, 1999: 573, Table 6).

settings of the developing world (Feder *et al.*, 1985; Christensen, 1989). Those determinants include: (i) market access (as a proxy for prices); (ii) physical factors (as a proxy for the technology set); (iii) capacity to invest; (iv) land-tenure security; (v) other socio-institutional factors (including community pressure and government services); and (vi) household demographic characteristics (including human capital).

Both our conceptual and our empirical models distinguish between those factors that trigger the decision to invest and those factors that determine the degree (intensity) of investment, based on a double-hurdle model linking the probability of adoption (as a probit regression) and, where terraces were adopted, the degree of adoption (density in metres of terrace per hectare) as a truncated regression[1] (Cragg, 1971). This analysis was applied to stone-terrace density, but not to soil bunds, which were present on only 1% of the fields studied. The analyses were applied separately to decisions on adoption of both stone terraces and soil bunds on 638 fields in Tigray in 1995/96.

The results of the analysis (Table 6.3) highlighted the importance of the institutional setting within which Ethiopian

farmers make conservation decisions (Gebremedhin, 1998; Gebremedhin and Swinton, 2000). Land-tenure security was a major determinant of the conservation-technology adoption. Farmers with secure land tenure who (i) expected to bequeath their fields to their children and (ii) lived in villages with no recent land redistribution were both more likely to build stone terraces and less likely to build soil bunds. Those who expected to operate the field in 5 years' time (but presumably not bequeath it to their children) were less likely to build terraces. In contrast, farmers with an immediate time horizon – those who currently operate a field – were more likely to adopt soil bunds.

Other government interventions influenced adoption as well. Public soil-conservation programmes had a substitution effect on fields where they had operated, making subsequent private conservation investments less likely. However, the existence of food-for-work programmes in the village increased the adoption of stone terraces, while decreasing the adoption of soil bunds, perhaps because of either a demonstration effect or a liquidity effect. Interestingly, the number of extension contacts did not affect adoption of either stone

Table 6.3. Statistically significant* determinants of adoption and intensity of use of conservation practices (signs in parentheses) (from Gebremedhin, 1998: 194–195, 198–199).

	Stone terrace		
Variable	Adoption (probit regression)	Intensity of use (truncated regression)	Soil bund adoption (probit regression)
Financial incentives to invest	No significant variables	Distance to market (+) Distance to road (+)	Distance to market (–)
Physical factors	Location in rainier upper highland[†] (–) Hilly topography (+) Distance from homestead (–) Loam soil[‡] (–) Plot on lower slope[§] (+) Slope (+) Slope squared (–) Concave slope[ǁ] (+) Plot area (+)	Location in rainier upper highland (+) Silt soil (+) Slope (+) Slope squared (–) Plot area (–) Plot age (+)	Loam soil (+) Number of plots cultivated (+) Plot on upper slope (–) Plot on middle slope (–) Slope (+) Slope squared (–) Mixed slope (–) Plot age (+)
Capacity to invest	Number of working-age household members (+)	No significant variables	No significant variables
Land tenure security perception	Up to 5 years (–) Bequeath land to children (+)	No significant variables	Bequeath land to children (–)
Socio-institutional factors	Beneficiary of public campaign for conservation (–) Food-for-work project available (+) Years since last land redistribution in village (+)	No significant variables	Owner operator (+) Beneficiary of public campaign for conservation (–) Food-for-work project available (–)
Demographic characteristics	No significant variables	Literate HH head (–)	Age of HH head (–) Literate HH head (–)
Chi-square	141.89	n/a	101.2
Prob. > chi-square	0.000	n/a	0.000
Pseudo R-squared	0.28	n/a	0.27
Predicted prob. at mean	0.219	n/a	0.013
N	638	139	638

*Significant at least at 10% level.
[†]Upper highland is defined as location at or above 2500 m above sea level.
[‡]Soil dummies are compared against clay soil.
[§]Location of plot dummies were compared against location at the flat land part of a catchment.
[ǁ]Slope dummies are compared against rectilinear slope.
HH, household; n/a, not applicable.

terraces or soil bunds, related, perhaps, to its significant negative effect on the perception of soil erosion. A variety of other physical factors also played fairly predictable roles in determining the adoption of conservation investments.

The intensity of adoption did, in fact, depend on different variables from those affecting adoption alone. Market-access factors proved especially relevant, as the density of terraces increased with distance to an all-weather road and to a regional market. This link suggests that off-farm labour opportunities may be fewer in more remote areas, reducing the opportunity cost of terrace construction. The majority of farmers in the study area are likely to be net buyers of food grains, thus rendering the price advantage of proximity to a road less important. As expected, terracing density was less on larger fields (suggesting economies of scale in terrace construction) and greater in the rainier upper highland areas (where erosion pressure is greater) (Gebremedhin, 1998; Gebremedhin and Swinton, 2000).

Inducements for Sustainable Management of Communal Lands

Apart from the direct effect of reducing yields on a given field, water-driven soil erosion on one field triggers further damage down-slope. It can induce gully formation and harm terraces and bunds on lower-slope fields, as well as contributing to the sedimentation of waterways. When impacts beyond one household's fields affect the welfare of others, these economic externalities may mean that private initiatives are inadequate to rectify resource-degradation problems, since external costs are not considered in private decision-making. Due to the rugged and mountainous topography, soil erosion and excessive runoff on uplands of the Tigray region result in significant public externalities.

Private incentives for conservation are also inadequate in common-property resources, where open access can make the rewards for good resource stewardship open to anyone, regardless of effort. Common-property resources, which have been important sources of fuel wood, timber and grazing lands in Tigray, have been severely degraded due to unrestricted access or ineffective use regulations.

In the following subsection, we draw on community-level data to analyse the nature, impact and determinants of collective action for community management of wood lots and grazing lands.

Managing common property resources: wood lots and grazing lands

Community management of common-property resources is increasingly recognized as a viable alternative to privatization, state ownership or environmentally regulated private or communal ownership (Rasmussen and Meinzen-Dick, 1995; Baland and Platteau, 1996). However, devolving to local communities the right to manage natural resources is a necessary but not sufficient condition for successful community-resource management. Sustainable resource management also requires that community rules and regulations be effectively observed (Turner *et al.*, 1994; Swallow and Bromley, 1995). Hence, the identification of factors that favour or retard the development and effectiveness of community institutions for resource management becomes important. In order to investigate the nature, impact and determinants of effectiveness of community wood-lot and grazing-land management in the region, we held group interviews with a stratified random sample of community leaders from 50 *tabias*[2] and 100 villages in Tigray during the 1998/99 cropping season.

How to measure collective action and its effectiveness raises the challenge of identifying measurable indicators suited to each natural resource. For wood-lot management, our indicators of collective action included the amount of collective labour input per hectare invested in managing the wood lot, whether the community paid for a guard to protect the wood lots, whether there were any violations of use restrictions of the wood lot, the number of trees planted per hectare on the wood lot since it was established and the survival rate of trees. The indicators of collective action for grazing-land management included whether the community practices use restrictions on its grazing land,[3] whether the community had established penalties for violations of use restrictions, whether there had been any violations of use restrictions in 1998 and whether those violations were penalized.

The analysis showed that in the highlands of Tigray 88% of *tabias* have wood lots and 89% of villages have restricted-grazing areas (Gebremedhin *et al.*, 2002). While most wood lots (96%) were promoted by external organizations, most restricted grazing lands (78%) were established by local communities, indicating the existence of local initiatives to develop use restrictions of grazing areas by rural communities in Tigray. While the establishment of community wood lots is a recent phenomenon in Tigray, especially since 1991, the establishment of restricted grazing areas has a long tradition in the region.

Most wood lots and all restricted grazing areas are managed at the village level. Hired guards are the dominant means of protection for both wood lots and grazing lands, and communities use cash penalties for violations of use restrictions for both resources. Compared with *tabias*, villages reported more intensive management of wood lots, with fewer problems and more benefits from wood lots. Despite the limited current benefits that communities receive from wood lots due to use restrictions, it was estimated that at one harvest wood lots can contribute more than US$600,000 to *tabia* (community) wealth in timber value. Communities tend to be more likely to enforce penalties when violations of use restrictions are more frequent. The communities perceived that community management of wood lots and grazing lands had resulted in significant regeneration of the resources. They also reported few problems as a result of the use restrictions of wood lots and grazing lands.

These descriptive differences between motives for collective management of wood lots and grazing lands prompted an econometric analysis of their roots. Following the literature on collective action and induced institutional innovation in managing common-property resources, we used population density, access to market, agricultural potential, the presence of external organizations, whether the wood lot was managed at the *tabia* or village level, area of the wood lot or grazing land, and age of the grazing land as determinants of collective action or its effectiveness (Boserup, 1965; Olson, 1965; Hayami and Ruttan, 1985; North, 1990; Rasmussen and Meinzen-Dick, 1995; Baland and Platteau, 1996; Otsuka and Place, 1999; Pender, 1999; Pender and Scherr, 1999).

The econometric analyses point to several key determinants of collective action for wood-lot and grazing-land management (Tables 6.4 and 6.5; Gebremedhin *et al.*, 2002). Intermediate population density (rather than high or low population density) generally favours community management of wood lots. This finding supports an inverted-U-shaped relationship between population density and collective action for resource management. Violations of use restrictions of grazing lands were also low at intermediate population density. Market access undermines collective action in wood-lot management, but favours collective action in grazing-land management. The effect was powerful, undermining not only collective labour input, but also tree-planting density and the survival rate of trees in wood lots. On the other hand, proximity to market appears to have increased the resource value of the grazing lands or returns from their use. Farmers who live closer to towns are more likely to sell dairy products, especially milk, thus perhaps increasing the return from sustainable use of grazing lands.

The presence of external organizations detracted from collective action in wood-lot management but failed to have a significant impact on grazing-land management, suggesting that external organizations displace the local effort of community wood-lot management. Although external organizations play an important role in promoting the establishment of wood lots, their role in managing the resource seems to be substituting or contradicting local efforts and/or preferences. Since most grazing lands were established by local communities themselves, the role of external organizations appears to be insignificant.

Table 6.4. Statistically significant* determinants of collective action and its effectiveness for community wood-lot management (signs in parentheses) (from Gebremedhin *et al.*, 2002).

	Indicators of collective action and its effectiveness				
	Collective labour input (person-days ha^{-1}) (Tobit regression)	Whether community pays for guard (probit regression)	Whether any violations of restrictions occurred (probit regression)	Number of trees planted ha^{-1} (OLS regression)	Survival rate of planted trees (Tobit regression)
Significant determinants	Central zone[†] (–) Eastern zone (–) Western zone Population density (+) Population density squared (–) Distance to district town (+)	Central zone (–) Eastern zone (+) Wood lot promoted by external organization[‡] (–)	Eastern zone (–)	Central zone (–) Population density (–) Population density squared (+) Distance to district town (+)	Central zone (+) Eastern zone (+) Distance to district town (+) Wood lot promoted by external organization (–)

*Significant at least at the 10% level.
[†]The study region is divided into four zones. The southern and western zones are considered relatively high-potential areas. Zonal dummies are compared against the southern zone.
[‡]External organizations are those organizations which are not locally constituted, such as the Bureau of Agriculture, NGOs, etc.

Table 6.5. Statistically significant* determinants of collective action and its effectiveness on grazing-land management (signs in parentheses) (from Gebremedhin *et al.*, 2000: 17).

	Indicators of collective action and its effectiveness			
	Whether village has restricted grazing area (probit regression)	Whether penalties for violations of use restriction were established (probit regression)	Whether violations of use restrictions occurred (probit regression)	Whether violations in 1998 were penalized (probit regression)
Significant determinants	Central zone[†] (–) Eastern zone (–) Western zone (–) Population density (–) Population density squared (+)	Population density (–) Area of restricted grazing land (–) Distance to district town (+)	Central zone (–) Eastern zone (–) Population density (–) Population density squared (+) Distance to district town (+)	Population density (–) Population density squared (+) Area of restricted-grazing area (–)

*Significant at least at 10% level.
[†]The study region is divided into four zones. The southern and western zones are considered relatively high-potential areas. Zonal dummies are compared against the southern zone.

Wood-lot size had no significant impact on wood-lot management, indicating that there are no economies of scale in wood-lot management. However, more extensive grazing lands reduced the need to set and enforce penalties for misuse. Perhaps the detection of violations of use restrictions was difficult in larger areas. Community experience in grazing-land management (as measured by the age of the restricted-grazing areas) did not matter for effective management, suggesting that there is little 'learning effect' in community grazing-land management.

Low-potential areas reduced collective labour input and planting density but increased the survival rate of trees planted in wood lots. Low-potential areas were also less likely to have restricted grazing lands but more likely to observe use restrictions once the grazing areas are established. These results suggest that community resource management tends to be more difficult to establish in low-potential areas but is more likely to be effective once the hurdle of establishment is overcome.

Conclusions and Implications

This synthesis of NRM adoption research in northern Ethiopia offers several lessons that may be extrapolated to other mountainous areas of sub-Saharan Africa. It appears that most farmers who live in a degraded, hilly and rugged environment are well aware that soil erosion is a problem. Most connect it with the physical conditions that aggravate erosion. Farmers are more likely to recognize erosion on plots that they have cultivated longer or which are closer to the homestead, suggesting that stable tenure systems may contribute to awareness of NRM problems. Literacy of farmers, as evidenced by younger farmers being more likely to perceive the erosion problem, appears to be one entry for public intervention to increase awareness of NRM problems. Although extension services are important communicators, they may need to change in order to succeed in raising farmer awareness of NRM problems.

Farmer perceptions of the severity of soil erosion and the need to treat it are a necessary but not sufficient condition for farmer investment in soil-conservation technologies. Conservation practices must also offer short-term benefits and be profitable. The profitability of conservation practices depends not only on biophysical factors but also on such institutional factors as the availability of credit and secure land tenure. These elements determine the length of the planning horizon and hence the expected return on investment. In a steeply sloped East African highland area like Tigray, the most effective soil-conservation investments are terraces.

Stone terraces increased yield substantially under farmer management. In the mountainous terrains such as those found in Tigray and in many other parts of Ethiopia, stone terraces can be important in the intensification process of agricultural production by: (i) conserving water; and (ii) preventing fertilizer from being washed away. The yield-stability advantage from stone terraces in an environment characterized by erratic and unreliable rainfall reduces the risk of crop failure. These combined effects are likely to contribute to food security in areas that are chronically food insecure.

Yet the high initial investment in terrace construction is practical only if a prolonged pay-off is expected. For poor farmers operating in an imperfect credit market like Tigray's, costly credit is likely to constrain conservation investment. Our investment analysis found that investment in stone terraces can yield a 50% internal rate of return. Impressive as that may sound, it is no more than equal to the prevailing rural discount rates (Shiferaw and Holden, 2000). So investment in stone terraces is merely a break-even proposition to private farmers. Although the yield-stability benefits offered could increase a private household's expected utility, the benefits from terrace construction that are pivotal to induce adoption are the social benefits that pertain beyond the farm's own fields. The value of these benefits has not been quantified, but would arise from reduced gullying, micro-dam sedimentation and consumer losses due to higher food prices resulting from production losses. Assessing the value of these benefits would be a first step to determining the justifiability of the added financial inducements needed to elicit more soil-conservation effort – for example, subsidized credit. But institutional innovations, such as enhancing land-tenure security, can yield comparable inducements without drawing on the public treasury.

Prior research on conservation adoption has considered that the determinants of adoption and the intensity of use are the same (Sureshwaran *et al.*, 1996; Pender and Kerr, 1998). Our results show that the determinants of both decisions can indeed be different. Land-tenure security was a key determinant of adoption of stone terraces, but not of how much terracing was done. The same was true of household labour availability. The opportunity cost of labour and the greater erosion threat due to higher rainfall were important determinants of the intensity of adoption of stone terraces, but not of the likelihood of adoption. These results imply that the cost of investment and returns to investment influence effective use of labour-intensive conservation practices.

Apart from private cultivated lands, communal lands, such as wood lots and grazing lands, are subject to degradation if utilized under unrestricted access or ineffective use regulations. Under such management institutions, resource economic theory suggests that each individual user of the resource tends to use the resource up to the level where his or her average revenue is equal to the marginal cost of utilizing the resource. These incentives tend to result in overexploitation of the resource and the dissipation of the scarcity rent. The effectiveness of public interventions to improve NRM also depends to a large extent on local-level institutions and organizations of resource management (Rasmussen and Meinzen-Dick, 1995).

Our research on collective action for resource management showed that community resource management tends to be more effective at intermediate population densities and if conducted by the most local of collective institutions. When population density is low, the need for collective action to manage resources may be low and the cost of organizing effective collective action may be high. Resource scarcity increases with population growth, raising the benefits of improved resource management. However, when population density becomes very high, the incentive to benefit from 'free riding' on the effort of others may outweigh the benefit from abiding by community rules.

External organizations have played important roles in establishing community management institutions of wood lots and grazing lands. However, external organizations can best promote community resource management by complementing local, demand-driven efforts, rather than displacing them. When the NRM practice is labour-intensive, community resource management can be more effective in remote areas, far from markets, where the opportunity cost of labour is low. In densely populated, well-connected areas, labour-intensive community NRM may not be effective, at least not for wood lots. When community NRM is less labour-intensive and the return from use of the resource is more directly integrated with the market, such as grazing lands in Tigray, market access can have a positive impact on collective action. Community NRM appears to be more difficult to establish in low-potential areas, but is more likely to be effective if established.

Overall, the Tigray experience suggests that the NRM adoption process hinges not only on the natural environment, but also the human institutional environment and the kind of decision-maker. The NRM practices relevant in Tigray are those of populous mountainous regions, where the leading natural-resource challenges relate to soil erosion on sloped lands and where impoverished populations have overexploited shared forests and pastures. But the institutional lessons can be extrapolated more widely. They suggest that public policies to foster NRM adoption should be attuned both to private and community incentives for action. For NRM investments that pay off over time, public intervention may be necessary if private decision-makers are to find NRM investments more attractive than alternatives. Where significant public benefits can be had that are unlikely to be captured by the private decision-maker, public subsidies are justifiable and can be effective if well administered.

Community-managed resources will require different policy incentives from individually managed ones. While guidelines for influencing individual action have been developed fairly well, further work is needed on the design and support of local institutions for NRM in sub-Saharan Africa. Communities that are neither too dispersed to organize shared natural-resource access rules nor too large to prevent free-riding have most effectively managed community wood lots. But, for this part of Africa's

eastern highlands, what policies will best facilitate collective NRM for other scales of community and what specific local institutional designs work best remain to be determined.

Notes

[1] A comparison between the Tobit and double-hurdle models showed that the double-hurdle model fits the data better (Gebremedhin, 1998: 187).

[2] *Tabia* is the lowest administrative unit in the region and usually consists of four or five villages.

[3] Every community has some kind of communal grazing land.

7 Poverty and Land Degradation: Peasants' Willingness to Pay to Sustain Land Productivity[1]

Stein T. Holden and Bekele Shiferaw

Department of Economics and Social Sciences, Agricultural University of Norway, PO Box 5033, N-1432 Ås, Norway

Land degradation may be the most serious environmental problem requiring prompt attention in sub-Saharan Africa (SSA) (Stoorvogel and Smaling, 1990; World Bank, 1996). Many SSA countries are among the poorest in the world and the farming populations constitute both the majority and the poorest segments of these societies. Land degradation is particularly severe in the densely populated areas of East Africa (Stoorvogel and Smaling, 1990; World Bank, 1996), including the Ethiopian highlands, where the majority of the population lives in acute poverty.

It is frequently claimed that poverty may inhibit investment in land conservation and induce myopic survival strategies that prove detrimental to the natural resource base. Few studies have tested these claims empirically, however. An exception is Holden *et al.* (1998), who used data from Indonesia, Zambia and Ethiopia to show that poverty may cause farm households to have high discount rates. High discount rates may induce households to underinvest and to mine their natural resource base. This may be seen as an intertemporal externality because of the discrepancy between private and social discount rates. This chapter uses the same data from Ethiopia but complements them with additional data

on willingness to pay (WTP) to sustain land productivity and WTP for access to improved technologies with different return and payment profiles. The additional data provide insights on the significance of intertemporal externalities due to land degradation. In places where environmental degradation is severe, as in the Ethiopian highlands, it is important to investigate farm households' interest, WTP and ability to pay (ATP)[2] to sustain the land productivity of their own land. Such information is a prerequisite for a sound analysis of whether there is a need for policy intervention and for establishing the appropriate sharing of intervention costs if action is warranted.

In this chapter, we test the hypotheses that market imperfections and poverty are important determinants of farm households' incentives to conserve their own land. We explore this question by asking peasants about their WTP to sustain the productivity of their own land and by analysing which factors are correlated with their stated WTP. We propose that poverty undermines conservation investment on private land even if peasants are fully aware of the problem and have secure rights to the land. Pervasive market imperfections are necessary for this argument to hold.

©CAB *International* 2002. *Natural Resources Management in African Agriculture*
(eds C.B. Barrett, F. Place and A.A. Aboud)

We show that peasants themselves are willing to pay only a small fraction (1.8–3.5%) of the external on-site costs of their own soil-degrading practices. This is largely due to the effects of households' liquidity constraint in increasing their discount rate. Spreading payments over several years would relax the liquidity constraint for the poorest households, thereby reducing the discount rate. Farm households would then be willing to pay a larger fraction of the costs of mitigating soil degradation.

In the next section, we outline the theoretical basis for the analysis and develop testable hypotheses. In the third section, we discuss the methodological approach and its limitations. The fourth section presents the study area, followed by econometric results and discussion in the fifth section, and a concluding section.

Theoretical Framework

High transaction costs, imperfect information, risk and subsistence constraints typically lead to market imperfections in resource-poor rural economies (Hoff et al., 1993). As land users are both producers and consumers, a farm-household perspective is most appropriate for analysing resource use and conservation decisions in these economies (Reardon and Vosti, 1992, 1995).

The contingent valuation method (CVM) may be useful when markets are missing and preferences cannot be revealed through market responses. Markets are more often found to be missing or imperfect in developing countries than in developed countries. This may render the CVM even more relevant in developing countries (less developed countries (LDCs)), although its validity and reliability have yet to be established clearly in the LDC context. We apply CVM to establish how the value of resources varies across households in the presence of market imperfections. In an economy where land sales are prohibited, we elicited households' WTP through sacrificing current production or income in order to sustain or increase the productivity of farm land in the future. The welfare loss due to a sacrifice of initial income may be represented by an expenditure function:

$$e(p, EU_0, F_0) \qquad (1)$$

where p is the vector of prices, EU_0 is the current expected utility level and F_0 is the old technology and farm characteristics. The expenditure function $e(.)$ represents the minimum expenditure level to reach the expected utility level EU_0. The WTP to sustain current productivity can be represented as

$$WTP = e(p, EU_0, F_0) - e(p, EU_0, F_1) \qquad (2)$$

where F_1 is the new technology and farm characteristics vector that sustains productivity and WTP is the Hicksian compensating surplus (Mitchell and Carson, 1989).

CVM has usually been applied in the context of pure consumers. Our context is different as we apply CVM to farm households, which are both producers and consumers, so as to value farm households' non-tradable land resource, the basis for their survival and farm income. This change in context has important theoretical and practical implications (Singh et al., 1986; deJanvry et al., 1991). Market imperfections lead to non-separable production and consumption decisions and cause farm investment decisions to depend on household characteristics and household wealth, unlike in the perfect-market case.

This structure generates the following testable hypotheses.

- H1: Market-imperfection hypothesis. Markets in resource-poor rural economies are imperfect, so investment in conservation depends not only on farm (land) characteristics but also on poverty and possibly other household characteristics. We test the market imperfection hypothesis in an economy which is favourably located in relation to major markets in Ethiopia, has high agricultural potential (inherently good soils and good and reliable rainfall) and produces a surplus of grains for the outside markets. If market imperfections are found to be significant here, they also probably exist in other, less favoured settings in the Ethiopian highlands.

• H2: Poverty hypothesis. Poverty reduces a household's willingness and ability to invest in conservation (poverty-environment trap). WTP per unit of land for sustaining land productivity increases with wealth and income when markets are imperfect (i.e. when H1 holds). A relaxation of the house-holds' liquidity constraint, by spreading payments over time, should then lower its discount rate and increase its total WTP.[3] We may also postulate the following sub-hypotheses.

• H2a: Poverty leads to an increase in the discount rate (rate of time preference) when credit markets are imperfect (H1 holds). Accordingly, a relaxation of the credit constraint should result in a lowering of the discount rate. An alternative cause of high discount rates may be good investment opportunities. If it is the consumption rate of interest (CRI) that is driving up the minimum internal rate of return (MIRR),[4] rather than the other way around, this should imply that poor households have higher CRIs and MIRRs than wealthier households. On the other hand, if investment opportunities drive the CRIs and higher wealth or income is correlated with better investment opportunities, then wealthier people may have higher CRIs and MIRRs. This may also indicate a non-linear (convex) relationship between CRI/MIRR and income/wealth variables. We use farm income, non-farm income, cash liquidity and debt as wealth indicators and test for the potential non-linear impact on WTP. We expect WTP to increase with farm income, non-farm income and cash liquidity. The marginal response in WTP may be higher at low levels of these variables, implying a typical concave response curve. For higher levels of non-farm income it is possible that households depend less on their land resource and therefore are willing to pay less to conserve it. The

poorest households may be rationed out of the credit market. Debt may therefore be a sign of wealth and may therefore be positively correlated with WTP. Credit is also typically obtained for farm inputs. Those with debt are therefore more farming-oriented and also likely to be more concerned with the productivity of their farms (have higher WTP).

• H2b: Poverty leads to a shortening of the planning horizon of poor people. WTP for benefits occurring beyond a household's planning horizon equals zero. We expect that the probability that households are willing to invest in projects the benefits of which occur far into the future increases with income and wealth.

Tenure insecurity could be another reason for farm households to adopt short planning horizons and to have low WTP for conservation. Tekie (1999) argued that the land policy in Ethiopia, which redistributed land based on family size, causes relatively land-rich households to feel more tenure-insecure than relatively land-poor households. He found that farm size was inversely related to investment in soil conservation in central and northern Ethiopia. Holden and Yohannes (2000) have argued that the same inverse relationship could result from Boserupian effects, wherein land scarcity increases incentives to conserve (Boserup, 1965). In a study covering 15 communities in central and southern Ethiopia, including our study area, they found that the probability of feeling tenure-insecure increased with farm size per capita overall, but, when this was tested more specifically in each community, tenure insecurity even decreased significantly with farm size in four of the communities and was significant and positive in only four of the remaining communities. Local power structures appear to counter the effect of the national land-redistribution policy.

We include farm size per consumer unit to control and test for tenure insecurity and Boserupian effects on WTP and MIRR. If these effects are important, the sign of the coefficient on this variable should be negative. A positive sign would imply that the wealth effect of farm size is more important, although we expect the wealth effect also to be captured by the income variables.

The WTP represents the subjective present-value equivalent of future productivity gains due to switching from the current land-degrading development path to a sustainable development path. This can be expressed as follows, assuming that individual households maximize expected intertemporal utility:

$$E_i \begin{bmatrix} [-U_{i0}(C_{i0})+U_{i0}(C_{i0}-WTP_i)+ \\ \sum_{t=1}^{\infty}(1+\delta_i)^{-t}U_{it}(C_{1it}-C_{0it})] \end{bmatrix} = 0 \quad (3)$$

WTP is the amount that leaves the household indifferent between the (expected) marginal utility of current WTP and the discounted expected marginal utility of the change in future incomes. We assume that preferences are intertemporally separable and that the individual's pure rate-of-time preference, δ_i, is constant. $U_{it}(C_{1it} - C_{0it})$ is the utility gained by household i from the difference in land productivity in period t when switching from the land-degradation regime to the sustainable regime, E_i is the mathematical expectation for individual i conditional on all information available to the individual at time $t = 0$. By taking expectations, we arrive at Equation (4).

$$U_{i0}(C_{i0})-U_{i0}(C_{i0}-WTP_i)= \sum_{t=1}^{\infty}(1+\delta_1)^{-t}EU_{it}(C_{1it}-C_{0it}) \quad (4)$$

This may also be reformulated into the Euler equation for intertemporal trade-offs:

$$U'_{i0}(C_{i0})(WTP_i)= \sum_{t=1}^{\infty}(1+\delta_i)^{-t}EU'(C_{it})dC_{it} \quad (5)$$

Assuming marginal changes[5] and no market imperfections,

$$dC_{it} = C_{1it} - C_{0it} = c[\eta(F_i, X_i), F_i, X_i = c(F_i, X_i)]$$

where η is the expected rate of soil productivity decline. We expect $\dfrac{\delta c}{\delta \eta} > 0$, implying a

larger difference on large farms and on vulnerable farms where a larger share of the farm is on upland.[6] We may also write Equation (5) as follows;

$$WTP_i = \qquad\qquad\qquad (6)$$
$$\sum_{t=1}^{\infty}\Big[(1+\delta_i)^{-t}\big(EU'(C_{it})/U'_{i0}(C_{i0})\big)dC_{it}\Big]$$

With market imperfections, this WTP measure will be subject to a set of income constraints and cash constraints. The income constraint is the sum of net farm income (FI) and non-farm income (NFI),

$$C_{1it} = FI_{it} + NFI_{it} \text{ and}$$
$$C_{i0} = FI_{i0} + NFI_{i0} - WTP_i \qquad (7)$$

Net farm income is a function of farm characteristics (F_{i0}) and household characteristics (X_{i0}) when markets are imperfect. Farm production and income are also driven by household food requirements (subsistence constraints) when food markets are imperfect, $FI_{it} = FI_{it}(F_{it}, X_{it})$.

Equation (6) implies the following reduced-form equation for WTP:

$$WTP_i = WTP_{i0}(FI_i(F_{i0}, X_{i0}), NFI_{i0}, CASH_{i0}, DEBT_{i0}, F_{i0}, X_{i0}) \quad (8)$$

where $CASH_{i0}$ is the amount of cash in hand and $DEBT_{i0}$ is the households' debt at the time the survey took place. With perfect markets and information, WTP would only depend on farm characteristics because the level of investment in conservation would be independent of household income and other household characteristics.[7] Therefore, if WTP per unit of land changes with household farm or non-farm income, we have evidence in favour of the H1 hypothesis. Non-separability causes the discount rate to become household-specific and to depend on both household and farm characteristics and market characteristics (e.g. credit constraints).

The Study Area and Data

This study was carried out in 1994 in three peasant associations (PAs) 20–30 km from the town of Debre Zeit, in Ada district, east

Shewa, in the Ethiopian highlands. The area is near the major Ethiopian markets and has high agricultural potential; it is at an elevation of 1900–2100 m above sea level and the average annual rainfall is 830 mm. An integrated crop–livestock system prevails, with oxen used as traction power. Agriculture is rain-fed and the risk of crop failure is low. Teff is the major staple as well as a cash crop. Wheat is the second most important crop. Most farm households are net sellers of food.

Basic characteristics of the 120 peasant households surveyed are given in Table 7.1. Households were stratified on the basis of traction power (number of oxen). Four categories of households were thus distinguished: households with no oxen, households with one ox, households with two oxen and households with more than two oxen. For each category a random sample of 30 households was drawn and interviewed. The land may be divided into two categories: sloping uplands, where soil degradation was significant, and flat lowlands, where degradation was much less severe. The average landholding is 2.25 ha, with 1.25 ha upland and 1 ha lowland. The number of oxen is the most important income and wealth indicator, due to oxen's important role in the farming system as a source of traction power. Without oxen, households are unable to farm their land and have to rent it out to others. Holden et al. (1998) show that oxen ownership is significantly and inversely related to the discount rate (see also Table 7.4). The number of oxen is one of the most significant determinants of farm income in the study area.

None of the farm households in the area have adopted land-conservation technologies (e.g. soil bunds, stone bunds, terraces, grass strips, tree hedgerows), even though they clearly see land-degradation as an important problem. This lack of investment in conservation implies that no part of the land-degradation externality is internalized. This is probably because private benefit–cost ratios are below 1 (Shiferaw and Holden, 1999, 2000). However, the fact that the large majority (111 out of 120 households) are willing to pay a positive amount to sustain land productivity on their farms shows that land degradation is widespread and serious in the area.

Methodology

The CVM has been subject to careful assessment by the National Oceanic and

Table 7.1. Basic farm-household characteristics in the survey area.

Variable	Household category: number of oxen				
	0	1	2	> 2	All
Share of population (%)	25	17	34	24	100
Number of households surveyed	30	30	30	30	120
% Female-headed households	27	7	7	3	11
Farm size (kert)*	4.48	7.4	8.18	11.25	7.83
Male workforce	0.71	1.34	1.57	2.84	1.62
Female workforce	0.91	1.01	1.12	1.68	1.18
Consumer units	2.47	3.76	4.12	6.47	4.2
Tropical livestock units	0.31	2.46	4.46	9.12	4.09
Farm income (Birr)[†]	2133	3767	3990	8129	4527
Non-farm income (Birr)[†]	899	1241	1665	4151	2002
Cash (Birr)[†]	30	27	30	131	55
Debt (Birr)[†]	50	68	45	52	53
Upland ratio	0.55	0.66	0.68	0.59	0.62
Education, years	1.65	1.1	1.05	1.05	1.2
Farm experience, years	44.3	41.7	45.1	45.4	44.2

*1 kert = 0.3 ha.
[†]6.5 Birr = US$1.

Atmospheric Administration (NOAA) Panel[8] (Arrow et al., 1993), which specified guidelines for contingent valuation (CV) surveys. These include a high response rate, in-person interviews, discrete choice, referendum questions with follow-ups and responsiveness of WTP to the scope or amount of what is offered. Smith (1996) points out the importance of the responsiveness to the scope (or amount) of the commodity offered to respondents. The validity test requires that WTP estimates be related to a set of economic variables hypothesized as important to observed choices and that CV choices involving objects generally taken as different should be significantly different. Full compliance with the NOAA Panel requirements would make the CVM too expensive for application in LDCs, however. Recent research has gone in the direction of identifying lower-cost, robust CVM approaches.

In our case, the stated WTP is related to respondents' individual farms and should therefore be very well known to the respondents, as should be the expected benefit streams from their farms. The variation is in accordance with economic theory. The WTP responses are clearly responsive to variation in farm size, technology characteristics and severity of land degradation. The validity of this CVM exercise is also supported by R^2 goodness-of-fit measures that are much higher than the minimum recommended value of 15% set by Mitchell and Carson (1989). The CVM questions were only a part of a large and general household survey, which also reduces the probability that responses were strategically biased, as no promises about projects or interventions were made in relation to the survey.

We used three questions of which the first (Q1) focused on WTP to sustain current land productivity on their farms, the second (Q2) asked for WTP in year 1 for an increase in production (due to a new technology) of 100 kg of teff year^{-1} from year 2 and onwards, and the third (Q3) asked for WTP year^{-1} over a 5-year period for an increase in production of 100 kg of teff from year 6 and onwards. Q2 is used to estimate the current

MIRRs of households based on Equation (6) above.

$$WTP_i = \frac{dC_{it}}{MIRR1_i\,(1 + MIRR1_i)} \rightarrow MIRR1_i =$$

$$-0.5 + \sqrt{\frac{100}{WTP_i} + 0.25} \qquad (9)$$

This assumes that there is no change in households' level of expected utility over time or in their MIRRs. Growth in the economy would also affect the elasticity of marginal utility and expected growth.

The differences in responses to Q2 and Q3 reveal the importance of the subsistence and liquidity constraints in limiting ATP for productivity increases in the short run and thus reveal MIRRs when the investment could be distributed over a period of 5 years. If subsistence and liquidity constraints are severe, Q3 (MIRR2) should signal a significantly lower MIRR than Q2 (MIRR1) (H2a holds when $MIRR1_i > MIRR2_i$). Unwillingness to pay anything in the case of Q3 is a sign that respondents have planning horizons shorter than 5–6 years (a test of H2b). MIRR2 may be derived from the following expression when expected consumption and expected utility are assumed to be constant over time:

$$\sum_{t=0}^{4} \left(1 + MIRR2_i\right)^{-t} WTP_{it} = \left(1 + MIRR2_i\right)^{-5}$$

$$\left[\sum_{t=6}^{\infty} \left(1 + MIRR2_i\right)^{-(t-6)} dC_{it}\right] \qquad (10)$$

The response to Q1 was used to deduce average (constant) rates of soil productivity decline and to analyse which factors affect household WTP to sustain land productivity.

In the econometric estimation, we use WTP as the dependent variable. WTP offers an instant measure while the independent variables are predetermined, allowing for time-recursive causality. There could, however, be a multicollinearity problem, because the independent variables are interrelated. We use instrumental variables to predict farm income.

The other independent variables include non-farm income, cash liquidity, debt, two dummy variables for PAs with different land characteristics, the upland share of farm land, farm size, gender and years of farm experience and education of household head. The non-farm income, cash and debt variables could also cause endogeneity problems, but we had no good instruments to predict these variables. Davidson and MacKinnon (1993) tests reject the hypothesis of endogeneity and inconsistency.

The first model (Table 7.2) includes eight censored observations (WTP = 0), too small a number to permit a reliable estimation of a censored model. There were no censored observations in the second model (Table 7.3). We also included squared terms for some of the variables that could be non-linearly related to the dependent variable.

Ultimately we removed two variables[9] that were highly collinear with other variables, thereby increasing the precision of the estimates of a number of variables' coefficients.

Results and Discussion

Table 7.2 presents the results for the first two-stage least-squares model of household WTP to sustain land productivity. Standard errors have been corrected using the White method. The R^2 of 0.62 indicates a good explanatory power of the variation in household WTP responses. The farm income, non-farm income squared and

Table 7.2. Determinants of willingness to pay to sustain land productivity.

Variables[†]	2SLS param. est.	2SLS robust t ratio	VIF[‡]	2SLS param. est.	2SLS robust t ratio	VIF
Constant	1.942	1.00		2.40	1.451	
Farm income, predicted[§]	0.0156	1.539	22.70	0.0126	1.741*[‖]	4.38
Farm inc., pred., sq.	−0.0000156	−0.948	47.24			
Non-farm income	0.0127	0.616	9.83			
Non-farm income, sq.	−2.96e-06	−1.312	2.64	−2.74e-06	−2.29**	1.74
Cash	0.346	1.016	20.64	0.425	2.108**	12.47
Cash, sq.	−0.002	−0.505	39.25	−0.004	−2.266**	12.54
Debt	0.309	0.983	1.17	0.0426	3.209***	1.08
PA dummy 1	−0.87	−1.241	2.13	−0.37	−0.546	2.02
PA dummy 2	−0.396	−0.891	1.75	−0.194	−0.421	1.71
Upland ratio	−1.574	−1.218	1.24	−1.968	−1.593	1.20
Farm size	−0.137	−1.776*	2.34	−0.091	−0.929	2.18
Sex	−0.2626	−0.18	1.53	−0.446	−0.269	1.51
Education	−0.1237	−0.46	2.61	0.09	0.35	1.85
Farm experience	0.0124	0.57	1.40	0.0071	0.41	1.33
R^2	0.62			0.6		
Mean VIF[2]			11.18			3.67
Number of observ.	119			119		

[†]All variables, except upland ratio, farm experience, sex and education, are per unit of land and per consumer unit in the household.
[‡]VIF is the variation inflation factor. For the farm income and non-farm income variables, we removed the term with highest VIF and lowest t value in the first regression. This also seemed to reduce the multicollinearity problem for the cash variable. If we removed the squared term for the cash variable, R^2 fell to 0.53. Therefore, we kept it in.
[§]Male workforce, female workforce, oxen, farm experience (years), farm size, purchased farm input expenditure, all linear and quadratic, PA dummies, education of head of household and upland ratio were used to estimate and predict farm income. Adjusted R^2 was 0.80 in the instrumenting equation. Hausman, and Davidson and MacKinnon tests were used (several specifications) to test for the endogeneity of the non-farm income, cash and debt variables. The Hausman test: $\chi_2 (8) = -4.28$. Davidson and MacKinnon test: $F(3, 100) = 0.70$, Prob. $> F = 0.5534$.
[‖]*, ** and *** indicate 10%, 5% and 1% levels of significance based on robust standard errors.
2SLS, 2-stage least squares.

Table 7.3. Determinants of peasants' minimum internal rates of return (MIRRs) measured as maximum WTP for a fixed productivity increase.[†]

Variables[‡]	2SLS param. est.	2SLS robust t ratio
Constant	5.895	0.721
Farm income, predicted[§]	0.0096	0.873
Farm inc., pred., sq.	−2.86e−06	−1.495
Non-farm income	0.0051	0.3
Non-farm income, sq.	−0.00001	−2.207**
Cash	0.382	4.293***
Debt	0.0407	1.426
PA dummy 1	5.479	1.114
PA dummy 2	3.623	1.052
Farm size	6.421	2.085**
Sex	−12.29	−1.061
Education	1.662	2.264**
Farm experience	0.0293	0.274
R^2	0.50	

[†]WTP has an inverse relationship with $MIRR1$.
[‡]The dependent variable is the response to question Q2. All variables except the WTP, upland ratio, farm experience, sex and education variables are on a per consumer unit basis. The data set contains 119 observations.
[§]The same instruments are used as in Table 7.2. The adjusted R^2 of the instrumenting equation is 0.44.
2SLS, 2-stage least squares.

cash liquidity variables are statistically significant, with signs supporting the market imperfections (H1) and poverty (H2) hypotheses. Cash liquidity has a significant non-linear effect that causes the marginal response to cash to fall. These results indicate that poverty and liquidity constraints reduce the WTP/ATP to invest in conservation. Only the non-linear (quadratic) effect of non-farm income is significant, but the linear effect has the expected sign. The non-linear effect reflects the hypothesized concave relationship between non-farm income and WTP. The debt variable is highly significant and positive, indicating that poor households are rationed out of credit markets, so that people with debt are able and willing to pay more for sustaining land productivity. The per capita farm-size variable proves to be negative but insignificant, indicating that local variation in tenure insecurity and Boserupian effects

have no significant effect on WTP. None of the other variables are significant.

Table 7.3 presents the results of the two-stage model analysing the factors correlated with peasants' MIRRs, represented by the maximum WTP for fixed productivity increase. WTP is inversely related to MIRR. We see that cash liquidity has a highly significant and positive impact on the WTP and thus a negative impact on MIRR. Many households are liquidity-constrained, so MIRRs are high at low levels of cash availability, in line with H2a. The farm-income and non-farm-income variables also have the expected signs, but only the non-linear element of the non-farm income is significant (5% level). Higher levels of non-farm income contribute less to increase WTP or lower MIRR. This suggests that people with high non-farm income also have access to better investment opportunities, which causes them to have higher MIRRs. The debt variable is insignificant, but its positive sign is consistent with the credit-rationing hypothesis (H1). The farm-size variable is positive and significant, reflecting the poverty effect (H2) and that tenure insecurity due to differences in relative farm size do not undermine incentives to invest. The education variable is also significant and positive, signalling that human capital increases WTP and reduces MIRR.

In Table 7.4 we present the calculated average normal CRIs and MIRRs of the surveyed farm households for the four wealth categories. Poorer households – those with no or one ox – have significantly higher CRIs and MIRRs than households with two or more oxen. There is larger variation in the estimated CRIs than in the MIRRs. This pattern indicates that the MIRRs are driven by the CRIs for the poorer households, consistent with H2a. If investments in productivity increases could be distributed over 5 years, MIRRs fall significantly, indicating the significance of the credit constraint in limiting ability and willingness to make larger investments. The share of households with a planning horizon of less than 6 years is as high as 33% for the poorest category (no oxen), while all households in the richest category (more than two

Table 7.4. Average CRIs and MIRRs of farm households of different wealth categories.

Oxen ownership category	CRI*	MIRR1[†]	MIRR2a[‡]	% with < 6 years planning horizon	MIRR2b[‡]
0 (n = 30)	1.20	0.89	0.40	33.3	0.32
1 (n = 30)	0.84	0.73	0.32	20	0.27
2 (n = 30)	0.69	0.65	0.30	3.3	0.29
> 2 (n = 30)	0.32	0.55	0.24	0	0.24
All	0.71	0.68	0.30	14.2	0.27

*From Holden *et al.* (1998), but these consumption rates of interest (CRI) are discrete-time nominal rates. Holden *et al.* reported only continuous-time discount rates.
[†]*MIRR1* is based on question Q2 and Equation (9). There were no zero responses to Q2.
[‡]*MIRR2a* is based on question Q3 and Equation (10) including all observations. *MIRR1* was used in place of *MIRR2* when the response to Q3 was zero. *MIRR2b* is based on the same data, but the observations with zero WTP (planning horizon < 6 years) have not been included in the calculation.

oxen) have a planning horizon of more than 6 years. Poverty appears to lead to short planning horizons (H2b), as well as higher discount rates (H2a).

Based on peasants' average WTP to sustain land productivity (Q1) and variation in their discount rates and planning horizons, we deduce their implicit subjective rates of land-productivity decline. Table 7.5 reports our sensitivity analysis on this variable. Peasants' implied, perceived rates of productivity decline are lower than the scientifically estimated rate of 1.1% – probably about half of this rate. The scientifically estimated rate is estimated based on the universal soil-loss equation, adapted to Ethiopia by Hurni (1985) and as related to this case-study area by Shiferaw and Holden (1999, 2000). Lack of good time-series data from the area makes this scientific estimate uncertain.

A short planning horizon of 3 years, indicating high tenure insecurity, and a discount rate of 70% would imply a lower perceived rate of productivity decline (1.04%) than the scientifically estimated rate. As we cannot be sure that the scientifically estimated rates are more accurate than the lower perceived rates that we have deduced from peasants' responses, we have carried out a sensitivity analysis for different rates of productivity decline with respect to the economic costs of this intertemporal externality.

Table 7.6 presents a sensitivity analysis of farm households' WTP to sustain land

Table 7.5. Perceived rates of productivity decline (% per year) based on peasants' WTP to sustain land productivity,* alternative discount rates and time horizons.

Planning horizon	Farmers' discount rate			
	30	50	70	100
Infinite	0.14	0.33	0.57	0.99
5 years	0.32	0.51	0.72	1.10
3 years	0.59	0.80	1.04	1.43

*The average WTP to sustain land productivity was 55 Ethiopian Birr. We assume the rates of productivity decline to be constant over time.

productivity at different discount rates, rates of land productivity decline and planning horizons. We compare these WTP rates with society's cost, and thus society's maximum WTP to internalize the intertemporal on-site externality of land degradation at a social discount rate of 10% (average inflation = 5%). Within the range of peasants' discount rates and perceived rates of productivity declines and even with infinite planning horizons, private WTP for sustaining land productivity is far below society's estimated WTP. This difference reflects a considerable intertemporal externality (on-site external costs). The probability of private investment in conservation increases with income or wealth as private discount rates fall. Policies that stimulate economic growth or redistribute income from rich to poor may thus reduce the externality by making conservation

Table 7.6. Willingness to pay to sustain land productivity.

WTP/NPV estimates in Ethiopian Birr	Rate of productivity decline (% per year)	Farmers' discount rate (%)				Society's discount rate, 10%*
		30	50	70	100	
Infinite planning horizon						
Farmers' perceived rate of productivity decline	0.277	112	47	27	16	836
Farmers' perceived rate of productivity decline	0.55	222	93	55	31	1629
Scientifically estimated rate of productivity decline[†]	1.1	436	184	107	62	3094
5 years planning horizon						
Farmers' perceived rate of productivity decline	0.277	47	30	21	14	836
Farmers' perceived rate of productivity decline	0.55	94	60	42	28	1629
Scientifically estimated rate of productivity decline[†]	1.1	187	120	84	55	3094
3 years planning horizon						
Farmers' perceived rate of productivity decline	0.277	26	19	15	11	836
Farmers' perceived rate of productivity decline	0.55	52	38	29	21	1629
Scientifically estimated rate of productivity decline[†]	1.1	103	76	59	43	3094

*Society is assumed to always have an infinite time horizon.
[†]Based on the universal soil-loss equation adapted to Ethiopia (Hurni, 1985) and in Shiferaw and Holden (1999, 2000).

investments privately profitable. At the time of the study, none of the surveyed households had sufficiently high incomes and low discount rates to undertake any investments to internalize the soil-degradation externality, suggesting a need for external intervention and/or collective action.

Average household WTP is 55 Birr ha^{-1}, the average discount rate about 70% and the average perceived rate of productivity decline is 0.55% per year. This WTP is only 1.8% of that of society's WTP at a discount rate of 10%[10] and the scientifically determined productivity-decline rate of 1.1% per year, underscoring the vast gap between private and social WTP. Relatively slow economic growth, combined with high population growth, suggests that relying on economic growth to solve Ethiopia's soil-degradation problem may be unrealistic. Private incentives established by (imperfect) market forces alone will probably take too long to avert serious land degradation in

an important surplus grain-producing area. A labour mobilization (taxation) policy for conservation investment, like that instituted in Tigray,[11] may be a good alternative, as it mobilizes local resources and requires little external capital. Interlinkage (also known as cross-compliance) policies, linking, for example, provision of subsidized credit and fertilizers to conservation behaviours, may be another promising approach (Holden and Shanmugaratnam, 1995; Shiferaw and Holden, 2000).

Shorter planning horizons may be caused not only by poverty but also by tenure insecurity. Tenure appears quite secure in our study area since the large majority of households were willing to pay for benefits occurring more than 5 years into the future. In another survey in the study area, only 14% of households considered their land tenure insecure (Holden and Yohannes, 2000). Table 7.5 also reflects the effects of shorter planning horizons on WTP at different discount rates and rates of productivity decline. Comparison with the infinite-horizon WTP estimates indicates that high discount rates caused by market imperfections and poverty explain much more of the intertemporal externality than do shorter planning horizons resulting from tenure insecurity.

Conclusions

We employ CVMs to estimate household WTP for soil conservation in the Ethiopian highlands. The CVM approach may also be useful for estimating shadow prices related to market imperfections, although more applied research is needed in order to develop cost-effective CVM approaches and to test their robustness in developing countries. Future studies should also study household WTP in terms of household labour, since much of the conservation investment is in the form of labour rather than cash. We find that market inperfections and poverty combine to undermine WTP to sustain land productivity, even in an area with relatively good market access and high agricultural potential in the Ethiopian highlands. Poverty and cash-liquidity constraints reduce farm households' willingness and ability to invest in farm-land conservation and drive up household CRIs and MIRRs. The poor's WTP for future productivity increases could be increased if the investments could be distributed over a number of years. Poverty also increases the probability that farm households will adopt short planning horizons and thereby neglect benefits accruing further in the future. Poverty appears to undermine much of the (Boserupian) population-pressure effect that many analysts hypothesize creates incentives to invest in conservation. Private WTP for soil conservation is so low as to not even lead to a partial internalization of the intertemporal on-site land-degradation externality. Peasants are willing to cover only 1.8–3.5% of the estimated social costs of soil degradation.

The core implication is that society may need to intervene in order to protect the interests of future generations. Poverty reduction may eventually lower households' discount rate, thereby stimulating increased private investment in soil conservation, but this alone is almost surely insufficient to approach socially optimal conservation investment in the short run. More direct, targeted interventions appear to be necessary. Labour mobilization and cross-compliance policies offer promising options, but require local participation and motivation to succeed (Holden and Binswanger, 1998; Shiferaw and Holden, 2000).

Appendix: the Formulation of the WTP Questions

Q1. 'If you are convinced that the productivity of your land is gradually declining every year due to soil erosion and the conservation technology would only help to sustain current levels of output starting from the second year of use but reduce production in the first year, how much reduction in the first year are you willing to sacrifice to sustain current levels of productivity from the use of the new technology?
a. I have nothing to sacrifice from current levels of output.
b. I can sacrifice less than 100 kg. Specify:
c. I can sacrifice 100 kg.
d. I can sacrifice more than 100 kg. Specify:'
To make them specify a quantity, a sequential procedure was followed, e.g. if you are not willing to sacrifice 100 kg, would you sacrifice 90 kg, if no: 80 kg, etc. These responses were used as estimates of the WTP per household for the different oxen categories of households.
Q2. 'Assume you have an option of using a new technology which will increase the production of teff by 100 kg starting from the second year of adoption. However, the new management practice is known to reduce your output of this crop in the first year: how much reduction in production in the first year could you sacrifice to be able to achieve an increment in production of 100 kg from

the second year onwards? (a, b, c, d as above).'

Q3. 'If the new technology is believed to reduce production in the first 5 years, but production improves by 100 kg from the sixth year, how much reduction in production could you sacrifice per year for the first 5 years? (a, b, c, d as above).'

Notes

[1]   We are grateful for comments from Chris Barrett, Frank Place and two anonymous reviewers on an earlier draft of this chapter. Funds for this research have been obtained from the Research Council of Norway.

[2]   We may see the 'ability to pay' (ATP) as the WTP when WTP is restricted by liquidity and subsistence constraints. Provision of credit would then increase the ATP/WTP or, if the payment could be spread out over a number of smaller payments, this could relax the liquidity constraint and thus lower the discount rate and increase the ATP/WTP. We use ATP and WTP as synonyms in the rest of the chapter.

[3]   Holden *et al.* (1998) derived an explicit relationship between the discount rate and the credit constraint: $-dC_1/dC_0|U = e^r + \lambda e^\delta/Ev'(C_1)$, showing that the pure rate of time preference and the shape of the utility function matter for the discount rate only when the credit/liquidity constraint is binding.

[4]   The minimum internal rate of return (MIRR) is the minimum interest required on a project for it

to be chosen instead of an extra unit of funds to be used for consumption.

[5]   We assume that the inter-period expected changes in consumption levels are small.

[6]   We included a variable for the share of the farm being on the upland and two dummies for peasant associations (PAs), as there was some systematic soil variation among PAs.

[7]   This rests on the assumption that all land-quality attributes are tradable. Non-tradable land attributes linked to heterogeneous household preferences for these land attributes may be a cause of non-separability.

[8]   The US National Oceanic and Atmospheric Administration's Expert Panel on the Contingent Valuation Method.

[9]   Farm income squared and non-farm income; these had lower *t* values than multicollinear variables.

[10]   Some would argue that the 10% social discount rate we use is too high. Those who believe in a lower social discount rate are more concerned about soil degradation and are therefore more supportive of intervention.

[11]   In Tigray, all able-bodied adult household members are required to work 20 days year^{-1} for the community without payment. This work takes place outside the busy agricultural seasons, when the opportunity cost of labour is low. Much of this labour has been invested in soil conservation in the communities, but other tasks, such as building and maintaining roads, schools or irrigation structures, are also common. Conservation investments under this scheme have taken place on both communal and private land.

8 Input Use and Conservation Investments among Farm Households in Rwanda: Patterns and Determinants[1]

Daniel C. Clay,[1] Valerie Kelly,[2] Edson Mpyisi[3] and
Thomas Reardon[4]

[1]Institute of International Agriculture, Office of International Programs,
College of Agriculture and Natural Resources, Michigan State University,
324 Agriculture Hall, East Lansing, MI 48824-1039, USA; [2]Department of
Agricultural Economics, Michigan State University, Agriculture Hall, East Lansing,
MI 48824-1039, USA; [3]Food Security Research Project, Rwanda Ministry of
Agriculture, Animal Resources and Forests, c/o USAID, BP 2848, Kigali, Rwanda;
[4]Department of Agricultural Economics, Michigan State University,
211F Agriculture Hall, East Lansing, MI 48824-1039, USA

While emergency relief and rehabilitation persist as a top concern in postwar Rwanda, the focus of both domestic and international attention has gradually turned to opportunities for long-term development. High on the list of priorities is the growing need to find ways to reverse a generation of decline in agricultural productivity. Increasingly poor farm yields have been well documented in Rwanda (Byiringiro, 1995; Clay *et al.*, 1995), as in sub-Saharan Africa more generally. It is also broadly recognized that prospects for ethnic harmony and political stability will be enhanced by concomitant growth and poverty alleviation in the rural economy, since 93% of Rwanda's population is rural and agriculture provides most of them with an important share of their income.

In a departure from the long-standing, low-external-input approach to Rwandan agricultural production (CNA, 1991), postwar policy debate has centred around the question of how to stimulate markets for commercial fertilizers and other improved inputs. Joint government and donor programmes have experimented with fertilizer sector reforms designed to stimulate full privatization of fertilizer import and marketing activities. In a recent 'Fertilizer Use and Marketing Policy' workshop, organized in the agriculture ministry, an action plan was developed that emphasizes fertilizer demonstrations, complemented by policy changes to expand the informal fertilizer import system through more private-sector entry, training programmes to enhance private-sector fertilizer distribution and import capacity and improvements in the credit availability to farmers, as well as fertilizer distributors and importers (Desai, 2001).

Though emphasis in recent policy discussions has been on increasing fertilizer use, Rwandan scientists, farmers and policy analysts recognize that the positive effects of fertilizer applications are enhanced by:

(i) improvements in farming practices, particularly soil conservation, such as bunding, mulching, agroforestry and alley cropping; and (ii) the use of animal and green manures, compost and other organic matter that are both nitrogen-fixing and beneficial to soil composition (Kelly and Murekezi, 2000). These observations are by no means specific to Rwanda; similar conclusions have been reached in other parts of Africa (Sanchez *et al.*, 1997b; Weight and Kelly, 1999).

This chapter contributes to the ongoing policy dialogue aimed at raising Rwanda's capacity to produce more food for a growing population through increasingly sustainable means. Working from a rich household-level database, we endeavour to identify the determinants of farmer investments in agricultural intensification and to examine how these determinants either constrain or enable farmer investment strategies. Particular attention is given to conservation investments, use of organic matter, purchase of chemical inputs and soil erosion associated with land-use patterns. Developing a better understanding of what factors influence these types of farm decisions and outcomes and the relative impact they have on investments is a first step toward developing better policies and programmes to promote agricultural modernization.

The following section presents our conceptual approach, explaining farm-level investment in terms of the incentives facing farm households and the capacity of households to undertake investments, and briefly describes the data and models developed. A brief description of the Rwandan farming context follows, with particular attention being given to levels of conservation and input investments and regression specifications. We then present econometric results, describing the relative importance of different determinants for the four different types of agricultural intensification discussed. We conclude with an examination of the implications of the model results for the design of policies and programmes to support the government's current agricultural-intensification objectives.

Conceptual Framework, Model and Data

The decision to use improved technologies or soil-conservation practices constitutes a major investment for farmers in Rwanda. Farmers must make difficult choices to allocate scarce resources between consumption (e.g. food, education, health, housing) and production ends. Because investing in agricultural intensification implies foregoing other consumption and/or investment opportunities (at least temporarily), farmers are likely to pose two basic questions before making such investments: Will it be profitable? Can I afford it?

Factors that influence profitability can be thought of as the 'incentives' to adopt a particular technology. It is useful to consider two key categories of incentives: monetary and physical. Monetary incentives are those associated with the agricultural market conditions in a zone (e.g. output prices, input prices, access to markets, prevailing wages for agricultural and non-agricultural activities). In general, higher output prices, lower input prices, better market access and lower wages/incomes from competing non-farm opportunities provide positive monetary incentives for agricultural investments. Physical incentives are those associated with farm and plot characteristics (size and location of plot, amount of fallow, fragmentation of plots, slope, rainfall, etc.). Our hypothesis is that farms are more likely to invest in soil conservation and improved inputs if they are under greater stress (more fragmented plots, less fallow) but possess land that can be improved (good location on slopes and/or slope not too steep, plots not too fragmented or far from the residence, and acceptable levels of rainfall).

Risk can alter farmers' perceptions of both monetary and physical incentives. For example, investments become riskier and incentives decline if farmers are not sure that they will be able to recover the full benefits of their investments (e.g. applying manure to a rented field). Similarly, volatile, unpredictable output prices can reduce incentives as farmers become uncertain as to their ability to recover their investment costs by selling surplus production.

Whether farmers can afford to invest in soil conservation or agricultural intensification depends on their capacity to acquire and correctly use improved technologies. Capacity improves when financial capital (fixed assets, cash and/or credit) increases, permitting farmers to invest more, and when levels of human capital (nutrition, health, education, management skills) are higher, as this enables farmers to use improved technologies efficiently. Thus, wealth, broadly defined to include cash for purchases, human capital and household labour resources, constitutes a major determinant of such investments. The level of financial and human capital is often reflected in farm characteristics, such as size of holdings, household size and location. In theory, household liquidity is especially important where the credit market is underdeveloped or absent, as is the case in Rwanda and elsewhere in the tropical highlands of East Africa.

Our general model for farm investments reflects the conceptual framework summarized above and the literature on firm- and farm-level investment (Feder *et al.*, 1985, 1992; Christensen, 1989). Farm investments are functions of five sets of variables:

Investment = *f* (financial incentives, physical incentives, risk, wealth, agrosocio-economic context)

The dependent variables examined (all at the parcel level) are: (i) land-conservation investments (represented by metres per hectare of grass strips, radical terraces, ditches, hedgerows, etc.); (ii) organic inputs (a dummy variable indicating use/non-use of composting, manure, green manure or mulch); (iii) chemical inputs (a dummy variable indicating use/non-use of fertilizer, pesticides or lime); and (iv) the land-use erosiveness index (represented by the C value).[2]

A relatively unique and advantageous aspect of the models developed below is that they model investment/adoption at the parcel level rather than at the household level. Most models designed to explain the determinants of investment/adoption use zone- and household-level data to try to explain why farm households invest in or adopt a particular technique. These types of models do not deal with the decision to adopt different technologies for parcels with different characteristics.

The core question we explore is: What explains investment/adoption for each parcel on the farm? The model design takes into account that adoption and investment decisions are not made uniformly for the entire farm. In attempting to answer this question, the models use zone and household characteristics, as well as the parcel-level characteristics that are seldom taken into account when the focus is on the household.[3]

Another unusual attribute of the present analysis is the degree of disaggregation made possible by an unusually rich database. Few adoption/investment analyses are able to cover all five categories of explanatory variables and the four outcome variables identified above. Such models require not only detailed information on farmers' input and conservation investments, but also a broader set of data needed to understand the farm-management and household-strategy context of these investments, including household farm and non-farm income, assets, demographic characteristics and plot ecological properties. Such multilevel data are rare.

The Rwandan data meet varied requirements. They derive principally from a nationwide, stratified random sample of 1240 farm households operating 6464 parcels interviewed in 1991 (prewar) by the Division des Statistiques Agricoles (DSA) of Rwanda's Ministry of Agriculture. Interviews with heads of households and/or their spouses were conducted over a 6-week period beginning in June 1991. We integrated these data with those on farm- and livestock-enterprise management from the Ministry's national longitudinal survey on the same sample of households.

The Rwandan Farming Context and the Specification of the Regressions

The purposes of this discussion are to describe the general farming context in Rwanda and to present descriptive statistics on the full range of variables used in the conservation-investment, input-use and land-use models. Table 8.1 presents summary statistics on metres of conservation investments per hectare, percentage of parcels treated with organic inputs, percentage treated with chemical inputs and land-use erosiveness (C value). These variables represent the four indicators of land use and investment activity for which we run regressions. Coefficients of variation (CVs) are reported to illustrate variability around the means.

The land-use index (C value) measures soil erosion associated with land use: the lower the C value, the less erosive the land use. Land use in Rwanda is fairly non-erosive on average (with a C value of 0.16), though variation across parcels is moderately high (the ratio ranges from a low of 0.01 to a high of 0.4; the CV is 0.55). Controlling for production techniques, the C value reflects crop mix. Erosiveness is higher for some crops than others – lowest for bananas (0.04) and increasing gradually for beans (0.19), manioc (0.25), sorghum (0.35) and maize (0.40).

The land-use model explicitly reflects choice of an outcome (level of soil erosion), but also reflects a choice between perennial cash crops and annual crops, which are either cash or subsistence crops. That decision reflects two sets of objectives (controlling for physical, cultural and economic constraints): (i) to reduce erosion, a long-term objective that requires short-term (crop) choices; and (ii) to maximize returns to land and labour, a short-term objective that requires a short-term choice of crops with high returns. We model this as a function of variables reflecting incentives related to the long-term objective of controlling erosion (e.g. steeper slopes of fields should spur investment in perennials to control runoff) and of variables that reflect short-term profitability (e.g. the price of bananas relative to that of sweet potatoes).

The parcel-level average of all land-conservation investments is 438 m ha^{-1}. There is, however, great variation across parcels (coefficient of variation of 2.12). Grass strips are most common, followed by anti-erosion ditches, then hedgerows, then radical terraces. Ditches and terraces are the most labour- and equipment-intensive to build and maintain, and grass strips the least. Hence, the abundance of grass strips can be explained by the relative ease of their installation. About half (49%) of the parcels receive organic matter (primarily animal manure but some farmers also use compost, green manure or mulch). Only 2% of parcels receive chemical inputs (some combination of fertilizer, lime and pesticides).[4]

To provide more detail on patterns of investment and input use, we calculate (not

Table 8.1. Descriptive statistics for dependent variables (estimated from survey data).

Descriptive statistics	Means	Standard deviation	Coefficient of variation
Conservation investments (m ha^{-1})	438	928	2.12
Grass strips	205	274	1.34
Anti-erosion ditches	161	270	1.68
Hedgerows	56	160	2.86
Radical terraces	1	29	25.20
Organic inputs (% of parcels benefiting)	49	50	1.02
Chemical inputs (% of parcels benefiting)	2	1.5	6.36
Land use (C value)	0.16	0.09	0.55

Summary statistics reported at the parcel level are for all holdings under cultivation or fallow (thus excluding pasture and wood lot).

shown in Table 8.1) the shares of farm land[5] receiving land-conservation measures, organic matter and chemical fertilizer. Only 15% of Rwandan farm land is cultivated without external inputs or conservation investments. Conversely, intensification using all three types of improvements (conservation investments, organic inputs, chemical inputs) accounts for only 4% of farm land. Most farm land falls between the two extremes.

Table 8.2 presents summary statistics on a wide range of variables thought to represent the five basic determinants of land use, input use and conservation investment decisions described above: (i) financial incentives; (ii) physical incentives; (iii) risk factors; (iv) wealth; and (v) the agrosocioeconomic context in which farmers are operating. The last column of the table indicates the level at which the data are applied in the models: parcel, household, sector or prefectural levels. Among the financial incentives, we report four prefecture-level indicators of prices and wages and two measures of market access. The agricultural profitability index is the average value product of labour per prefecture, calculated using aggregated sample household data, valued at market prices. Across all prefectures the average is 96 RWF, with a minimum of 60 and a maximum of 179 RWF. The distance variables represent the transaction costs of getting products to markets; these costs are not reflected in market prices. On average, farmers in all sectors covered live 4.4 km from the nearest market; the nearest paved road is reported to be an average of 21 min away on foot (with a relatively high CV of 1.06).

The non-agricultural wage can be thought of as the opportunity cost of working on one's own farm. The average value of the non-farm wage (205 RWF) is about double that of the profitability index for agriculture – suggesting that those who are able to earn non-agricultural incomes are receiving a higher return to their labour. The impact that the non-agricultural wage might have on soil quality and investment is ambiguous. Better returns off farm will compete for both labour and investment capital that could be used in agriculture. This is not necessarily bad, as labour and cash diverted to off-farm uses might reduce pressure on the land by providing cash to purchase food or encouraging less intensive land-use patterns requiring less labour (perennial crops, fallow, pasture). Greater off-farm income could also promote investments to improve soil quality because more cash is available.

Crop prices are hypothesized to affect land quality and investments through the incentives they create for soil-conserving crops (e.g. perennials, such as coffee and bananas) versus more erosive crops (such as cereals and beans). Given the cross-sectional nature of the database, we are only able to look at how differences in banana and sweet-potato prices across prefectures affect investment, but changes in the relative prices of crops across time would also be expected to affect investment behaviour.

The physical characteristics reported at the parcel level include average years operated (18), average parcel size (0.18 ha), average distance from residence measured in walking time (11 min) and average slope (13°). Farm/household-level characteristics include average share of land in fallow (14%), wood lots (7%) and pasture (2%). For each of these variables, the CV is extremely large – reflecting the highly variable capacity of farmers to leave some part of their holdings uncultivated. Figure 8.1 shows that differences in shares of uncultivated land are highly correlated with farm size. The quartile of smallest farms (ranked by arable land per adult equivalent) cultivates 86% of their arable land (with 14% in fallow, wood lot or pasture), whereas the quartile of largest farms cultivates only 57% (with 43% allocated to other uses). The Simpson farm-fragmentation index (0.64), a quantitative indicator of farm fragmentation, combines the number of parcels in a farm and their relative size. Annual rainfall, measured at the sector level, is high (1140 mm) and should therefore provide an incentive for conservation investments (to reduce erosion from runoff) and input use.

Risk factors likely to influence incentives are the percentage of parcels rented in (20% of all parcels are rented) and the percentage of price variation during the 1986–1992 period (19%). Use of the rent/own dummy to reflect risk is a common practice, but it can result in misleading results if, at the time a rental agreement is entered, the renter knows already that he/she will only have access for a fixed period, in which case there is little risk involved

Table 8.2. Explanatory variables: descriptive statistics (estimated from survey data).

Model variables	Overall mean or percentage	Coefficient of variation	Level of observation (Parcel* = 5460 HH = 1146 Sector = 78 Pref. = 10)
Monetary incentive to invest			
Agricultural profitability index (FRW)	96.3	0.40	Prefecture
Non-agricultural wage in prefecture (FRW)	205	0.34	Prefecture
Price of banana (FRW)	23.5	0.12	Prefecture
Price of sweet potato (FRW)	14.7	0.221	Prefecture
Distance to nearest market (km)	4.4	0.32	Sector
Distance to paved road (min)	20.8	1.06	Sector
Physical incentive to invest			
Share of holdings under fallow	0.14	1.16	Household
Share of holdings under wood lot	0.07	1.55	Household
Share of holdings under pasture	0.02	3.571	Household
Slope (degrees)	13	0.73	Parcel
Location on slope (1 = summit, 5 = valley)	3.27	0.36	Parcel
Farm fragmentation (Simpson)	0.64	0.33	Household
Size of parcel (ha)	0.18	1.79	Parcel
Distance from residence (min)	11	1.6	Parcel
Years operated	17.7	0.86	Parcel
Annual rainfall (mm)	1,140	0.27	Sector
Risk of investment			
Dummy for rent/own land (% rented in)	20%	–	Parcel
Price variation (1986–1992)	0.19	0.27	Prefecture
Wealth and liquidity sources			
Non-cropping income (FRW)	17,606	2.38	Household
Cash-crop income (FRW)	12,317	1.62	Household
Value of livestock (FRW)	12,582	1.68	Household
Landholdings owned (ha)	0.91	0.93	Household
Human capital			
Number of adults (aged 15–65)	2.77	0.52	Household
Dependency ratio	1.23	0.74	Household
Literacy of head of household (% literate)	49%	–	Household
Knowledge of conserv./prod. technologies	2.85	0.85	Household
Age of head of household (years)	45.9	0.332	Household
Sex of head of household (% male)	80%	–	Household
Sector-level variables			
Sector land-use patterns (C value)	0.14	0.90	Sector
Sector conservation investments (m ha^{-1})	436	1.12	Sector
Sector use of organic inputs (av. % area using)	0.71	1.01	Sector
Sector use of chemical inputs (av. % area using)	0.05	2.25	Sector

*Summary statistics reported at the parcel level are for all holdings under cultivation or fallow (thus excluding pasture and wood lot).

(Place and Hazell, 1993). The data used do not provide the information necessary to make this distinction; hence results must be interpreted with some caution.

Variables reflecting assets and capacity to invest show that the average household owns 0.91 ha and livestock valued at about 12,600 RWF and it earns about 12,300 RWF from crop sales and 17,600 RWF from off-farm activities. The distribution of landholdings is uneven, with a sevenfold difference in land per person between highest and lowest landholder quartiles. Non-farm income (wages from hired agricultural and non-agricultural work, plus own-business income) constitutes about one-third of total income; about two-thirds of households earn some non-farm income (breakdown not shown in Table 8.2). Most households own a few small ruminants; less than a quarter own cattle. Households have on average 2.77 adults of working age (15–65) and a dependency ratio of 1.23, indicating that, for each working-age adult (15–65 years) in the household, there are 1.23 non-working-age dependents (< 15 or > 65 years). Household heads are 46 years old on average, 80% male and 49% literate. The variable representing knowledge of conservation/production practices is a summated index across a set of practices such as knowledge of recommended fertilizer-use packages and measures to intensify livestock production.

The average value of the knowledge index (2.85 of a possible 9 points) is relatively low and varies significantly across households.

Sector-level variables (*secteurs*, the primary sampling units, are relatively low-level administrative districts, numbering approximately 1500 nationally) reflect the general context in which farmers are making their land-use and investment decisions. The nationwide sample of 1240 households is stratified by sector and falls into 78 sector-level clusters of 16 each (there are eight missing cases). Household observations for each of the four land-use and investment variables were averaged across households in each sector to create sector-level variables.[6] These sector-level variables can be used to represent: (i) social and administrative conditions in the immediate area; (ii) 'imitation effects'; and (iii) positive externalities of neighbours' undertaking land-protection measures. Kerr and Sanghi (1992) argue, using examples from watersheds in India, that these types of sector effects should have a positive impact on a given household's investments. The average values of the sector-level variables do not differ substantially from the overall averages based on parcel-level data. The extremely high CV (2.25) on the chemical-inputs variable, which has an average of 0.05, suggests that the share of parcels receiving chemical inputs is highly variable across sectors.

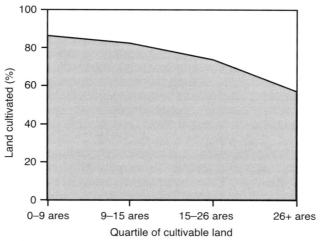

Fig. 8.1. Proportion of land under cultivation by farm size.

Determinants of Land Use and Conservation and Input Investments: Model Results

Overview of results

Table 8.3 lists regression results for each of the four independently estimated models.[7] Columns represent the different models estimated; dependent variables are used as column headings. The explanatory variables are listed in rows below the major determinants they represent. These models were estimated with observations for 1077 households and 5159 parcels that had complete data for all the variables in each of the four models.

All four models explain a great deal of the variation observed in the dependent variables as reflected in high χ^2 statistics. The Wald test statistic is largest for the land-use model (820), somewhat smaller for the conservation-investment and organic-input models (636 and 641) and much smaller for the chemical-input model (270). The smaller statistic for the chemical-input model is not surprising, given that only 2% of the parcels in the data set received chemical inputs, making it difficult to model accurately. The complex nature of household decision-making is confirmed by the large number of individually significant coefficients and the joint tests of significance for the five sets of determinants. The only instance of a non-significant joint test (< 0.05) was the 'risk of investment' category for the chemical-input model (see discussion below).

As noted previously, most models of input and land-use decisions rely on household-level data rather than a combination of parcel and household data. Results reported in Table 8.3 suggest that this could be a serious omission in a country such as Rwanda where there is substantial variation in parcel characteristics, both within zones and within households. For example, a parcel's distance from the residence is significant in two of the four models. The parcel-level explanatory variables of location on slope and ownership status are significant in three of the four models and

the parcel size and slope variables are significant in all four models.

The sector-level variables are all jointly significant, suggesting that the market context has an impact on plot-level adoption and investment decisions. The key explanatory variable in each model tends to be the sector-level variable associated with that model (e.g. the sector average for conservation investments is individually significant in the conservation-investment model, the organic-input average in the organic-input model, etc.).

Individual model results

We turn now to a more detailed discussion of the results for each of the four models, focusing on those variables that are likely to have the greatest impact on decisions to invest or adopt particular farming practices. For the conservation-investment model, the location of a parcel on the toposequence (1 = summit, 2 = just below summit, 3 = middle, 4 = below middle and 5 = lowlands/marshes) appears to be the most important determinant of conservation investments. Metres of conservation investments are highest at the summit (estimated level of 614 m ha^{-1}) and decline dramatically as one moves down the hillside (e.g. 246 m ha^{-1} at level 2). This pattern is what one would anticipate given that the need for conservation investments is greater towards the summit and very low in the lowlands. Other variables examined for their potential impact on conservation investments are the sector-level effect and the total landholdings owned; a 10% change in the sector variable stimulates an equivalent 10% change in the dependent variable, while a similar change in the landholding variable stimulates only a 5% change in the dependent variable.

The organic-input model predicts 50% adoption, but only 72% of the predictions are correct. The ownership variable is very important here. If the parcel is rented, the probability of using organic inputs declines to 22%. As the application of organic inputs is labour-intensive and the benefits of these

Table 8.3. Estimated coefficients and statistical significance (estimated from survey data).

Independent variables by group	Conservation investments (m ha^{-1}) (Tobit)	Organic inputs (probit)	Chemical inputs (probit)	Land use (C value) (linear)
Monetary incentive to invest	Joint test **	Joint test **	Joint test **	Joint test **
Agricultural profitability index	0.45	−0.00	−0.00	−0.00
Non-agricultural wage	−0.82	−0.00 **	−0.01	0.00
Price of banana	−2.04	−0.02	0.00	0.00 **
Price of sweet potato	−18.20	−0.00	−0.06 *	−0.00 **
Distance to nearest market	−4.76	0.04 *	0.11 **	0.00
Distance to paved road	−2.01	−0.00 **	−0.01 **	0.00 **
Physical incentive to invest	Joint test **	Joint test **	Joint test	Joint test **
Share of holdings under fallow	−357.96 *	−0.67 **	−0.50	−0.04 **
Share of holdings under wood lot	−28.81	0.32	−0.82	−0.03 **
Share of holdings under pasture	−990.14 **	−1.53 **	−0.57 **	−0.05 **
Slope (degrees)	15.22 **	−0.01 **	−0.03	−0.00 *
Location on slope (1 = summit, 5 = valley)	−368.62 **	−0.20 **	0.02	0.01 **
Farm fragmentation (Simpson index)	221.39	−0.13	0.15 **	−0.01
Size of parcel	3.32 **	0.01 **	0.01	−0.00 **
Distance from residence	−5.66	−0.02 **	−0.00	0.00 *
Years operated	3.37	0.01 **	−0.00	−0.01
Annual rainfall	0.04	−0.00	−0.00 NS	0.00 **
Risk of investment	Joint test **	Joint test **	Joint test	Joint test **
Land-rental dummy (0 = own, 1 = rent)	−287.04 **	−1.10 **	−0.02	0.06 **
Price variation (1986–1992)	−383.04	−2.02 **	−3.10 **	−0.08 *
Wealth/liquidity sources and human capital	Joint test **	Joint test **	Joint test	Joint test **
Non-cropping income	0.00	0.00 *	−0.00 *	−0.00
Cash-crop income	0.00	−0.00	0.00 **	−0.00
Value of livestock	0.00	0.00 *	0.00	0.00 **
Landholdings owned (ha)	−1.80 **	−0.00 *	0.00	−0.00
Human capital				
Number of adults (aged 15–65)	−11.06	0.01	−0.05	0.00
Dependency ratio	0.04	0.00	0.00	0.00
Literacy of head of household (0 = no, 1 = yes)	13.86 *	−0.04	−0.11	0.00
Knowledge of conserv./prod. technologies	22.12	0.04 **	0.02	−0.00 **
Age of head of household (years)	−0.52	−0.01 **	−0.01	−0.00
Sex of head of household (0 = male, 1 = female)	−25.84	0.07	−0.02 *	−0.00
Sector-level variables	Joint test **	Joint test **	Joint test	Joint test **
Sector land use (C value)	−749.53	−3.47 **	−2.17	0.30 **
Sector conservation investments (m ha^{-1})	0.54 **	−0.00	−0.00	0.00
Sector use of organic inputs	−56.83	0.62 **	−0.04	−0.05 **
Sector use of chemical inputs	245.33	−0.24	3.02 **	−0.02 **
Constant	1512.00 **	2.77 **	0.75	0.11
Statistics				
Wald chi-square statistic	636.23	641.31	269.96	820.12
Probability of chi-square	0.00	0.00	0.00	0.00
% Observations correctly predicted (probit only)	−	73.00	92.00	−

*Sig. T ≤ 0.05; **Sig. T ≤ 0.01.
Positive coefficients are desirable for models 1–3 and negative coefficients for model 4.

inputs are realized over time rather than entirely in the year of application, it is logical that farmers would prefer to apply organic inputs to land they were sure to have access to in the future. Distance of a parcel from the homestead influences use of organic inputs, but the size of the impact is relatively small. The sample average for parcel distance from the homestead is 11 min walking time. If this is decreased to 5 min, the probability of input use increases from 50 to 53%. Location of a parcel on the toposequence is important: at the summit (where many homesteads are located and most organic matter is collected) the probability of organic-input use is 65%; midway down the hill it is 52% and in the lowlands only 38%. This use pattern not only follows a logic of using organic matter in places closest to where it is collected but also a logic of using it where it is most needed (lowlands accumulate topsoils washed down from the hillsides and thus tend to have better fertility than higher locations).

The chemical-input model has a high rate of accurate prediction (92%) and many of the hypothesized determinants (slope, size of parcel and variables reflecting access to cash from crop and livestock income) are statistically significant. Nevertheless, changes in these explanatory variables offer little promise for large increases in chemical-input use. Ten per cent changes in the value of the slope, parcel-size and cash-access variables stimulate extremely small changes (<1%) in the probability of adoption. There is also the surprising result that the further farmers are from a market, the more likely they are to adopt chemical inputs; one would expect easier access to markets to stimulate use rather than the opposite. The lack of significance for the risk-of-investment variable represented by the rent/own dummy is not surprising. The rental status of land seldom affects chemical-fertilizer use (the main component of the chemical-input variable) because fertilizer does not have residual effects over several years, as is the case with organic fertilizers. Although this model provides some insights about factors affecting chemical-input use, until a larger number of farmers are using

chemical inputs in Rwanda or such farmers are statistically oversampled, it will remain very difficult to model the adoption decision compellingly.

The land-use model again highlights the importance of parcel location on the hillside. As one moves from the summit to just below it, there is a 6% increase in the erosiveness of land-use practices; this increases by another 5% for each move down the hillside until arriving at the lowlands. This is a logical pattern in Rwanda, as previously noted by Clay and Lewis (1990), illustrating that farmers reserve the most erosive practices for the lowlands, where erosion is less likely to occur, and the least erosive practices for the more vulnerable locations on the hillside. This attention to using more erosive practices on less vulnerable land changes on land that is rented; moving from owned land to rented land changes the estimated index from 0.14 to 0.20 – a 40% increase in the erosiveness of land-use practices (all other factors being equal). Ownership of more fallow land is also associated with an increase in the erosiveness of land-use practices, but the impact is relatively small; a 10% increase in fallow land results in a 3% increase in the land-use index (i.e. more erosive practices).

Policy and Research Implications

The focus of government agricultural policy in the postwar period has centred on agricultural modernization through intensification and commercialization. As this policy is pursued, research identifying the determinants of farmer investments in conservation practices and soil-enhancing inputs can provide policy-makers with valuable information, which they can use to design better agricultural policies and programmes. The research reported here confirms that all five hypothesized determinants of adoption – financial incentives, physical incentives, risk, wealth and agrosocioeconomic context – play a role in shaping farmers' investment and input-use behaviour.

An examination of the specific variables used to operationalize each of these five determinants reveals that conservation investment and input use are strongly affected by parcel characteristics, such as the position of the parcel on the hillside, its slope, parcel size and the time it takes to walk from the homestead to the parcel. The finding that larger parcels and those closer to the homestead are more likely to benefit from conservation investments and input use suggests that policies preventing further deterioration in the average size of parcels or policies to promote land and housing patterns permitting farmers to live close to their fields are likely to stimulate intensification. A case in point is the Rwandan government's current villagization policy, which is intended to provide housing for those displaced by the war in a way that will cut the cost of social services (water, electricity, health care, schools, etc.) and simultaneously stimulate the growth of commercial centres in rural areas. However, the location of households in villages, where they reside much further from their fields than in the past, may have the unanticipated consequence of limiting conservation investments and particularly the use of organic fertilizers. Similarly, the knowledge that conservation investments and input use are much less likely to be applied on certain parts of the hillside can be used by extension agents for targeting their activities.

Rented land is less likely to receive conservation investments and inputs than owner-operated parcels. This implies that Rwandan farmers need confidence in the longer term through secure land tenure. In turn, this means not only helping farmers to farm their own rather than rented land by legalizing land transactions, but also reducing the perceived risk of appropriation – which has probably been unusually high during this postwar transition period. Enhancing farmer access to land markets will require reform of existing and antiquated land laws.

Recent developments on land-related issues in Rwanda have been encouraging. The importance of land issues to the government is clearly evidenced by the formation of a new Ministry of Lands and Settlement, created in February 1999. The government has a proposal for new land regulations, which will be going before cabinet/parliament in the near future. The new regulations are aimed at ensuring land tenure to individual farmers and facilitating land transactions.

Another lesson from all four models is that sector-level patterns of conservation and input use can serve as a stimulus for the further promotion of adoption. If the general context in which farmers are making their land-use and investment decisions favours conservation and investment, individual farmers will be more likely to move in the same direction. This implies that sector-level investments in extension, markets and infrastructure that get a few farmers moving in the right direction will have a multiplier effect in helping to spread adoption to others.

The underlying message of this analysis for policy-makers is that conservation and input-investment decision-making are multidimensional and complex. These characteristics are highlighted by the large number of statistically significant variables in the estimated models, each offering a small contribution to the overall decision to invest or not to invest. This implies that major changes in conservation and input investments will require attention to the broader package of determinants, as no one factor can be singled out and thus no single policy instrument offers great leverage to induce greater soil conservation.

Notes

[1] We thank the Division des Statistiques Agricoles (DSA) of the Rwandan Ministry of Agriculture, Animal Resources, and Forests (MINAGRI) for provision of the data. We thank USAID/AFR/SD/PSGE (FSP and NRM), USAID/Kigali, and AID/Global Bureau, Office of Agriculture and Food Security, for funding via the Food Security II Cooperative Agreement.
[2] The C-value index reflects the overall protective quality of crops. It is defined as 'the ratio of soil loss from an area with a specific cover and tillage practice to that from an identical area in tilled continuous fallow'; a higher C value indicates

more erosive land-use practices (Wischmeier and Smith, 1978).

[3]   However, there have been previous efforts to model investment decisions at the parcel level in Rwanda (Blarel *et al.*, 1992; Place and Hazell, 1993; Blarel, 1994).

[4]   Only 0.08 kg ha^{-1} of fertilizer were used in rural Rwanda in 1991 – substantially less than is used in cash-cropping areas of highland Kenya and Uganda (Byiringiro, 1995). Current estimates show levels in the 4–6 kg ha^{-1} range (Kelly *et al.*, 2001).

[5]   We used share of farm land rather than share of households, because many households use inputs on only a small share of their land, thereby rendering figures on the share of households misleading.

[6]   The sector average for each household excludes the observation on that particular household, thus the means obtained are 'non-self' means.

[7]   The models were estimated using a random-effects (population averaged) procedure, described in Liang and Zeger (1986), to control for possible bias due to correlations among the multiple parcel-level observations for each household. A Breusch and Pagan Lagrange multiplier test for the linear model corroborated the superiority of the random-effects procedure over simple OLS estimation.

9 Agroforestry Adoption Decisions, Structural Adjustment and Gender in Africa

Christina H. Gladwin,[1] Jennifer S. Peterson,[2] Donald Phiri[3] and Robert Uttaro[4]

[1]Department of Food and Resource Economics, University of Florida, Box 110240 IFAS, Gainesville, FL 32611-0500, USA; [2]Consultant, Niamey, Niger; [3]World Vision, Zambia; [4]Department of Political Science, University of Florida, Gainesville, FL 32611, USA

In Africa, as in the rest of the world, women are traditionally in charge of reproductive activities in the household (i.e. care of the children, adult family members and the home), as well as some productive activities. What makes Africa unique is the large extent of women's involvement in food-crop production. In most African societies – if one can generalize at all across the myriad diversity of African societies – women provide half of the labour force in agriculture and produce most of the subsistence food crops consumed by the family, while men produce export and cash crops. Women's yields are generally low, however – too low by green-revolution standards and much lower than men's yields in societies where a comparison can be made, e.g. where women and men grow the same crops on different fields or yields of female-headed households (FHHs) can be compared to those of male-headed households (MHHs). In these situations, gender differences in productivity have been shown to be due to differences in the intensity of use of productive inputs, such as fertilizer, manure, land and labour, credit, extension training and education, rather than to differences in the efficiency or management styles of men

and women (Quisumbing, 1996). Because women farmers lack access to yield-increasing inputs of production, they tend to produce less and more of their crops are consumed within the family (Due and Gladwin, 1991; Gladwin, 1996, 1997a,b). Estimates show that, if productive inputs like fertilizer, manure and labour could only be reallocated within the African household from men's to women's crops, in some societies the value of household output could increase by 10–20% (Udry, 1996).

Although the literature on African women in development shows that separate income streams of men and women in African households give some autonomy to African women, women's incomes do not necessarily give them power, which usually accrues to the household head, most often a male relative. The relative powerlessness of African women as compared with men is symbolized by the long hours they spend head-loading water and firewood and by their devotion to subsistence crops rather than cash crops, as well as their lack of political voice.

Yet women are often *de facto* female household heads for some period in their lives, so that 25% of African households are

FHHs, where women have relatively more autonomy and decision-making power in the household than women in MHHs (Due, 1991). They are generally poorer, however, and therefore less powerful than MHHs in their rural communities: Due's (1991) data from Zambia and Tanzania show that FHHs have less adult labour, less access to credit and smaller incomes than MHHs, plant smaller crop acreages and more subsistence crops relative to cash crops and are not as productive as MHHs. Quisumbing (1996) notes that this is not directly due to their gender but rather to their low incomes, which prevent their purchase of modern, yield-increasing inputs, such as fertilizer, hired labour, etc. Gladwin *et al.* (2001) point out that, due to their greater poverty, FHHs have a greater tendency to be chronically food-insecure than do women in MHHs.

Impacts of Structural Adjustment Programmes

Due to their lower incomes, rural women and especially rural FHHs are considered a vulnerable group. As such, they are the first to suffer when a macroeconomic downturn or recession hits and the last to recover from it (Elabor-Idemudia, 1991). Women in particular have borne the social costs of structural adjustment programmes (SAPs) in sub-Saharan Africa (SSA). Because women are in charge of reproduction in the household, they suffer when the costs of food, education, health care and medicines rise due to government budget cutbacks mandated by SAPs (Meena, 1991). Because women are in charge of the household granary, they suffer when currency devaluations and the removal of fertilizer subsidies result in the rise of fertilizer prices to such an extent that its use on hybrid varieties is unprofitable and unaffordable (Gladwin, 1991; Bumb *et al.*, 1996) and they have to switch back to local unfertilized varieties with lower yields and watch their granaries empty earlier in the hungry season. Because gender ideologies tell women they are the ones responsible for feeding the family

(Goheen, 1991), they especially suffer when the hungry season lengthens as a result of structural adjustment reforms (Uttaro, 1998).

This is not to say that the 'bitter pill' of structural reform in SSA was unnecessary. By now, most observers realize that globalization demands changes in the way open economies formulate macroeconomic policies and finance budget expenditures. No longer can individual countries hang on to an overvalued exchange rate and negative current-account balance for very long. Since the early 1980s, African governments have thus been forced to learn the rhetoric of stabilization, fiscal and monetary policies and market-liberalization measures.

The deflationary measures mandated by structural reform, however, have had the most severe impact on women in African households, especially on FHHs, which tend to be poorer and more vulnerable. Because most FHHs are net buyers, not net sellers, of food crops and sell little, if any, export crops or tradables encouraged by SAPs, they are unable to benefit from increased price incentives under market-liberalization programmes (Mehra, 1991). FHHs thus suffer when the price of food is allowed to rise so that fertilizer use on food and cash crops once again becomes profitable, especially if the government has no safety-net programme in place to ameliorate the negative impacts of SAPs. They also suffer when the government's safety-net programmes do not treat them as producers but only as consumers of food, and make them more dependent on government hand-outs of subsistence crops that they can grow themselves.

Agroforestry Innovations

It is within such a setting that we should examine the potential of recent agroforestry innovations extended to farmers in Africa and note their usefulness to women farmers as well as men. If women farmers as well as men adopt agroforestry innovations, then we may conclude that they are indeed gender-neutral innovations worth diffusing on a continent where women produce most of the subsistence food crops. If the poorest quintiles of the population, including FHHs, also adopt agroforestry innovations, then we may hope that these innovations may make Africa more food-secure, while also avoiding the 'second-generational' income-distribution problems associated with the green revolutions of Asia in the 1960s and Latin America in the 1970s (Falcon, 1970; deJanvry, 1981). It is the purpose of this chapter to show that, under particular conditions, certain agroforestry innovations – namely, improved fallows – have the potential of being gender-neutral, scale-neutral soil-fertility technologies adoptable by women as well as men, by the poor and food-insecure as well as the food-secure.

Yet the original rationale for introducing agroforestry innovations was not to provide a gender-neutral, scale-neutral soil-fertility technology. Originally, the sole aim was to stop soil-fertility depletion on smallholder farms, which Sanchez *et al.* (1997b) claimed to be 'the biophysical root cause of declining per-capita food production in Africa'. Smaling *et al.* (1997) estimated that soils in SSA were being depleted at annual rates of 22 kg ha^{-1} for nitrogen (N), 2.5 kg ha^{-1} for phosphorus (P) and 15 kg ha^{-1} for potassium (K). Soil-fertility depletion was all the more alarming, given that recurring devaluations and removal of fertilizer subsidies due to SAPs had made inorganic fertilizer unaffordable in most Africa countries by the mid-1990s (Bumb *et al.*, 1996).

Sanchez and colleagues (1997b) recommend a two-pronged strategy to stop this mining, the first to replenish phosphorus nutrients and the second to replenish nitrogen. The first strategy involves the high-phosphorus-fixing soils of Africa, an estimated 530 million ha where phosphorus fixation is now considered an asset, and not a liability as previously thought. Here, inorganic phosphorus fertilizers are necessary to overcome phosphorus depletion (Jama *et al.*, 1997). They become 'phosphorus capital' as sorbed or fixed phosphorus, almost like a

savings account, because most phosphorus sorbed is slowly desorbed back into the soil solution over 5–10 years. Phosphate rock, moreover, can be helped to desorb by the decomposition of organic inputs that produce organic acids to help acidify the phosphate rock, e.g. the organic acids in tithonia (*Tithonia diversifolia*), a common shrub in western Kenya.

To reverse nutrient depletion of nitrogen, Africa needs a second strategy, amounting to an increased use of organic sources of nitrogen from animal manures and compost, biological nitrogen-fixation technologies, biomass transfers of organic matter into the field and more efficient use of trees and shrubs, whose deep roots capture nutrients from subsoil depths beyond the reach of crop roots and transfer them to the topsoil via decomposition of tree litter. One proposed solution to soil-fertility depletion is thus to import Minjingu phosphate rock from Tanzania and to replace nitrogen via agroforestry innovations, such as hedgerow intercropping with leucaena (*Leucaena leucocephala* (Lam.) De Wit), biomass transfer with tithonia, manures improved with calliandra (*Calliandra calothyrsus* Meissner) and improved fallow (IF) systems using nitrogen-fixing shrubs like sesbania (*Sesbania sesban*) or tephrosia (*Tephrosia vogelii*).

Questions persist about this innovative approach to Africa's soil-degradation crisis, however, and centre on the question of whether the nitrogen demands of food crops can be met in full with only organic sources of nutrients. International Centre for Research in Agroforestry (ICRAF) scientists claim that, biophysically, organic sources can produce mid-range level yields of 4 tons ha^{-1} but not 6 tons ha^{-1}, where combinations of organic and inorganic fertilizers are needed because recovery of nitrogen by the crop from leaves of leguminous plants is lower (10–30%) than recovery from nitrogen inorganic fertilizers (20–40%). To reach these higher crop yields, more research is needed on the synergistic effects of combining the different kinds of organic and inorganic fertilizers (Palm *et al.*, 1997).

Agroforestry innovations: a solution for African women farmers?

More research on the social-science side is also needed to answer questions such as: who is adopting agroforestry innovations? Do women farmers adopt them as much as men do? Or do they face constraints to adoption more severe than those facing men farmers, thus limiting adoption? Do women have different motivations and reasons for adopting from those of men? Finally, do women in FHHs differ from women in MHHs and, because of their poverty, do they adopt less than women and men in MHHs?

Previous ethnographic and policy research suggests that women have more limiting factors to adoption than men. Rocheleau (1995) found an interaction between gendered property relations and gendered resource uses, user groups, landscapes and ecosystems in western Kenya, a region of high population density and small farms. Before colonization, the decisions about where sons would cultivate was left to the son's mother to decide; wives and daughters had only usufruct rights to the land and its products. Under the land-reform laws of 1956, men aged 18 and over were automatically entitled to land title from the colonial government (Pala-Okeyo, 1980). This policy change lowered women's status in the lineage system, since sons no longer had to go through their mothers to acquire land. Scherr (1995) subsequently found that gender differences in agroforestry practices were quite significant. In one study, men had 50% more trees on their farms and an almost 30% higher tree density. They also tended to plant trees in crop land while women's farms had more trees used primarily for fuel wood. Women were also subjected to the authority of a man before making most decisions, whereas men were more free to take risks and experiment with new technologies. As we shall see below, this power differential between men and women lays the foundation for gender bias from household-level decisions to policy-level decisions.

Agroforesty Adoption Decision-tree Models

In order to definitively answer questions about whether or not factors like land availability and power or authority influence agroforestry adoption decisions, researchers should propose a testable model of the adoption decision process and test it on a gender-disaggregated sample of both adopters and non-adopters. This has been done with respect to soil-fertility amendments in general and adoption of agroforestry innovations in particular, using 'ethnographic decision trees' or hierarchical decision models,[1] whose usefulness comes from their relatively high prediction rate (at least 80%). For example, decision-tree models have been generated of farmers' decisions to use chemical fertilizer versus manure in Guatemala and Malawi (Gladwin, 1989, 1991), to increase fertilizer use in Mexico (Gladwin, 1975), to use credit for fertilizer in Mexico, Malawi and Cameroon (Gladwin, 1992) and repay it in Malawi (D'Arcy, 1998; Uttaro, 1998), to adopt other agroforestry technologies, such as hedgerow intercropping, in Kenya and Malawi (Swinkels and Franzel, 1997; Williams, 1997) and to use grain legumes as soil-fertility amendments in Malawi and eastern Zambia (D'Arcy, 1998; Uttaro, 1998; Peterson, 1999; Peterson *et al.*, 1999).

Decision trees predict because they are cognitive-science models, which aim to process information in the same way humans do (Simon, 1979), as opposed to artificial-intelligence methods which are not so concerned with modelling the exact process that humans use but seek some alternative processing technique that approximates the human solution, e.g. linear programming models or probit analysis. Because cognitive-science models aim to represent psychological reality and to mimic the mental processes people use, they should be better descriptions of human information processing and better predictors of human choice.

The decision trees are relatively simple to design and test, as Uttaro's model of the decision to adopt IFs in southern Malawi shows (Fig. 9.1). Read from top to bottom, they have alternatives in set notation ({Use Improved Fallows; Don't Use}) at the top of the tree, decision outcomes in boxes ([Don't plant an Improved Fallow]) at the end of the paths of the tree, and decision criteria in angle brackets (<Farm large enough to leave a portion of it fallow?>) at the nodes of the tree. There are only two alternatives or decision outcomes in this set – [Plant an Improved Fallow now] and [Don't] – and they are mutually exclusive. The only trick to the trees is eliciting the decision criteria from the decision-makers themselves, who are the experts in making their decisions. They alone know how they make their choices, and so their decision criteria should be elicited from them in ethnographic interviews or by participant observation and other participatory methods (e.g. role playing). Uttaro (1998) elicited this particular tree from 20 decision-makers and then tested it on another sample of 60 farmers via questions on a formal questionnaire designed after he had elicited these decision criteria.

Given a sample of data from decision-makers, e.g. Uttaro's (1998) 60 farmers interviewed in southern Malawi in 1997/98, one can test the tree easily by putting the data from each individual choice (as a separate, independent Bernoulli trial) down the tree and counting the errors in prediction on each path. This model is a simple one, because most individuals in the sample say 'No' to the first criterion, 'Farm large enough to leave a portion of it fallow', and so 52 out of 60 farmers go to the outcome [Don't plant an Improved Fallow now]. The results of testing this simple tree show no errors, meaning that 60 cases were sent to the outcome [Don't plant an Improved Fallow], and in fact no one did plant an improved fallow. With more variability in outcomes in the sample, however, the researcher should expect to find more prediction errors.

When a decision tree is correctly specified, it allows the research team to identify the main factors limiting adoption at a specified time and, if possible, to

recommend policy interventions to alleviate these constraints and speed up adoption (Gladwin, 1975, 1979). These limiting factors may change or disappear over time, however. The model is assumed to be valid only for the time period during which it is tested and should be retested at later times. Given low adoption rates, the research team may gradually conclude that the chances of much future adoption of the technology are not good, if there are a number of structural factors persistently blocking adoption (e.g. lack of land) that are not amenable to policy intervention (as opposed to limiting factors that are changeable, e.g. lack of knowledge or seeds or credit). In this case, the usefulness of the adoption decision-tree model lies in sending the designers of the technology, the biophysical scientists, back to the drawing-board to redesign.

Applications of decision trees to agroforestry adoption choices

Much adoption work has been done by ICRAF social scientists using ethnographic decision-tree modelling on the adoption and expansion of hedgerow intercropping or alley cropping (David, 1992; Shepherd *et al.*, 1997; Swinkels and Franzel, 1997). Their work in western Kenya showed that women farmers' constraints of lack of knowledge, labour and land did not allow many of them to plant hedges of leucaena or calliandra in between rows of maize, the subsistence crop. Their conclusions were corroborated by Williams (1997), who interviewed 40 women farmers in Maseno, western Kenya, and found less than 20% adoption (Gladwin *et al.*, 1997b), due to both structural factors (e.g. lack of land (five cases) and labour (four cases)) and limiting factors more amenable to policy change (e.g. lack of knowledge (15 cases) and seeds (two cases) and termite problems (two cases)). Williams (1997) concluded that the future prospects for women's adoption of hedgerow intercropping in western Kenya were not good.

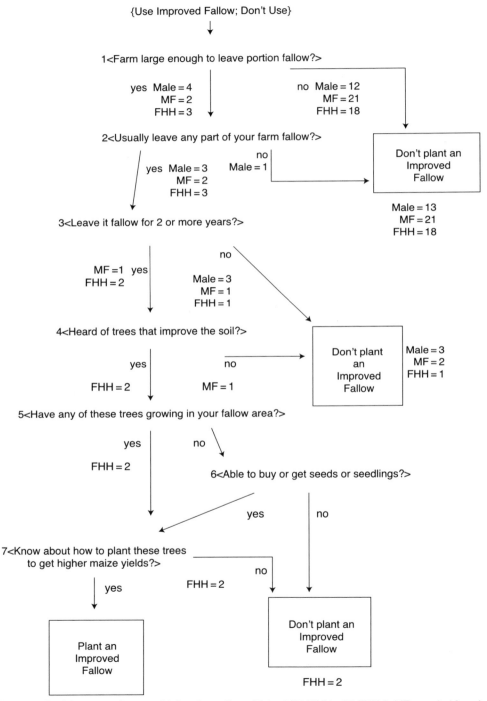

Fig. 9.1. Decision to use improved fallow in southern Malawi (39 MHHs, 21 FHHs). MF, married female.

Adoption of biomass-transfer technologies

Williams (1997) also modelled women farmers' decisions to adopt or not adopt biomass transfer innovations with a subsample of 23 women farmers in the same region.[2] Biomass transfer involves the use of leaves and stems from shrubs (*T. diversifolia* and *Lantana camara*) for mulch. These shrubs are homestead border markers found everywhere in rural western Kenya and are under the control of women, but they are traditionally used for goat fodder and medicine for stomach ailments and not for mulch. Williams's (1997) model of women farmers' decisions not to use tithonia leaves and cuttings as mulch on their food crops suggested that most women lacked access to tithonia or lantana shrubs growing nearby or did not know about the technology. Labour was also limiting, because many female heads of households and women with small children felt they did not have the time themselves or access to the labour required to cut and carry enough biomass from these shrubs to mulch their crops adequately. The amount of biomass required to produce significant soil-fertility benefits is large – by some estimates, 7 tons ha^{-1} of leafy dry matter (and triple that for fresh biomass) (Jama *et al.*, 1997). Other women had problems with termites and no practice, like applying ash, to help with this problem. Still others felt they needed the tree or shrub more for fodder or medicine than for soil improvement. The cumulative result of all these constraints was that only three of the 23 women in the sample used tithonia for soil improvement.

The question of whether these results could be replicated in other parts of Africa was then taken up by Robin D'Arcy (1998), who modelled women farmers' decisions to adopt biomass transfer using *Faidherbia albida* (called the *msangu msangu* tree), an indigenous tree common in the study area in Dowa, Malawi. *Faidherbia albida* fixes nitrogen with its roots while young and later drops leaves to transfer nitrogen and biomass in the rainy season, thereby allowing more sunlight to pass down to crops. In the dry season, it restores its canopy and is therefore not a labour-intensive technology. Yet its effect is limited, because only one or two trees per field grow to maturity in 10–20 years.

How do women farmers in Dowa assess the agroforestry potential of *F. albida*? Its adoption decision tree (for brevity, not presented here), built during 20 initial interviews with women farmers and tested in interviews with another 60 women (41 women in MHHs, 19 in FHHs), shows women's use of it is basically to be a question of access to seeds or seedlings of *msangu* or having a mature *msangu* tree already growing on one's fields. Sixteen respondents had mature *msangu* trees on their land and 22 had additional seedlings given to them. In all, 29 of 60 informants had *F. albida* on their fields. Most women farmers knew of the tree because *msangu* is an indigenous species. Of the whole sample, 68% of respondents knew how to plant *F. albida* or felt they could find out how, and 98% of all informants said *msangu* leaves helped the soil. Most did not have problems with pests, such as termites. No women farmers reported land or labour constraints with *msangu*. At face value, therefore, agroforestry technologies using *msangu* trees seem promising with women farmers in Dowa.

Improved-fallow adoption decision trees

Uttaro (1998) modelled and tested 60 farmers' decisions to use IF technologies in Zomba, southern Malawi.[3] Uttaro's decision tree (Fig. 9.1) shows that lack of land was the most serious constraint to IF adoption in the Zomba region; most of the households engaged in continuous cropping. In his sample, nine informants (15%) had farms large enough to leave part of it fallow and eight (13%) usually left part of the farm fallow (criteria 1 and 2). Only three of the eight farmers left their land fallow for 2 or more years (criterion 3). Of the three informants left, only two FHHs had any trees or shrubs that improve soil fertility in fallow areas (criterion 5). They both lacked the knowledge of how to plant an IF in order to get higher yields after returning the land to maize production (criterion 7). In short, no farmer of the 60 used IFs.

The prospects for IF systems in southern Malawi thus appear poor. Even if information were disseminated about the use and management of trees and bushes in fallow systems, farmers would still need to have land available to place into fallow. And, with a population growth rate among the highest in Africa, that is something unlikely to occur in southern Malawi.

Eastern Zambia: the exception that proves the rule?

Our initial ethnographic results in three African locations prior to 1998 were discouraging, as they showed that women farmers tend not to adopt agroforestry innovations, such as biomass transfers, hedgerow intercropping and IFs. Why? The main limiting factors were lack of knowledge of the new technology, lack of access to seeds or seedlings and lack of cash or credit to acquire them. Yet structural factors – lack of land and labour – were also limiting women's adoption and posed more serious problems to adoption prospects than did factors more amenable to policy intervention, such as lack of knowledge or seedlings. Moreover, structural constraints were much more severe for women than for men, and even more severe for FHHs. We were therefore discouraged about the chances of agroforestry innovations replacing inorganic fertilizers as women's soil-fertility management technique of choice in the near future.

But could we extend these results to all of sub-Saharan Africa? In a word, no. Conditions in Africa are so diverse, location-specific, dynamic and dependent on historical contingencies and socio-economic specificities that results which held in western Kenya in 1996 and in Malawi in 1998 could not be generalized to later times and other locations in Africa.

More recent research results from on-farm trials of IF systems with *S. sesban* in eastern Zambia seem to agree (Franzel *et al.*, 1997; Kwesiga *et al.*, 1997; Peterson, 1999;

Peterson *et al.*, 1999). In 1988, ICRAF began to test IF technologies at Msekera Research Station, eastern Zambia, and in 1992/93 some on-farm trials of the IFs began. IF plots, ranging in size from 10 m by 10 m to 30 m by 20 m, are planted for 2 years with nitrogen-fixing tree species (*S. sesban* or *Gliricidia* seedlings or direct-seeded *T. vogelii* or *Cajanus cajan* (pigeon pea)), then followed by 2–3 years of maize. *S. sesban* is by far the most promising, although it may look like a 'dinky little tree'. *Sesbania* is grown in a nursery for 3–6 weeks before the rainy season. Results over the 5-year cycle showed that IFs increased total maize production by 87% over unfertilized maize (even without any yield in years 1 and 2), although estimates varied about the advantage of IFs over maize fertilized with 112 kg N ha^{-1}. Kwesiga and Beniest (1998) found that maize yields following 2-year IFs approach those of fully fertilized fields, but Franzel *et al.* (1997) found that fully fertilized maize yields were 2.5 times more than IFs over 5 years. The differing estimates did not matter for farmers, however, because, with the rising prices of fertilizer in the Eastern Province, fully fertilized maize was no longer an option. In many cases, even partially fertilized maize was not an option because farmers had neither the cash nor the access to credit to purchase fertilizer. By 1997, therefore, the multi-year trials of IF technologies were a major success story. Over 3000 farmers had participated, 49% of whom were women farmers (Franzel *et al.*, 1997).

Yet the question still unanswered is why are IFs being adopted so readily in eastern Zambia, especially by women, and not in central and southern Malawi? Is their success due to the fact that eastern Zambia is a region of lower population density than the other regions so that women farmers have enough land to put some of it in fallow? Or is it just a delayed reaction to structural adjustment policies that have raised inorganic-fertilizer prices to levels so high that women farmers have finally adjusted by deciding to 'grow their own fertilizer' and adopt a substitute soil-fertility amendment? To answer this question, Jen Scheffee

Peterson interviewed women farmers who both were and were not testing and expanding their on-farm trials of IFs.

In 1998, men and women adopters and non-adopters were interviewed in each of the four villages targeted by ICRAF with on-farm trials of IFs since 1992/93 (Peterson, 1998). After an initial composite model was built and refined, we designed a questionnaire to test the revised composite decision model (Figs 9.2 and 9.3) during personal interviews with another sample of 81

women farmers and 40 men farmers in the camps surrounding the four villages (Peterson, 1999). Women in both FHHs and MHHs were interviewed. The samples were chosen so that half the sample of each gender would be testers who planted at least one IF plot, and half non-testers, who did not plant even one IF plot. Half of the sample of testers would be testers–expanders, who planted at least two IF plots, and half testers–non-expanders, who planted only one IF plot.[4] The model in Figs 9.2–9.3 has 'descriptive

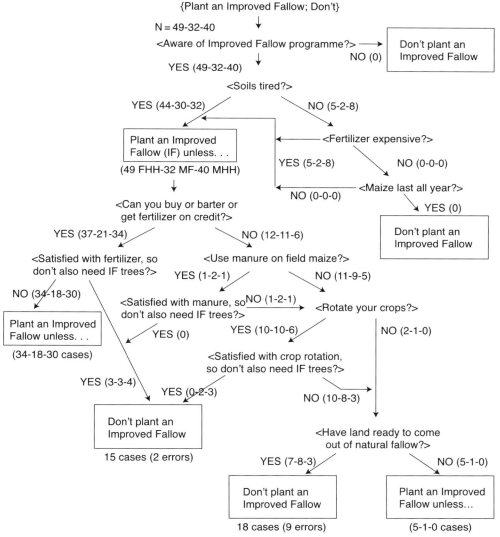

Fig. 9.2. Model 1, eastern Zambia, 1998, 121 cases (49 FHHs, 32 MFs, 40 MHHs).

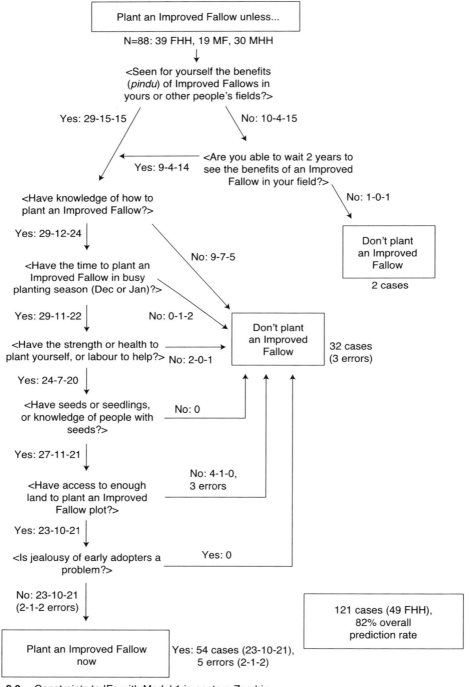

Fig. 9.3. Constraints to IFs with Model 1 in eastern Zambia.

adequacy', meaning that it matches informants' statements about how they decided to plant an IF.

Motivations to plant trees

Nearly all women say they plant an IF because their soils are tired (*nthaka yosira/ yoguga*), fertilizer is too expensive (*wodula ngako*) or their maize harvest does not last all year until the next harvest. Figure 9.2 shows that any one of these reasons is enough for a farmer to consider planting an IF and thus sends them (i.e. their data) to the outcome 'Plant an Improved Fallow unless . . .'. In the eastern Zambian sample, every farmer had at least one of these reasons to plant an IF and thus the whole sample passed on to the first set of 'unless conditions', constraints which will block a farmer from planting an IF even though she or he has a good reason to plant.

Constraints to planting an improved fallow

Figure 9.2 also lists the first set of constraints. If farmers are already satisfied with their current soil-fertility amendments, they do not also need to plant an IF. Farmers are therefore sent to the outcome [Don't plant an improved fallow] if they can buy or barter for fertilizer, they have used manure on field maize in the recent past, they rotate crops in the field (e.g. groundnuts with maize with cotton) or they have land ready to come out of a natural fallow now and they are satisfied with this extant technique.

Most farmers can either buy or barter or get some fertilizer on credit. Whereas men mostly buy fertilizer, women (especially in FHHs) mostly barter for fertilizer. In this post-SAP era, almost no one gets credit for fertilizer in eastern Zambia. Probably as a consequence, almost no one is satisfied with the amount of fertilizer acquired, as usually it is much less than what they used pre-SAP. In addition, almost no one uses manure on maize. Manure is saved for garden

vegetables grown in the *dimba* in the dry season and is not usually used on field maize. Finally, almost all farmers rotate their crops as a soil-fertility measure, but that does not satisfy their need for more soil nutrients. Results further show errors with the fallow criterion 'Have land ready to come out of fallow now?' but, when we omit this fallow criterion, there are more errors in the model (29 vs. 21), so, with these data, the fallow variable clearly helps the prediction rate.[5]

If the farmer is satisfied, he or she moves to the outcome [Don't plant an IF now]. If the farmer is not satisfied, and also feels a need for the soil-fertility amendment by IF trees, he or she is sent to the outcome [Plant an IF plot unless . . .], meaning that the farmer must pass another set of constraints in order to go to the outcome [Plant an IF]. These constraints (Fig. 9.3) start with a benefits criterion ('Have you seen for yourself enough benefits of IF?'). If yes, farmers are asked if they can wait for 2 years to see the benefits. Because of ICRAF's intense work in these four villages, most farmers have either seen the benefits of IF plots on their or their neighbours' land, so most are willing to wait the 2 years until the maize harvest after the IF.

Most (86) farmers in this sample proceed to the other constraints: lack of technical knowledge of how to plant the IFs (planting the nursery, transplanting the seedlings or direct-seeding tephrosia), lack of time to plant an IF during the busy rainy season, lack of strength and health, lack of access to seeds or seedlings and lack of land. In addition, farmers were asked if their only access to land was to borrowed land (so they would not plant an IF) or if villagers' jealousy of early adopters of IF might be a problem.[6] Results show that only 54 of 86 farmers pass all these latter constraints and are predicted to adopt. The most important limiting factor (for 21 farmers) is lack of technical knowledge of how to plant an IF. Of the 86 farmers who make it down the tree to this constraint, lack of technical knowledge is a limiting factor for more married women (37%) than FHHs (24%) than men (17%). This gender difference is expected, based on previous literature, and it affects adoption. This model predicts adoption for only 31% of the married women in MHHs, compared with 47% of the FHHs and 52% of the men in MHHs. There are 22 total errors in the model, for an overall 82% success rate.

Conclusion

This chapter shows that, under certain conditions, namely land availability and unaffordable fertilizer, IFs are potentially gender-neutral, scale-neutral soil-fertility technologies adoptable by women as well as men, by the poor and food-insecure as well as the food-secure. The evidence comes mainly from eastern Zambia, where in the four ICRAF villages an almost equal amount of FHHs (47%) and men in MHHs (52%) have adopted improved fallows. Indeed, the laggard group seems to be married women in MHHs (31% adoption), some of whom say they lack 'authority' to decide to plant IFs for themselves.

Why are IFs being adopted so readily in eastern Zambia? Their success is partly due to greater land availability and lower population densities than in southern Malawi, where no surveyed farmer adopted IF. Figure 9.2 also suggests that the relatively high adoption rates in eastern Zambia are a delayed reaction to structural adjustment policies that have increased the price of inorganic fertilizers to levels so high that farmers, especially poorer FHHs, have finally decided to 'grow their own fertilizer', regardless of the labour requirements of planting and transplanting trees during the busy rainy season. Decision-tree results (Fig. 9.2) show that few women farmers now have cash and none have credit for fertilizer. Most can only barter for inorganic fertilizer, often receiving poor terms of trade. So they are now open to finding a substitute for fertilizer, even if it means 'growing their own'.

Both the pull factor of land availability and the push factor of unaffordable fertilizer are causing women and men, FHHs and MHHs, to adopt IFs in eastern Zambia. Evidence from Malawi (Fig. 9.1) shows that

farmers who do not have enough land to consider planting a natural fallow also do not consider planting an improved fallow. Results of testing the Zambian model are not as clear. They do not clearly show whether the effect of farmers' use of natural fallows has a positive or negative impact on farmers' planting an IF, because there are too many errors at the fallow criterion 'Have land ready to come out of natural fallow', where a natural fallow is assumed to be a substitute for an IF. Further research needs to clarify this question of whether natural fallows are substitutes or complements for IFs.[7]

How does gender affect these decisions? Unexpectedly, we found in eastern Zambia that married women have lower adoption rates and more limiting factors to adoption than do women in FHHs. This seems to be because they face an authority constraint (*malamuno* or *mphavu*) that household heads of either gender do not face, i.e. they lack the authority or mandate to decide for themselves whether or not to plant an IF and must ask their spouses for permission. This authority criterion was further tested (with success) with another sample in eastern Zambia, but for brevity's sake those results are not presented here. Finally and unexpectedly, where FHHs were directly targeted by ICRAF staff in eastern Zambia with new IF technologies, we did not find FHHs to be more severely constrained than either men or women in MHHs. We found FHHs to be adopting at rates almost as high as men.

Do our results imply that researchers have to reinvent the wheel and elicit all new decision criteria every time agroforestry research is done in a new location or at a later time? Not entirely. In similar decision contexts (e.g. western Kenya, Malawi), we would expect some criteria to be widely shared by women farmers, including: access to seeds or seedlings or cash to acquire them, knowledge of the new technology, previous beneficial experience with it, problems with termites, preferred use of the trees (for fodder, fuel wood or soil-fertility improvement) and women's lack of land and labour. These decision criteria are amazingly similar, although the order and phrasing of the shared criteria are different. Given these similarities, researchers should expect to find similar criteria in other locations in Africa and should thus ask informants specifically about them. In different decision contexts (e.g. eastern Zambia), the context framing choice is completely different and therefore we would expect to elicit some different decision criteria and even new branches of the decision tree. Researchers starting fieldwork in a new location should therefore keep their options open and use open-ended questions of the type traditionally used by ethnographers, as well as practising tried and true ethnographic methods, such as immersion techniques, participant observation and field notes, utilizing techniques to elicit new and unexpected decision criteria.

Notes

[1] The term hierarchical decision models distinguishes decision trees from linear additive models, such as linear regression analysis, probit analysis or logit analysis. The term 'hierarchical' refers to the fact that the decision criteria or dimensions are mentally processed in a certain order such that alternatives are compared on each dimension or criterion separately, and criteria or dimensions are ordered so that all of them may not be processed by all individuals. This simplifies the decision process considerably and saves the individual cognitive energy. A linear-additive model, in contrast, assumes that all the criteria or dimensions of each alternative are weighed by the decision-maker, each alternative is assigned a composite score and the alternative with the highest score is chosen. Much debate about these two types of models of the search-for-information process has occurred between psychologists (Rachlin, 1990: 76–77).

[2] Williams used two samples of women, one to build the adoption models and one to test them. Both samples included (*de jure* and *de facto*) FHHs, members and non-members of women's groups (high to low resource, newly and well established), and women generally considered to be of above-average, average and below-average wealth according to such socio-economic criteria as farm size, house type, numbers and types of livestock, etc. The sample of women used to build the models consisted of 25 Luo women, while the sample used to test the models was made up of both Luo and Luhya women (10 and 13, respectively).

[3] D'Arcy (1998) in Dowa, Malawi, also tested Williams's IF decision model with 60 farmers. He found IF systems were unknown in his sample of 60 women. No one had ever tried an IF system, probably because the closest site of the US Agency for International Development (USAID)-funded Malawi Agroforestry Extension (MAFE) Project or the European Union (EU)-funded PROSCARP project was 50 km away at Mponela. No respondents planted any trees or shrubs that did not occur naturally on their fallow.

[4] We first planned to find 40 women who began testing IFs before 1995/96. As it turned out, however, only 28 women tested IFs before 1995/96, because most of the early testers were men. In many instances, however, farmers were so convinced of the success of the technology (especially after having visited farmers in other camps as part of field-days or farmer-to-farmer visits) that they did not wait until they harvested their first IF before they planted another. Of the 81 women in the ICRAF sample, Peterson interviewed 40 non-testers, 23 tester–expanders and 18 tester–non-expanders. Of the 40 men, she interviewed 15 non-testers, 16 tester–expanders and 9 tester–non-expanders.

[5] The question used to test this criterion may have been wrongly specified. There is a taxonomy of native terms for 'cleared land' that was not elicited before designing the questionnaire. We used the Chewa words *chisala* and *tsala*, meaning 'cleared land put back in fallow and now ready to be taken out again', but should have used the word *mphanje*, meaning 'land that was in woodland (*tengo*) and is now ready to be cleared'. Further, we should have added an additional constraint, 'Do you have time and strength to clear this land?' The single fallow criterion here should be replaced with a more complicated fallow-land subroutine in the future.

[6] In other versions of the model, we tested a 'risk of loss of the IF plot' due to livestock eating the leaves during the dry season. Unfortunately, there were too many errors due to this criterion. Many farmers were indeed afraid of animals getting into their IFs, but had adopted a strategy to reduce the risk. They had trees that resprouted, had the IF plot farm away from the village, sent children to watch the field, informed the village leaders and got new rules about animals in the village or used a barbed-wire fence around the IF plot. We also elicited other sources of risks to the IF plots: children burning fields to catch mice during the hungry season, and damage to the trees due to beetles and termites. Unfortunately, we did

not ask every farmer if they perceived these sources of risk or had a risk-reduction method, and so could not properly test a 'risk of the loss of the IF plot' subroutine with these data.

[7] In another paper, we present logit and ordered-probit results that shed more light on this question (Gladwin *et al.*, 2000).

10 Liquidity and Soil Management: Evidence from Madagascar and Niger[1]

T J Wyatt*

*International Crops Research Institute for the Semi-Arid Tropics,
BP 12404, Niamey, Niger*

A major question confronting both researchers and aid programmes is the often low adoption rate of natural resource management techniques, even in the face of continued land degradation. Rates of erosion from agricultural land in the tropics have been a source of concern for some time (Brown and Wolf, 1984). Bationo *et al.* (1998) relate concerns that soil nutrients are being depleted rapidly in many areas of semi-arid and subhumid Africa, where use of chemical fertilizers is far below rates observed in other parts of the developing world. Land degradation, whether direct soil loss or mining of soil nutrients, could jeopardize future agricultural production. Given renewed concerns about population growth, particularly in areas where food production is not keeping pace, sustainable and enhanced production systems would appear to be vital.

Research has been effective in developing new technologies or identifying and improving indigenous technologies that can be effective in combating soil degradation. In the area of soil conservation and water harvesting, various methods can be used, including terraces, stone bunds and half-moons, all of which slow water flow, reduce the amount of sediment carried away and increase infiltration. Soil-fertility management practices include use of chemical and organic fertilizer, green manures and fallows. None of these practices come without a cost, of course. Construction of barriers to the flow of water can be very labour-intensive, purchased chemical fertilizers require an outlay of cash, and fallows and green-manure crops reduce immediate production in favour of future harvests. In fact, all of these practices essentially imply an investment of resources in the present to ensure production in the future. Yet, despite reports that these measures are frequently cost-effective, many producers fail to adopt them. This suggests that there are other social, institutional or economic factors that inhibit investments in soil conservation or soil-fertility management.

A number of hypotheses seek to explain why farmers pass up profitable investments. It may be that the economic analysis fails to properly account for the impact of risk on farmers' perceptions of what is a good investment. It may be that informational linkages between the farmer and the technology provider are weak or non-existent, in which case the farmer fails to adopt because

* Current mailing address: Mail code 7503C, US Environmental Protection Agency, 1200 Pennsylvania Ave, NW, Washington, DC 20460, USA.

©CAB *International* 2002. *Natural Resources Management in African Agriculture*
(eds C.B. Barrett, F. Place and A.A. Aboud)

he or she is unaware of the benefits available. Land-tenure regimes are frequently blamed for failing to provide sufficient security for the future benefits of a current investment. Another frequently cited reason for non-adoption is that credit markets are virtually non-functional and farmers lack the means to self-finance many of the investments. Projects are therefore proposed to address these potential problems, including credit schemes.

This chapter proposes a corollary to the credit- or liquidity-constraint hypothesis. A common characteristic of low-income households practising agriculture in high-risk environments is that they diversify their income sources. In the face of poorly functioning markets, households rely on their own resources for use in alternative activities; for example, crop residues can supply organic matter as soil amendments or can be used for feed. Similarly, cash resources must be allocated across different activities in such a way as to meet the objectives of the household to maximize return at an acceptable risk.

Heterogeneity between farmers in their resource endowments can lead to wide divergences in the opportunity costs of resources, which, in turn, lead some producers to value certain investments more than others. The implication is that we should not expect all households to avail themselves of any given technology. Further, if sustainable development is our objective, not some narrowly defined criterion of soil quality, then we should look for ways to enhance returns to investments preferred by producers, not attempt to sway them toward investments preferred by researchers and aid workers.

The rest of the chapter is organized as follows. In the next section, a brief analytical model is proposed that guides the analyses of two case-studies. One case-study is of investments in soil conservation in the highlands of Madagascar and the other is of soil-fertility management in the northern Sahel of West Africa. Results suggest that alternative investment options have a large impact on producer decisions. The final section summarizes the main conclusions and offers some policy recommendations.

Analytical Framework

Generally, an investment is said to be profitable if the net present value (NPV) of the benefit stream is greater than the NPV of the costs. It is important, of course, to accurately represent the benefits and costs as perceived by the investor. In this and in all further discussions, price refers to the price actually received or paid by the household, net of transaction costs, or it refers to the implicit price paid in terms of household resources for production consumed at home. Prices observed in the market, particularly a distant market, may not be reflective of those implicitly faced by the decision-makers.

In the case of soil-conservation measures, such as the construction of terraces, the benefits are the difference between the value of production through time with and without terraces. In addition to the construction costs of the terraces, which may be in terms of the value of family labour, terraces can have recurring costs, in that they must be maintained, and their construction may reduce the area that can be cultivated. In the long run, however, the benefits of conserving the soil against erosion may outweigh the costs. If the investment is profitable and if the necessary capital (or labour) is available, we would expect to see the investment take place. One of the key parameters, and one which is not readily observable, is the household discount rate – that is, how future costs and benefits are weighed relative to immediate costs and benefits. A higher discount rate reduces the attractiveness of an investment with a long period of small returns compared with a short period of higher returns, even if the total returns are the same. The discount rate can be influenced by a number of factors, including family size, age, gender and education of the household head and wealth.

The decision to use chemical fertilizer to maintain soil fertility, however, may consider a much shorter time horizon, especially if there is little or no residual effect of fertilizer on following campaigns. Returns to fertilizer depend on the relative prices of fertilizer and output levels (which are usually not known at the beginning of the season). Yet there is a lag between the time of purchase of the input and the benefits received from it, so the decision is essentially similar to an investment decision. To the extent that chemical fertilizer does have a residual impact on future production, the similarities are increased. But these decisions are rarely made in isolation. Instead, the household may confront a number of investment options, both short- and long-term, from which to choose. If capital is limited, then the household may forego investments with a positive return in favour of another investment that is more attractive, i.e. offers a higher return.

To see this more clearly, consider a household engaged in agricultural production. The household exists over a period of time, which may in fact be infinite if the current generation cares about future generations. If the household is concerned about maintaining production, then it will be concerned about maintaining soil fertility. Maintaining soil fertility, however, will have some cost associated with it. The household's problem is to maximize the benefits from production, which again includes the value of production consumed by the household. Let us distinguish between purchased inputs or expenditures, x, and household-supplied inputs, v. Further, let us consider alternative income-generating activities, R, that also use family-supplied and purchased inputs. Then we could represent the household's problem as:

$$\max J = \int_0^T \left\{ \begin{bmatrix} P\,Q(v_1, x_1, s) - x_1 \end{bmatrix} + \\ \begin{bmatrix} R(v_2, x_2) - x_2 \end{bmatrix} \right\} e^{-\rho t}\, dt \quad (1)$$

Subject to

$$s_{t-1} = s_t + G(v_{1t}, X_{1t})$$
$$v_1 + v_2 \le b$$
$$(x_1 + x_2) \le \overline{C}$$

where the t subscripts have been suppressed except where necessary for clarity. The G function describes how inputs influence soil fertility. Cultivation depletes nutrients and, to the extent that inputs increase production, they would also contribute to that depletion. However, some inputs will also contribute to the amount of nutrients in the soil and therefore the overall impact may be

positive. Thus, decisions today will affect future production through the effect on soil fertility. Similarly, purchases of inputs for the alternative activity can influence future returns – for example, the purchase of a cow that will produce milk and calves. The b vector represents the household resource endowments and includes land and labour. The use of purchased inputs is constrained by available cash, $\overline{C}$, which could be a function of previous production decisions. Assume that the functions Q, R and G are all concave in their arguments – that is, marginal products are declining.

If the staple crop is unavailable on the market, the household may be obligated to produce a sufficient amount for home consumption. This is likely to be a rare situation, however. On the other hand, where prices are highly variable, the household may want to produce a certain amount to insure itself against high purchase prices. Finkelshtain and Chalfant (1991) show that peasant farmers who produce at least a portion of the crop for home consumption will tend to increase production in the face of price risk, rather than reducing production as is the case of the prototypical firm. For ease in modelling, we shall simply assume that this is reflected in the value of production placed on it by the household. That is, the price of the output, P, may include a premium for production destined within the household.

Let x simply represent cash expenditures; x_1 is expenditures in agriculture and x_2 expenditures in the alternative activity. Solving this problem for the best use of purchased inputs shows that the returns in each activity must be at least equal to the opportunity cost of capital (μ) or no investment will occur.

$$\frac{\partial J}{\partial x_1, t=0} = \int_0^T P \frac{\partial Q}{\partial x_1} e^{-\rho t}\, dt + \\ \lambda \frac{\partial G}{\partial x_1} - (1 + \mu) \le 0 \quad (2)$$

$$x_1 \left[\int_0^T P \frac{\partial Q}{\partial x_1} e^{-\rho t}\, dt + \lambda \frac{\partial G}{\partial x_1} - (1 + \mu) \right] = 0$$

$$\frac{\partial J}{\partial x_2, t=0} = \int_0^T \frac{\partial R_t}{\partial x_{2, t=0}} e^{-\rho t}\, dt - (1 + \mu) \le 0$$

$$x_2 \left[\int_0^T \frac{\partial R}{\partial x_{2,t=0}} e^{-\rho t} \, dt - (1+\mu) \right] = 0$$

All variables are defined as above and λ is interpreted as the value of soil fertility so that the value of a fertility-enhancing input includes the benefits to future production. In the case of terraces, where a reduction in area cultivated may cause an immediate decrease in production, $\partial Q/\partial x$ may be negative. Marginal returns are discounted. For example, if we buy a heifer that will not enter production for 2 years, we value the additional milk and calves she produces in the future less than the immediate production obtained from buying an adult cow. If capital is limited, investment will be made in the activity that provides the highest return, which sets the total opportunity cost of capital. If sufficient capital is available to drive down the marginal return on that activity, it may become worthwhile to invest in a second activity. At that point, returns should be equalized across the different activities, that is:

$$\int_0^T P \frac{\partial Q}{\partial x_1} e^{-\rho t} \, dt + \lambda \frac{\partial G}{\partial x_1} =$$
$$\int_0^T \frac{\partial R_t}{\partial x_{2,t=0}} e^{-\rho t} \, dt \qquad (3)$$

Note that this does not include any risk premium on the activities. Where returns are risky, due to variability in prices and/or production, investment decisions depend on personal valuation of the trade-off between risk and return. For a given return, we would assume that preference is given to the less risky investment. If returns to different investments are uncorrelated or negatively correlated, there are benefits to investment in multiple activities in order to reduce the household's overall exposure to risk, even if returns in some activities are low. The variability and correlation of returns to different options cannot be stated a priori. Moreover, we cannot easily define the risk–return trade-off acceptable to various households. Since the aim of this chapter is to examine the impact of limited cash availability, we abstract away from risk considerations.

We can see that a liquidity constraint to the adoption of a technology is not merely a question of a complete lack of capital. Even if some capital is available, it may not be used on a profitable activity if returns are higher in another activity. Failure to recognize this may lead to erroneous conclusions on the part of researchers or policy-makers. They may assume that farmers are ignorant of the benefits of a new technology or that farmers put more weight on personal status than their family's livelihood. This in turn could lead to ineffective policies. Specifically, relaxing the liquidity constraint through the supply of credit may not encourage the intended investments in soil-fertility management. The next sections use two case-studies to examine more closely the issue of liquidity constraints and investment alternatives.

Soil Conservation in the Malagasy Highlands

This section studies adoption and investment in the construction of terraces by farmers of the highlands of Madagascar. This is an example of a situation in which high initial costs may keep farmers from undertaking a profitable investment in the protection of their soil resources. Credit would be the most logical means for farmers to transfer future returns to the present in order to make the necessary expenditures. In the absence of credit, farmers must have a means of self-financing their investments or they will be forced to forego it. If farmers are liquidity-constrained in the short term, however, credit may not necessarily be used for the construction of terraces, but could be used for investments in alternative activities. Thus, some indication of the effectiveness of credit as an instrument for increasing investments in sustainable practices would be useful.

The highlands of Madagascar are characterized by sharp relief and intense rainfall, which contribute to a dramatic rate of erosion that can be as high as 250 kg ha^{-1} year^{-1} (Randrianarijoana, 1983). Small valleys are cultivated in rice, the main staple; fields are flood-irrigated during the rainy season. Supplemental irrigation can be utilized and depends on a network of small canals. Hillside fields are cultivated in cassava, maize, beans and sweet potato. Households generally have several, widely scattered fields. Villages are fairly isolated. Access is difficult, especially during the rainy season. Besides crop production, households engage in a number of different activities, including petty commerce, wage labour and livestock husbandry. The livestock system is fairly intensive, with animals usually stall-fed. The crop and livestock production systems are complementary, with manure a main input into crop production and crop residues an important source of forage. Animal traction is primarily for transport. Use of ploughs on hillside fields is difficult because of the slope. Lowlands are sometimes 'tilled' by running livestock through the flood fields to work the soil, but only occasionally are they ploughed.

As in the analytical model above, we would assume that the value of terraces would be a primary factor in motivating their construction for soil-conservation purposes. The value of the terraces is basically the contribution of soil, which would otherwise be lost, to continued production. The amount of soil that would be lost is determined by the rate of erosion, which is itself a function of the physical characteristics of the field, including the length and steepness of the slope. The value of the soil depends to some extent on the value of the crop grown on the field and to the amount of soil currently present. A very deep soil might withstand considerable erosion before there is any impact of erosion on yields.

In addition to the value of terraces to crop production, other household and field-specific factors might influence farmers' decisions to engage in soil conservation. The tenure status of the field is one factor frequently suggested (see, for example, Southgate *et al.*, 1984). Others include family size, which may indicate higher immediate consumption needs, age and education level of the household head, which may influence his or her willingness to try new techniques, and the level of

wealth or assets, which may influence the household's willingness or ability to accept higher levels of risk. As we are interested in the effect of a liquidity constraint, we want to test if construction costs deter investment because capital is unavailable. We also want to look at the impact of credit on the investment decision, as well as the influence of other potential options.

Data for this analysis were gathered by means of a detailed household-farm survey in 1995. A total of 195 households were surveyed at random in 18 villages near the eastern edge of the central highlands. Estimation for the adoption decision was based on 130 households for which there was complete information. The estimation covers a total of 415 fields, of which 138 were recently terraced. Fields terraced more than 5 years previously were not included in the estimation, as conditions could have changed during the intervening years. In fact, major changes had occurred in terms of market liberalization and the withdrawal of state enterprises. Input prices had increased dramatically while output prices had risen much less.

The value of the terraces is calculated as the NPV of the investment (discounted value of future production with no change in soil depth, less the costs of construction) minus the NPV of the discounted value of production, given continued erosion. Production functions were estimated using the cross-sectional data to determine the influence of soil depth, as measured from the subsoil, on output (Wyatt, 1998). Average annual erosion rates were calculated using parameters of the universal soil-loss equation for slope length and angle, calibrated to measured rates of erosion to account for local climate conditions. Production under declining soil depth was then forecast to calculate future returns. Prices were those actually paid or received by the household. An arbitrary discount rate of 25% per annum was used. Interest rates for formal loans ranged from 10% (supplied by a Catholic Relief Services project) to 18% at the bank. Labour requirements for terrace construction were estimated as a function of field characteristics, particularly area and slope. The opportunity cost of labour was calculated from off-farm household activities at the village level (Wyatt, 1998).

Table 10.1 presents summary statistics of these field-level variables. Tenure regimes varied from field to field within the same household and are listed here. A small percentage were under government title, but the largest number were held under traditional tenure – that is, privately owned but without government registration. A few fields were owned by a parent of the cultivator and over one-fifth were owned by the extended family. Fields in this latter category could be allocated to another user within the family from year to year. Only two fields in the sample were owned by someone outside the family of the cultivator and they had to be dropped from the analysis.[2]

Household-specific variables are presented in Table 10.2. Female-headed households (*de facto* or *de jure*) were rare. Education represents years of formal education. The average was low, but the range extended to post-secondary education. Non-crop income was estimated as a function of human and physical assets (buildings, livestock and ox carts) and the predicted values were used in the adoption equation. Credit participation represents either borrowing or lending. Formal loans are rare, but informal networks are quite common and, within the preceding year, some households both borrowed from friends or family and loaned money as well. Participation was estimated as a function of household variables, but did not present an adequate instrument to use in the adoption equation. In any case, no one explicitly stated that credit was used for the construction of terraces. The vast majority, in fact, were for consumption. A few farmers even admitted that formal loans taken ostensibly for input purchases were actually for consumption. Lowland areas are for irrigated rice cultivation (rain-fed hillside rice was rare). Total land includes land left fallow and, in a few places, grazing area. Village associations play a number of different roles, including guarantors of loans from the formal sector and the primary means through which government and non-governmental extension services operate.

They also provide a ready source of labour under reciprocal arrangements.

Results of a probit estimation are shown in Table 10.3. The dependent variable is the construction of terraces on the field: 1 if terraced, 0 otherwise. The marginal effect of the explanatory variables was calculated at the sample mean. Clearly, the economic value of terraces plays an important role. An increase of 1000 Malagasy francs (Fmg) – about US$0.25 or about half the daily wage rate – in the calculated value increases the probability of investment by almost

Table 10.1. Summary statistics of field-specific variables, Madagascar (from Wyatt, 1998).

Variable	Mean (SD)
Terrace (1 = yes)	33.25%
Value (1000 Fmg)	27.50 (48.28)
Construction costs (1000 Fmg)	36.16 (33.87)
Tenure (1 = yes)	
Titled	4.1%
Traditional tenure	69.9%
Parent-owned	3.6%
Extended-family ownership	22.4%
n = 415	

US$1.00 = 4000 Fmg.
SD, standard deviation.

Table 10.2. Summary statistics of household-specific variables, Madagascar (from Wyatt, 1998).

Variable	Mean (SD)
Female-headed household (1 = yes)	4.6%
Age of household head (years)	45.3 (13.1)
Education of head (years formal schooling)	4.47 (3.16)
Family size (persons)	6.15 (2.73)
Number of men (persons)	2.03 (1.23)
Non-crop income (1,000,000 Fmg)	0.706 (0.692)
Credit participation (1 = yes)	48.9%
Lowland (ha)	0.598 (0.920)
Cultivated hillside land (ha)	0.769 (0.696)
Total land (ha)	2.153 (2.746)
Village association member (1 = yes)	58.9%
n = 130	

two-tenths of 1% and is highly significant. Variables included to reflect individual preferences or constraints have the expected signs. A larger family size decreases the probability of investment, as might be expected if consumption requirements lead the household to emphasize immediate needs. Controlling for credit participation and membership in village associations, both of which are biased against female-headed households, gender has no significant impact on the decision to invest in terraces. Non-crop income and total land both have positive effects on the decision to construct terraces. These represent wealth variables, as well as providing a source of investment capital.

Interestingly, the effect of titled land is negative and significant, but this is based on very few observations that are highly correlated with a single location. In this group

Table 10.3. Marginal effects on the decision to adopt terraces (from Wyatt, 1998).

Variable	Marginal effect	*t* Statistic
Constant	−0.420	−2.816***
Value of terrace	0.002	2.505**
Construction costs	0.003	3.138***
Female-headed household	−0.005	−0.041
Age of head	−0.0002	−0.066
Education of head	−0.036	−2.071**
Men	0.071	2.495**
Family size	−0.022	−1.685*
Non-crop income	0.085	1.646*
Credit participation	0.977	1.867*
Lowland	−0.212	−3.185***
Cultivated hillside land	−0.041	−0.832
Total land	0.034	2.31**
Village association membership	0.232	4.239***
Titled land	−0.522	−2.809***
Parent-owned	0.034	0.244
Extended-family-owned	0.138	2.090**
n = 415	Correct predictions	
Observations at 1 = 138	= 73.7%	
	McFadden's R^2 = 0.1572	

*, significant at below the 0.10 level; **, significant at below the 0.05 level; ***, significant at below the 0.01 level.

of villages, terraces had been constructed during the colonial period (although they had been allowed to degrade during the 1980s), and today agroforestry techniques appear to be preferred by farmers as a means of controlling erosion. More interesting is the finding that land owned by the extended family, which can be reallocated to other members and therefore would appear to offer less secure rights, is significantly more likely to be terraced than privately held fields. This underscores the potential bidirectional causality, as several farmers indicated that, by making improvements, they strengthen their claims and can eventually privatize the land.

Indications of a liquidity constraint are somewhat ambiguous. Construction costs do not appear to be an impediment to the adoption of terraces. The total impact of construction costs is β_{NPV} plus β_I (because the NPV of the terrace is benefits less costs). The result is positive, small (0.001) and insignificant (*t* statistic of 0.833). This could indicate that farmers do not face a liquidity constraint, but could also be related to risk factors. Construction costs increase with slope, but steeper slopes are more vulnerable to erosion, not merely on average, but also to sudden extreme events (slippage). Farmers may therefore be motivated by risk avoidance to preferentially terrace steeper slopes, as found in the Philippines (Shively, 1997). Results of a regression model replacing construction cost with slope (not shown) are very similar.

The positive signs on both non-crop income and credit participation suggest, however, that households that have better means of amassing the necessary capital are more likely to be able to make the investment. Credit participation, in particular, has a major influence, despite the fact that most loans are for small amounts and are of short duration. Viewed in the aggregate, the implication is that increasing participation in the credit system by 10%, from the current sample average of 48.9%, would increase the overall terracing rate by almost 10%, from the current sample average of 33.3% of fields. It may be that the advantage of these loans is to help households to overcome temporary shortfalls in consumption that might result from withdrawing labour from activities that provide immediate returns, in order to use it for the long-term investment in soil conservation.

But while credit is clearly important in aiding farmers to make investments that have a future pay-off, it does not necessarily follow that providing credit will induce all farmers to invest in terraces. The results also support the contention that alternative investment opportunities influence farmers' willingness to adopt natural resource management practices. Each additional hectare of lowland rice-fields reduces the likelihood of constructing terraces on a hillside field by over 20%. Rice is the main staple and many farmers in this region are net sellers. Thus, higher returns in the production of rice draw capital (and labour resources) from investments in hillside agriculture towards purchases of fertilizer and pesticides and construction and maintenance of canals. As investments rise in one activity, we might assume that marginal returns decrease. Therefore, additional capital, including that obtained through credit, might encourage the household to invest in hillside activities.

The education of the household head is also a negative factor in the adoption decision. Opportunities for unskilled labour are few, with the exception of temporary migration. Education, however, opens up alternative avenues and may even be a means of leaving agriculture entirely. If households see greater opportunities elsewhere, they may be expected to exploit their soil resources as a means of financing alternative activities (Pagiola, 1995). Inducing households with other investment opportunities to focus their cash resources on hillside agriculture may be difficult, costly and ultimately ineffective. However, targeting credit towards households with fewer lowland resources and lower education and towards female-headed households – essentially those households frequently unable to obtain credit through formal sources – will provide those with the incentives to invest in soil conservation the means to do so.

Soil-fertility Management in the West African Sahel

While the use of terraces is clearly an investment decision, the use of chemical fertilizer is less obviously so. Unlike terraces, fertilizer is a divisible input and has a relatively short period before benefits are received. Given variability in output price and the potential that the impact of fertilizer depends on rainfall, risk concerns are likely to play a large role in farmers' decisions regarding the use of fertilizer. However, failure to add soil amendments, including chemical fertilizer, results in a steady exportation of nutrients from the soil by means of the grain and stover. This loss of nutrients has been reported throughout the semi-arid regions of West Africa (Bationo *et al.*, 1998). In this respect, use of fertilizer to maintain or restore soil fertility can be seen to have long-term impacts, beyond that of residual effects from a single dose. It is this element of long-term fertility management that takes on the characteristics of an investment and upon which we focus.

The Sahel of West Africa is an arid zone stretching the breadth of the continent from Senegal to Sudan. The northern area is characterized by low rainfall (600 mm or less on average), which falls during a brief (July–September) rainy season, following a long, hot dry season. Rainfall is highly variable from year to year (coefficient of variation in excess of 20% (International Crops Research Institute for the Semi-Arid Tropics (ICRISAT) data)) and is spatially and temporally dispersed within the season. Millet is the staple crop. Cowpea is often cultivated in association with millet and small plots of groundnut and other speciality crops are sometimes cultivated, especially by women, mostly for home consumption. Since the major droughts of the 1970s and 1980s, mixed crop–livestock systems have become dominant, with former herders becoming more sedentary and cultivating millet fields and sedentary agriculturalists investing in small ruminants and cattle, which they often entrust to herders to range in search of forage.

The sandy soils are generally of poor quality, being low in organic matter, nitrogen and phosphorus. The former long-term fallow system, which was used to restore nutrients depleted through cultivation, has shortened, largely as a result of population pressure, to a duration of about 3 years (Samaké *et al.*, 1999). Nutrient budgets suggest that soils are being rapidly depleted as a result. However, during village meetings conducted by ICRISAT and national agricultural research systems (NARS) researchers at benchmark sites, farmers were more concerned about lack of rainfall and only mentioned soil-fertility problems when the issue was brought up by researchers. On-station and on-farm experiments confirm the beneficial effects of fertilizer on yields, and financial analysis suggests they are profitable (Ndjeunga and Bationo, 2000). Use of chemical fertilizer by the farmer is very low and well below recommended levels. A major factor in this lack of application may be limited cash availability to purchase inputs. The corollary is that what cash resources are available to the producer, be they traditionally pastoralists or traditionally agriculturalists, tend to be invested in livestock. One question is whether this allocation of resources represents an investment strategy aimed at maximizing returns. If so, will producers eventually use returns from livestock to make investments in agriculture and can credit programmes provide the means for producers to invest immediately in soil-fertility management through purchased fertilizer?

To address this question, a mathematical programming model was developed at the household level. Data for the model come from characterization reports describing the physical and socio-economic environments of benchmark sites in Mali, Burkina Faso and Niger and from on-station and on-farm agronomic and animal nutrition trials of NARS, ICRISAT, the International Fertilizer Development Centre (IFDC) and the International Livestock Research Institute (ILRI). The model is meant to represent a typical household in the Sahelian zone of Niger that depends on rain-fed agriculture, animal husbandry and temporary migration

to urban or coastal areas for income. Millet and cowpea are the cropping possibilities, while livestock consists of cattle and small ruminants. Millet and cowpea are both consumed, but can be marketed as well. A consumption constraint must be met, either through own production or by purchases in the market. Transaction costs drive a wedge between sale and purchase price. Crop residues can also be sold. Livestock income is derived from sales of live animals and of milk. Minor amounts of wage-labour opportunities are available or labour can be employed if family labour is insufficient.

Crop yields depend on soil fertility and the use of variable inputs, including manure, crop residues and fertilizer. Fertility is measured by an index, where 100 represents soils newly placed in cultivation. Fertility declines linearly with production and increases with soil amendments and the amount of land left fallow. Composite fertilizer, the most widely available type, is specified in the model (nitrogen–phosphorus–potassium (NPK) 15–15–15). Studies have shown that phosphorus is often the limiting nutrient in the northern Sahel and much emphasis is currently placed on local sources of rock phosphate – availability of which remains limited, however. Cattle management within the village can involve extensive grazing, intensive stall feeding or transhumant migration. Calf and milk production, availability for pulling ploughs and carts, and fodder and cash requirements vary accordingly. Sales are mostly cull cattle or bulls going to the urban or coastal market for slaughter. Breeder stock carries a very high premium and is usually unavailable in the market. Milk is consumed within the household and must be purchased if production is insufficient to meet the needs. Milk sales are possible, but transaction costs of fresh milk are high because of the difficulty in transporting and the lack of refrigeration. Small ruminants are primarily sold for slaughter and are frequently fattened for sale in the period immediately following grain harvest when forage supplies are at their peak. In the absence of a rural banking system, livestock also form the principal means of short- and long-term savings.

The model maximizes the NPV of a 10-year income stream plus the value of assets held at the end of the time horizons. Assets are agricultural equipment (plough, weeder, cart), livestock and land. Equipment and livestock are valued at their sale prices, while the value of land is calculated as the NPV of net benefits from production over an infinite time horizon. In the final year of the time horizon, soil fertility is not permitted to change – that is, a steady state is presumed – but the soil quality level to be maintained is chosen by the model. The household must utilize a combination of fallow, organic material and inorganic fertilizer to maintain this chosen level, weighing the costs against the benefits, which include home consumption. We compare the strategies of two households, one relatively better endowed with land and labour resources (and with initial livestock holdings) than the other (Table 10.4). If different households behave in a similar fashion, similar technologies or policies could be used without distinguishing a target group.

Selected results for the better-endowed household are shown in Table 10.5. Throughout the planning horizon, even the well-endowed household 'mines' its soil resources, although fertility appears to be stabilizing in the final years. The household mainly relies on crop residues (CR) and manure to maintain soil fertility, but begins

Table 10.4. Initial conditions for programming model of Niger households.

	Better endowed	Lesser endowed
Population (persons)	15	10
Labour (persons)	8	5
Land (ha)	15	8
Land per labourer (ha)	1.875	1.6
Pasture (ha)	20	10
Cattle (head)	5	0
Small ruminants (head)	6	2
Ox cart	1	0

Table 10.5. Model results for better-endowed household.

Year	NPK (kg ha⁻¹)	CR (Mt ha⁻¹)	Manure (Mt ha⁻¹)	Fertility	Expenditures (1000 FCFA)
1	0.0	1.034	0.125	82.2	0.000
2	0.0	1.074	0.154	80.7	0.000
3	0.0	1.197	0.200	79.3	0.000
4	0.0	1.012	0.183	78.3	0.000
5	0.0	1.286	0.228	77.0	0.000
6	13.6	1.542	0.347	75.9	38.885
7	0.0	1.098	0.265	75.5	0.000
8	18.6	1.794	0.476	74.0	47.880
9	12.2	1.286	0.415	74.3	35.819
10	14.5	0.861	0.336	73.9	54.849

Year	Purchases, small rum.	Total, small ruminants	Purchases, cattle	Stock, cattle	Expenditures (1000 FCFA)
1	0.0	6.0	0.0	5.0	0.000
2	0.0	7.2	0.0	5.9	0.000
3	0.0	8.2	0.0	7.5	0.000
4	0.0	3.8	0.0	9.0	10.882
5	0.0	0.0	0.0	10.7	12.869
6	0.0	0.0	0.0	13.1	8.340
7	0.0	0.0	0.0	14.0	15.212
8	0.0	0.0	0.0	16.2	8.624
9	0.0	0.0	0.0	18.2	43.188
10	12.9	2.4	0.0	21.5	59.177

NPK, chemical fertilizer (nitrogen–phosphorus–potassium); CR, crop residues.

to use chemical fertilizer regularly in the last 3 years of the horizon, when it begins to substitute fertilizer for residues. The extensive nature of the livestock system prohibits amassing large quantities of manure for use on the fields, but growth of the herd permits increasing applications, although it also demands some feeding of crop residues. The emphasis in expenditures is on livestock to pay transhumant herders, vaccinations and feed supplements. Purchases and stocks are shown. Small ruminants are liquidated within the first 3 years and the large purchase in the final year is actually for fattening; only two animals are kept in stock. The model predicts the sale of bulls at about maturity, as weight gain slows; cows are sold only as they advance in age. Expenditures during the initial years of the time horizon are on investments complementary to livestock: equipment purchases of a plough and a weeder, both for personal use and for rent.

In contrast, the lesser-endowed household (Table 10.6) invests somewhat more in purchased inputs for crop production, despite the fact that it would be assumed to be more liquidity-constrained. This is particularly true in the later years of the time horizon, as soil fertility levels continue to decline. Usage of fertilizer is far below recommended levels (250–300 kg ha^{-1}) but corresponds roughly to levels actually observed. Hopkins and Berry (1994) report average fertilizer use between 6 and 8 kg ha^{-1} in 1990, prior to the devaluation of the Communauté financière d'Afrique franc (FCFA). More recently, an ICRISAT study has found application rates of about 15 kg ha^{-1} (Hima Amadou, 2000). With a lower land-to-person ratio, this household needs to produce more cereal per unit of land to meet its food needs or must pay the market price for cereal. Thus, over time, they are more willing to intensify and more willing to maintain the resource base. This is consistent with Pagiola's (1995) contention that subsistence households, especially those that lack alternative income sources,

Table 10.6. Model results for lesser-endowed household.

Year	NPK (kg ha^{-1})	CR (Mt ha^{-1})	Manure (Mt ha^{-1})	Fertility	Expenditures (1000 FCFA)
1	0.0	1.350	0.031	81.4	0.000
2	4.0	1.218	0.088	79.8	7.617
3	0.0	1.214	0.125	78.3	0.000
4	0.2	1.176	0.173	76.8	0.465
5	0.0	1.169	0.241	75.4	0.000
6	0.0	1.175	0.293	74.0	0.000
7	0.0	1.153	0.327	72.6	0.000
8	33.5	1.144	0.357	71.3	63.369
9	56.8	1.500	0.402	70.0	98.495
10	74.5	1.639	0.462	69.0	127.112

Year	Purchases, small rum.	Total, small ruminants	Purchases, cattle	Stock, cattle	Expenditures
1	5.6	2.0	0.00	0.0	24.183
2	3.4	12.0	0.00	0.0	13.199
3	1.8	20.9	0.00	0.0	40.164
4	7.4	24.0	0.05	0.0	61.261
5	0.0	38.6	0.05	0.1	30.027
6	0.6	47.9	0.02	0.1	12.319
7	0.0	56.0	0.00	0.1	0.000
8	0.3	61.7	0.16	0.3	48.375
9	0.0	61.4	0.02	0.3	5.890
10	0.0	69.3	0.00	0.3	0.000

will be more likely to maintain their land base.

Unlike the better-endowed household, the lesser-endowed household tends to make short-term investments. The household is shut out of the cattle market by high purchase prices for young cows and the need for complementary investments in equipment to obtain quicker returns from draught oxen. (The model predicts purchases of less than a tenth of an animal; since livestock comes in discrete units, in actuality these investments would have to be delayed or forgone.) Small ruminants are actively traded and form a sort of liquid savings. Even while purchasing small ruminants in order to build a stock of capital, the household also makes frequent sales of males after the animals have been fattened. Livestock is also valued for its manure.

Despite the differences in livestock species, both households place the emphasis on livestock investments over those in the cropping system. As demonstrated by the model, this behaviour can be explained by differences in returns to the relative investments where capital is limited. But would alleviating the capital constraint lead to investment in soil-fertility management, or would it simply permit the households greater freedom to pursue livestock husbandry?

We ran a simple experiment, enabling the household to take a loan at the beginning of the agricultural season to be paid back at harvest, with interest payment of 10%. The loan could be used for any purpose and the programme lasted 5 years, beginning in the second year of our time horizon. The results are instructive. The better-endowed household did not take a loan. The short payback period did not fit within their strategy of cattle husbandry and using the loan for short-term investments would have meant drawing other resources, primarily labour, from this activity. In contrast, the lesser-endowed household did avail itself of the credit programme, although at an amount below the borrowing limit of 50,000 FCFA (approximately US$70.00). The credit programme induced the household to increase usage of chemical fertilizer in the initial

years of the programme (from 8000 to 42,000 FCFA over 3 years) and to reduce investments on small ruminants (from 115,000 to 45,000 FCFA). However, in the last 2 years of the programme, investments shifted back to small ruminants (from 42,000 to 52,000 FCFA), compared with zero expenditures on fertilizer. Given the importance of manure in improving soil fertility, investments in livestock do not mean neglect of soil management. Rather, this experiment suggests that there is demand for chemical fertilizer on the part of some households who are unable to make their desired purchases due to capital constraints. Not all households are alike, however. An advantage of a credit programme seems to be that households would self-select into the programme on the basis of the terms, including the period of the loan. In this case, a credit programme would be effective not only in assisting farmers to make investments in chemical fertilizers, but also in reaching those most likely to use them.

Conclusions and Policy Recommendations

The main contention of this chapter is that the adoption of natural resource management techniques that require an investment in cash or other resources depends not simply on the absolute profitability of the investment, but also on its relative profitability within the diverse activities undertaken by the typical producer in rural areas. Analyses that focus on a single sector may fail to adequately account for the opportunity cost of scarce resources and misidentify the constraint to adoption. Whole-farm analysis, however, may help to better target the diffusion of existing technologies and highlight constraints that research must overcome in developing new technologies. It should also be recognized that households, because they have different endowments of resources and personal preferences, will choose different investment options, that is, technologies that best suit their needs, even within the same agroecological zone. A portfolio of resource management options is therefore better than a single one.

In the case of soil-conservation investment in the highlands of Madagascar, it was seen that farmers do respond to economic signals: an increase in the value of terraces increases the likelihood that the investment is made. Credit and off-farm income facilitate investments in terracing for erosion control by providing the household with available capital, even if it is only short-term. However, farms with greater opportunities in rice production and with skills that may command a higher return off the farm are less likely to put their resources into hillside production.

The implication is that credit by itself may not be sufficient to increase investment in soil conservation, since funds can be used for other activities. However, targeting credit towards farmers with fewer resources, and particularly women farmers, who are likely to be excluded from existing programmes, may increase the capacity of those producers most likely to make investments in hillside agriculture. This could be done by reducing transaction costs by providing banking services during local market days. Policies or programmes that increase returns to hillside agriculture could also be effective in reallocating household investments to soil conservation. Introduction of higher-value crops and/or facilitating transport to urban areas for hillside products would be possibilities. Farmer associations are an important factor, principally because they facilitate the organization of labour parties. Again, ensuring that female-headed households are included in these organizations and benefit from the additional labour would help to alleviate the constraints of those most likely to want to make investments in hillside production.

Results from a programming model of the northern Sahel in West Africa suggest similar conclusions. Farmers with more resources, irrespective of cultural background, prefer investments in livestock and even lesser-endowed farmers may use additional credit to supplement their own funds to purchase stock or supplement their feed, rather than purchasing fertilizer to augment crop production. From a research perspective, this suggests several avenues. First, if farmers are only going to use very minimal amounts of fertilizer, research should provide them with recommendations as to how to use it best, instead of insisting on the current levels recommended. Secondly, research should address crop diversification. Since a key means of inducing investment is to increase returns, higher-value crops, such as sesame, could play a key role in the development of arid zones. Thirdly, and related, agricultural research in the northern Sahel should view crop production as a key input into the livestock system, rather than the reverse. This means looking towards forage as alternative crops and, by helping to intensify livestock production, increasing manure availability for soil-fertility management.

Note that this model is deterministic, while the reality of the Sahel is highly variable rainfall. Under poor rainfall conditions, livestock mobility makes it less risky than crop production. Under extreme conditions or droughts lasting several years, however, much of the asset accumulated by the household could be lost. If the cycle of reinvesting in livestock then begins again, but from a weakened land base, a continual negative spiral might result. Since producers seem convinced that livestock offers the best return for their investment, policies should be tailored to support the livestock sector rather than some notion of self-sufficiency in grain production. This might include investments in the marketing system, which could facilitate the exchange of livestock and livestock products for grain. Linking wider areas of livestock production through markets could also facilitate sales during droughts, without a complete collapse in prices, and purchases for rebuilding herds in the aftermath.

In the final analysis, we must take our cues from the producers themselves, who are much better positioned to determine the best investments to make with their limited financial resources. Improving producers' ability to amass assets, whether in the form of land quality or livestock, will ultimately improve their ability to make investments in resource management.

Notes

[1]   Research for this chapter was conducted while the author was a graduate student at the University of California, Davis, and while working at the International Crops Research Centre for the Semi-Arid Tropics (ICRISAT) Niger research station. The author would like to thank Jim Wilen, Garth Holloway and Art Havenner of the University of California, Davis, Jean-Pierre Tiendrebeogo of Institut National de l'Environment et Recherche Agricole (INERA) Burkina Faso, Odiaba Samaké of Institut de l'Economie Rurale (IER) Mali, Niek van Duivenbooden of Creative Consultancy, and Tim Williams of the International Livestock Research Institute (ILRI) for their assistance and insights. All opinions and errors are solely those of the author.

[2]   Both fields were 'borrowed', that is, no rental fee charged or share of the output requested. In contrast, rental and sharecropping of rice-fields was common.

11 Smallholder Farmers' Use of Integrated Nutrient-management Strategies: Patterns and Possibilities in Machakos District of Eastern Kenya[1]

H. Ade Freeman[1,*] and Richard Coe[2]

[1]International Crops Research Institute for the Semi-Arid Tropics, PO Box 39063, Nairobi, Kenya; [2]International Centre for Research in Agroforestry, PO Box 30677, Nairobi, Kenya

Depletion of soil nutrients is a major constraint to crop productivity in semi-arid cropping systems characterized by low and highly variable rainfall and soils that are deficient in inherent soil nutrients, water-holding capacity and organic-matter content. In these areas crop yields vary considerably between on-station experiments or researcher-managed on-farm trials and those that farmers achieve on their plots (KARI, 1995). Crop improvement and better natural resource management (NRM) practices – particularly soil, water and nutrient management – are therefore essential for increasing productivity and sustaining the resource base. Yet, although farmers frequently use new seed varieties, they consistently ignore extension recommendations on improved NRM practices (Rukadema et al., 1981; Muhammad and Parton, 1992; Tiffen et al., 1994). Consequently, the quality of soil resources and its productive capacity remains low and many farmers realize a small proportion of the potential productivity gains possible from adoption of new crop varieties. This chapter examines the factors that condition farmers' choice of soil-fertility maintenance practices in a semi-arid cropping system. It describes recent data on farmers' soil-fertility maintenance practices in a semi-arid area of Kenya and draws implications for testing and developing practical soil-fertility management options with farmers.

Study Area and Sampling Methods

The study was conducted in Machakos district in eastern Kenya. This area has a bimodal rainfall distribution pattern, with the first rains, known locally as the long rains, falling between mid-March and the end of May and the second rains, known as the short rains, falling between mid-October and the end of December. Average seasonal rainfall varies between 250 and 400 mm, with large interseasonal variation and a coefficient of variation ranging between 45 and 58% (KARI, 1995). Average annual

* Corresponding author.

temperatures fall between 17 and 24°C. The study was conducted in three agroecological zones – semi-humid tropics, the transitional zone and the semi-arid zone – thus capturing the variability in agroecological conditions that span the district. The major soils in Machakos district are classified as sandy loams to loamy sands, with a brown to reddish-brown colour and characterized by low inherent fertility, low water-holding capacity, low organic-matter content and high erodability (KARI, 1995). The production system in the district is a mixed crop–livestock system, with the relative importance of crops and livestock varying across agroecological zones. The main food crops include maize, beans and cowpeas in the semi-humid tropic and transitional zones, while sorghum and millets are important in the semi-arid zone. Legumes, such as beans, pigeon pea and cowpea, are important in the transitional and semi-arid zones. Major cash crops include coffee, horticultural crops and fruit-trees in the semi-humid zone, while cotton, sunflower and fruit-trees are important in the transition zone. Intensive dairy activities, involving cross-bred cows and zero or semi-zero grazing, are also important in the semi-humid and transitional zones, in contrast to the local animal breeds and extensive

communal grazing that prevail in the semi-arid zone.

Machakos district is characterized as a medium-potential area in the dry transitional agroclimatic zone (Hassan, 1998). Table 11.1 shows that these areas receive more rainfall than those in the lowland tropics but less than areas in the moist transitional zone and highland tropics. Population density in the dry transitional zone is slightly higher than in the lowland tropics, but it is much lower compared with all other agroclimatic zones. Crop productivity estimated by average maize yields is lowest in the dry transitional zone and less than half of those in the highland tropics, even though average levels of total nutrients applied per hectare are not substantially lower than those in the highland tropics and the area reports the greatest proportion of farmers applying animal manure. The region has better access to input suppliers and local markets compared with the lowland tropics and mid-altitude zone and the access is comparable to that in the moist transitional and highland tropics.

Machakos district provides an interesting case-study site for examining the determinants of farmers' choice of soil-fertility maintenance practices. The seminal work of Tiffen *et al.* (1994) in this district

Table 11.1. Basic data by agroclimatic zone (from Hassan, 1998; Government of Kenya, 1999, 2000).

	Lowland tropics	Moist mid-altitude zone	Dry transitional	Moist transitional	Highland tropics
Altitude (masl)	< 400–700	1110–1500	1100–1700	1200–2000	1600–2900
Total precipitation (mm) (March–August)	300–550	> 500	< 500	> 500	500–1000
Average temperature (°C)	25.4	22.1	19.7	19.7	16.6
Population density (persons km^{-2})*	60	172	65	228	232
Maize yield (t ha^{-1})	1.36	1.44	1.21	2.76	2.91
Total nutrients of inorganic fertilizer applied (kg ha^{-1})	n/a	19.5	34.7	58.5	42
Animal manure (% used)	22	52	89	45	50
No input supplier in village (% reported)	80	75	50	41	39
Distance to nearest market (km)	9	18	10	7	10
All-weather road to nearest market (% sites)	0	12	40	23	44

*Population density estimated for representative districts: Kwale in lowland tropics, Kakamega in moist altitude, Machakos in dry transitional, Kissi in moist transitional and Trans Nzoia in highland tropics. masl, metres above sea level; n/a, not available.

highlighted the role of population growth in agricultural intensification and incentives to adopt improved NRM practices. This chapter provides fresh insights into the conclusions reached by Tiffen *et al.* (1994) by examining the underlying factors that condition farmers' technology choices in alternative soil-fertility maintenance practices.

Data for the study were collected in a random survey of households from May 1997 to December 1997. The sample was designed to maximize spatial coverage, so villages and households were randomly selected in all administrative divisions in the district. A multi-stage sampling procedure was used to select the sample within random populations of locations, villages and households. The total sample comprised 399 households, after eliminating five households that did not provide complete information. Table 11.2 shows selected characteristics of households included in the survey.

Fertility Management Practices

An initial step in testing and developing practical soil-fertility management options for semi-arid cropping systems involves understanding farmers' decisions on current soil-fertility maintenance practices. Table 11.3 shows that almost all farmers reported using some soil-nutrient input to maintain soil fertility, with many frequently

Table 11.2. Household characteristics by agroecological zones (from CARMASAK project; Omiti *et al.*, 1999).

| Characteristic | Agroecological zone | | | |
	Semi-humid tropics (*n* = 84)	Transitional (*n* = 206)	Semi-arid (*n* = 109)	Total (*n* = 399)
Sociodemographic				
Average age of household head	54 (19)	47 (15)	46 (16)	48 (17)
Average family size	6 (3)	8 (3)	7 (4)	7 (3)
Percentage of household heads				
Female	55	64	63	62
Male	45	36	37	38
At least primary education	73	80	72	76
Average distance to closest market (km)	3.1 (4.1)	4.1 (4.8)	5.6 (6.7)	4.3 (5.3)
Landholdings				
Average cultivated area (ha)	2.2 (9.3)	1.5 (1.6)	2.1 (2.6)	1.8 (4.6)

Figures in parentheses are standard deviations.

Table 11.3. Sources of nutrients applied in long rains, 1997 (from CARMASAK project; Omiti *et al.*, 1999).

Nutrient source	Semi-humid zone (*n* = 84)	Transitional zone (*n* = 206)	Semi-arid zone (*n* = 109)	Total (*n* = 399)
Percentage of farmers using				
Nothing	1	3	7	4
Animal manure	91	88	86	88
Inorganic fertilizer	81	35	13	38
Compost	46	35	13	32
Green manure	1	1	3	2
Legume intercropping	10	13	11	12
Legume rotation	10	8	12	9

combining organic and inorganic sources of nutrients.

Animal manure

Animal manure, reported by 88% of farmers, was the most widely used soil-fertility input in all agroecological zones. There was no significant difference in the proportion of farmers applying animal manure across agroecological zones, even though the size of livestock herds varied significantly. Disaggregating the data by gender suggested no significant difference in animal-manure use between male and female respondents both across and within agroecological zones.

Farmers reported declining availability of animal manure over time. This reflected steady reduction in livestock herds as a result of drought, disease and increasing human population density, which, in turn, puts pressure on grazing land. The most important source of animal manure was the household's livestock herd (Table 11.4). Thus, application of animal manure was closely associated with ownership of livestock and the average quantity of animal manure available significantly correlated with the size of the livestock herd. Only 15% of farmers in the survey reported buying animal manure from other farmers, while less than 1% bought animal manure from the market. The low frequency of market transactions in this input implied that animal-manure markets were relatively thin, with very small quantities traded.

The survey did not collect data on quantities of animal manure available to farmers, but estimates of farm-level availability based on the composition and size of farmers' livestock herds indicated that the average quantity of animal manure available per hectare of cultivable land ranged from 5 tons in the semi-arid zone to 3 tons in the semi-humid zone (Omiti et al., 1999). Farmers cited inadequate quantities of animal manure in relation to farm requirements and high labour demand as the most important constraints to use of animal manure in all agroecological zones.

Compost and green manure

Compost and green manure were less important sources of organic nutrients than animal manure. Almost one-third of farmers in the survey used compost as an important soil-fertility input. Green manure was not cited as an important soil-nutrient input in any agroecological zone. The size of a household's livestock holdings was strongly correlated with farmers' decisions to use animal manure or compost. There was significant variation in average stock units between animal-manure and non-animal-manure users as well as between compost and non-compost users. In general, animal-manure users tended to have larger livestock herds than those using compost. Although the frequency of animal-manure use did not differ significantly across agroecological zones, the observed differences in the size of livestock herds among animal-manure and compost users explains, in part, the skewed spatial distribution in the use of these inputs. Compost tended to be used more frequently in the semi-humid and transitional zones, where herds were smaller.

Table 11.4.   Percentage of farmers by source of animal manure (from CARMASAK project; Omiti et al., 1999).

	Semi-humid zone (n = 84)	Transitional zone (n = 206)	Semi-arid zone (n = 109)	Total (n = 399)
Source of animal manure				
Own livestock	81	81	80	80
Purchased from farmers	30	10	7	13
Purchased from market	1	0	0	0.3
Given by farmers	6	6	10	7

Legumes

About 90% of all farmers in the survey reported growing a cereal–legume intercrop and yet only 12% of respondents cited legume intercropping and 9% cited cereal legume rotations as important soil-fertility maintenance practices. The proportion reporting legumes as important soil-fertility inputs did not vary significantly across zones. Farmers cited maize–legume intercrops as important strategies for maximizing utilization of cultivable land and managing risk through diversification of food supplies on small landholdings. About 70% of farmers were aware that a legume–cereal rotation leads to higher yields of the cereal in the following season, but many could not use grain–legume rotations because of their small farm sizes and short-term pressures to maximize production of the staple maize crop.

The dominance of grain–legume intercrops in farmers' cropping system and the fact that many farmers do not perceive legume cultivation as an important strategy for maintaining soil fertility remain a major conundrum in soil-fertility management research. This raises the question of whether farmers underestimate the importance of the nitrogen contribution from legumes or whether the actual levels of nitrogen fixed through legumes under smallholder farmers' conditions are small and therefore have a negligible impact on maintaining soil fertility. Experimental trial data from southern and eastern Africa indicated that residues from a legume–maize rotation contributed the inorganic-fertilizer equivalent of about 30 kg ha^{-1} and a pigeon pea–maize intercrop contributed up to 50 kg ha^{-1} of nitrogen in Malawi (Snapp, 1999) and about 30 kg ha^{-1} in a year in semi-arid Kenya (Rao and Mathuva, 2000). While these data suggested impressive levels of nitrogen fixation at relatively low cost, such levels of nutrient supply are not normally translated into higher levels of soil fertility on farmers' fields. There are several reasons for this. First, the contribution to soil fertility in a legume rotation depends on crop-residue management practices. Under a smallholder

system where there is competition for the crop residue as livestock feed, seed and fuel, a substantial amount of residue is removed from the field and there is little net nitrogen contributed by the legume to soil fertility. In legume–cereal intercrops farmers normally use low plant densities of legumes, which result in relatively small quantities of nitrogen and organic matter being contributed to the soil. Secondly, the contribution of legumes to soil fertility critically depends on the growth habit of the legume and its adaptation to nutrient deficiency. Research on the intensification of legumes in low-nutrient cropping systems in Kenya and Malawi identified long-duration legume varieties with indeterminate growth habit and high biomass production as having the best prospects for producing high-quality residues and improving soil fertility compared with short-duration indeterminate varieties (Snapp and Silim, 2002). However, indeterminate growth habit is positively correlated with late maturity and moderate yields, characteristics that are increasingly becoming incompatible with smallholder preferences for early-maturing and high-yielding varieties. It is also likely that the observed low level of uptake of grain–legume technologies for maintaining soil fertility is due to constraints in existing mechanisms for disseminating information on legume-based technologies to farmers. The survey indicated that farmer-to-farmer informal communication was the dominant means for disseminating information on improved soil-fertility practices. But, whereas about one-third of farmers cited extension as an important source of information for inorganic-fertilizer-based technology, almost no farmer reported receiving advice on improved legume-based technologies from extension. This suggests significant constraints on the flow of information on legume technologies from extension systems to farmers.

Inorganic fertilizer

Farmers realized the importance of inorganic fertilizer in maintaining soil fertility.

Almost 40% of farmers in the survey used inorganic fertilizer as an input in soil-fertility maintenance. There was, however, significant variation across zones, with higher adoption of inorganic fertilizer in the relatively high-rainfall semi-humid zone. Only 13% of farmers in the semi-arid zone reported using inorganic fertilizer, compared with 35% in the transitional zone and over 80% in the semi-humid zone. These estimates, compared with the 8% of farmers who reported using inorganic fertilizer in the transition zone in 1980 (Rukandema et al., 1981) and the 18% reporting inorganic-fertilizer use in 1990 (Muhammad and Parton, 1992), suggested that inorganic-fertilizer use has been increasing, even though the grain–nitrogen inorganic-fertilizer price ratio declined by over 40% between 1988 and 1997. The upsurge in inorganic-fertilizer use has been stimulated, in part, by fertilizer market reforms, which shifted responsibility for fertilizer distribution and marketing from parastatal marketing organizations to the private sector.

Over half of the farmers who reported using inorganic fertilizer in the long rains of 1997 started using the input after fertilizer markets were liberalized (Omiti et al., 1999). This proportion is highest in the transitional zone, followed by the semi-arid zone, areas where fertilizer distribution systems were not well developed and supply-side constraints on the use of inorganic fertilizer were most binding prior to market reforms. The increased level of private-sector participation in fertilizer distribution and marketing following market reforms led to improvements in the efficiency of marketing and distribution systems at the retail level (Freeman, 2001; Omamo and Mose, 2001). Under liberalized fertilizer markets, traders timed fertilizer purchases to coincide with the planting season, a wider range of fertilizer types were available in local markets and there was greater competition in the fertilizer retail trade (Argwings-Kodhek, 1996; Omamo, 1996; Mose, 1998; Mwaura and Woomer, 1999; Freeman and Kaguongo, 2001). In addition, almost all traders repacked fertilizer into smaller packages, which many farmers found to be more

affordable. On the demand side, increased use of inorganic fertilizer was stimulated by sale of the input in smaller packages and increased availability in rural markets (Omiti et al., 1999). However, despite the increase in use of inorganic fertilizer, many farmers applied very small quantities of the input, with median levels of total nutrients applied on maize ranging from 24 kg ha^{-1} to 50 kg ha^{-1} (Omiti et al., 1999).

Farmers prioritized application of inorganic fertilizer, applying it selectively on crops they perceived to have a high relative response or return. Almost all farmers in the survey targeted the input to maize, regardless of the agroecological zone, because of its importance in satisfying household food-security objectives. Coffee was the second most important crop to which inorganic fertilizer was applied. Half of the farmers in the survey applied inorganic fertilizer on coffee, although this proportion declined from 71% of inorganic-fertilizer users in the semi-humid zone to 36% in the transitional and semi-arid zones. The relatively high proportion of coffee farmers applying inorganic fertilizer in the semi-humid zone is due, in part, to the organization of coffee marketing in these areas, where coffee markets were vertically integrated, with cooperatives providing inputs, output marketing facilities and access to credit. This market structure increased access to inorganic fertilizer and provided reliable market outlets, which stimulated use of the input on the crop. Another 25% of inorganic-fertilizer users applied the input to horticultural crops, which are frequently cultivated under irrigated conditions. Legumes, particularly beans, also benefited from inorganic fertilizer applied to maize in the traditional maize–bean intercrop. Table 11.5, showing the application of inorganic fertilizer by crop among farmers applying the input, indicates that the intensity of inorganic fertilizer use is highest on coffee, followed by maize and horticultural crops.

About two-thirds of inorganic-fertilizer users cited lack of knowledge how to use the input effectively as the most important constraint to its use. The reasons for non-adoption of inorganic fertilizer varied, depending on whether the farmer had ever tried the input. In general, many farmers did not use inorganic fertilizer because of lack of money to purchase the input, its cost and perceptions about production risk and benefits. Lack of money was the most frequent response, but the proportion of farmers citing it as an important constraint was highest among those farmers who had previous experience with the input. Production risk was more important among farmers who had never used inorganic fertilizer, with 40% of non-users in the semi-humid zone and less than 10% of non-users in the transitional and semi-arid zones citing it as a reason for not applying the input. In contrast, among farmers who had some experience with the input, production risk was less frequently cited as a constraint on its use. Only 15% of these farmers in the transitional and semi-arid zones cited risk of crop failure as an important constraint.

There were no significant gender-related differences in inorganic-fertilizer use among farmers who applied the input, suggesting that both male and female respondents were equally likely to be using the input. This pattern is different in the semi-humid zone, where male respondents were more likely to be using the input in comparison with female respondents.

Combining organic and inorganic sources of nutrients

Farmers in semi-arid Kenya used multiple sources of soil nutrients. Many farmers in the survey reported using combinations of inorganic fertilizer and organic nutrient sources, especially in the semi-humid zone, where a combination of inorganic fertilizer and animal manure was most frequent (Table 11.6). In about half of the cases where nutrient sources were combined, inorganic fertilizer and animal manure were used as substitutes rather than as complementary soil-fertility inputs, because farmers rarely had adequate quantities of either

Table 11.5. Fertilizer application rates (kg ha^{-1}) (from CARMASAK project; Omiti *et al.*, 1999).

	Semi-humid zone (*n* = 68)	Transitional zone (*n* = 71)	Semi-arid zone (*n* = 14)	Total (*n* = 153)
Maize				
Total nutrients (kg ha^{-1})	64	39	42	50
Standard deviation	42	20	22	69
Horticultural crops				
Total nutrients (kg ha^{-1})	19	11	11	15
Standard deviation	48	33	22	40
Coffee				
Total nutrients (kg ha^{-1})	167	77	127	121
Standard deviation	215	246	273	238

Levels of fertilizer application refer to averages for all farmers using fertilizer.

Table 11.6. Percentage of farmers combining soil-fertility inputs (from CARMASAK project; Omiti *et al.*, 1999).

	Semi-humid (*n* = 84)	Transitional (*n* = 206)	Semi-arid (*n* = 109)	Total (*n* = 399)
Inorganic fertilizer and animal manure	45	23	10	24
Inorganic fertilizer and compost	5	3	0	3
Animal manure and compost	5	13	4	9
Inorganic fertilizer, animal manure and compost	26	6	3	10

input to apply on the entire farm (Omiti et al., 1999).

Empirical Model of Farmers' Technology Choice

A series of binary choice models were fitted to test hypotheses about the factors conditioning farmers' choice of soil-fertility maintenance practices. The models analysed five soil-fertility management options – no fertility amendment, animal manure only, animal manure plus inorganic fertilizer, compost plus inorganic fertilizer and compost only. Use of inorganic fertilizer as a strategy was eliminated from the analysis because the eight farmers who reported using inorganic fertilizer as the only soil-fertility maintenance strategy did not provide sufficient information to fit the model. The aim of the analysis is to identify factors associated with farmers' use of these soil-fertility maintenance options. This is achieved by considering four binary choices: (i) use of a fertility maintenance practice versus none; (ii) use of compost versus use of animal manure; (iii) addition of inorganic fertilizer to animal manure versus no addition to animal manure; and (iv) addition of inorganic fertilizer to compost versus no addition to compost.

For each of these a logistic regression model is used to find the explanatory factors or variables that separate farmers using the first option from those using the second. The effects of significant ($P < 0.10$) factors are presented by using the model to estimate the probability of a farmer using the first option, conditional on using either the first or second, for each level of the factor. The estimates are calculated by averaging over other factors in the model and at the average level of variates in the model. The effect of significant ($P < 0.10$) continuous variates in the model are presented in a similar way, by estimating the probability of using the first choice when the variate takes its value for the lower quartile of the sample and for the upper quartile of the sample.

Results for the logistic regressions are shown in Tables 11.7–11.11. The result from the logistic regression models provides additional insights on the factors that condition farmers' technology decisions. Table 11.7 shows the sign and significance levels of factors that are hypothesized to be related with the soil-fertility maintenance practice farmers are faced with. Different factors appear to be important in each choice. The model presented is more of a predictive model rather than an explanatory model, since three explanatory variables are choice variables of farmers (coffee, livestock units,

Table 11.7. Factors and variables associated with each of the five choices.

	Sign and significance level (P) for effect							
	Choice 1		Choice 2		Choice 3		Choice 4	
	Sign	P	Sign	P	Sign	P	Sign	P
Agroecological zone (AEZ)*		0.100		0.005		0.004		0.009
Herd size (TLU)	+	0.001	–	0.005	–	0.880	+	0.052
Farm size (FARMHA)	–	0.006	+	0.807	+	0.979	–	0.078
Education (EDUCT)	+	0.316	+	0.024	+	0.058	–	0.870
Coffee farmer (COFFEE)	+	0.133	+	0.712	+	< 0.001	+	< 0.001
Gender (GENDER)	+	0.888	+	0.064	+	0.080	+	0.700
Off-farm income (OFFFARM)	+	0.397	+	0.025	+	0.007	–	0.611
Family size (TTALRES)	–	0.911	–	0.679	–	0.060	–	0.237
Age of household head	–	0.270	+	0.125	+	0.560	–	0.378
Distance to market	–	0.409	–	0.854	–	0.101	–	0.333
Sample size		391		223		276		92

*The sign for the agroecological zone variable is not reported because it varies between the two levels.

off-farm income). However, rerunning the regressions without these choice variables did not significantly influence the strength and significance of the remaining exogenous variables. The significant factors $(P < 0.1)$ influencing farmers' discrete choices are used in estimating predicted probabilities.

Table 11.8 shows that a farmer's decision to use some soil-fertility maintenance practice in comparison to not applying any soil-fertility input is strongly conditioned by agroecological factors and the size of farmer's livestock herd. The probability of a farmer using soil-nutrient inputs progressively increases from the drier semi-arid zone to the wetter semi-humid zone. But the significance of the agroecological variable is extremely high across all zones, confirming the survey result that most farmers in Machakos district are likely to respond to declining soil fertility by using some soil-fertility amendment practice. Farmers with larger livestock holdings were more likely to be applying some soil-fertility input. This is not surprising, given that use of animal manure is the most common soil-fertility maintenance strategy. Smaller farm size, probably due to increased population pressure, also influenced the choice to use

soil-fertility inputs but the magnitude of the predicted effect is relatively small compared with the effects of agroecological factors and the size of the livestock herd. Once farmers have made the decision to use soil-fertility inputs, the actual choice of strategy seems to be influenced by different factors.

Table 11.9 indicates that the feasibility of using compost rather than animal-manure strategies is strongly conditioned by agroecological factors. Compost use is most common in the wetter semi-humid zones but not in the semi-arid zone. This is probably due to the relatively high levels of biomass in the wetter areas compared with the semi-arid zone. Compost is also more likely to be used by educated farmers, because it is a relatively new practice in these areas and is a knowledge-based and management-intensive technology. Thus, education and learning become important conditioning factors influencing a farmer's decision to use the practice. Farmers with smaller livestock herds are also more likely to be using compost rather than animal manure because, under conditions of thin manure markets, farmers rely overwhelmingly on their herds for manure supplies. Hence, farmers with smaller herds are more likely to have

Table 11.8. Choice 1: predicted probability of using some soil-fertility maintenance practice versus no amendment.

Effect of significant variable	Level			Probability		
	1	2	3	1	2	3
AEZ	SH	T	SA	0.990	0.973	0.942
FARMHA	0.81	3.6	4	0.977	0.967	
TLU	0.5	5.4		0.891	0.980	

SH, semi-humid; T, transitional; SA, semi-arid.
For other abbreviations, see Table 11.7.

Table 11.9. Choice 2: predicted probability of using compost rather than animal manure.

Effect of significant variable	Level			Probability		
	1	2	3	1	2	3
AEZ	SH	T	SA	0.281	0.204	0.061
TLU	0.5	5.4		0.296	0.143	
EDUCT	No	Yes		0.090	0.206	
GENDER	F	M		0.123	0.220	

For abbreviations, see Tables 11.7 and 11.8.

inadequate quantities of manure to apply on their fields. Male farmers are more likely to be using compost compared with female farmers. This might be due to the knowledge intensity of compost use and gender biases that lead to discriminatory access to information for male and female farmers. Because compost and animal manure are both labour-intensive technologies, variation in family size does not seem to influence the choice of one practice over the other. Distance to markets similarly does not significantly influence the decision to use either of these organic technologies, reflecting the fact that both compost and animal manure are thinly traded soil-fertility inputs. The finding that different factors condition the use of compost and animal manure implies that, even though both inputs are usually lumped together as organic sources of soil nutrients, they are not homogeneous inputs.

The results provide useful insights into the factors that drive farmers to intensify from organic soil-fertility management strategies to integrated nutrient management strategies that combine organic and inorganic sources of nutrients. Such strategies are perceived to provide promising opportunities for soil-nutrient replenishment that will lead to adequate and sustained crop-productivity growth (Lynam et al., 1998). Predicted probabilities for the decision to add inorganic fertilizer to animal manure and compost are shown in Tables 11.10 and 11.11, respectively. Agroecological factors seem to drive the decisions to add inorganic fertilizer to animal manure and compost, although the ecological effect appears to be stronger with integrated nutrient strategies involving compost. Farmers in the semi-humid zones are more likely to intensify into integrated nutrient management strategies, while those in the semi-arid zones are least likely to intensify. The effect of coffee cultivation is a strong conditioning factor for intensifying and adding inorganic fertilizer to organic sources of nutrients. Coffee cultivation has the greatest effect on the decision to adopt integrated nutrient management strategies, irrespective of whether the strategy is an animal-manure-based or compost-based strategy. This probably reflects the incentives offered by a vertically integrated coffee sector, which increases farmers' access to fertilizer, credit and reliable market outlets. Smaller farm size

Table 11.10. Choice 3: predicted probability of adding inorganic fertilizer to animal manure.

Effect of significant variable	Level			Probability		
	1	2	3	1	2	3
AEZ	SH	T	SA	0.528	0.363	0.221
EDUCT	No	Yes		0.277	0.398	
COFFEE	No	Yes		0.237	0.737	
GENDER	F	M		0.310	0.415	
OFFFARM	No	Yes		0.260	0.413	
TTALRES	5.0	9.0		0.389	0.316	

For abbreviations, see Tables 11.7 and 11.8.

Table 11.11. Choice 4: predicted probability of adding inorganic fertilizer to compost.

Effect of significant variable	Level			Probability		
	1	2	3	1	2	3
AEZ	SH	T	SA	0.752	0.439	0.302
TLU	0.5	5.4		0.458	0.649	
FARMHA	0.81	3.64		0.598	0.460	
COFFEE	No	Yes		0.331	0.842	

For abbreviations, see Tables 11.7 and 11.8.

influences the decision to intensify into integrated strategies, but its effect is relatively weak in comparison with the coffee and agroecological effects. In addition to these factors, male farmers, being educated and households with smaller family sizes increase the probability that farmers will intensify from animal manure to integrated use of animal manure and inorganic fertilizer. The weak significance of the distance-to-market variables in intensifying from organic to integrated nutrient strategies might reflect the fact that fertilizer market reforms resulted in the establishment of several fertilizer retail outlets in rural areas, which reduced supply constraints on use of the input.

Policy and Research Implications

The description and quantitative analysis of farmers' choice of soil-fertility maintenance practices in this chapter identify several implications for the testing and development of improved soil-fertility technology options that many smallholder farmers are likely to use and benefit from.

The study does not provide clear evidence to corroborate the results from Tiffen *et al.* (1994) that population pressure, reflected in smaller farm sizes, endogenously drives the process of agricultural intensification and use of improved soil-fertility maintenance practices. The study suggests that agroecological factors, *ex ante* endowments and access to liquidity appeared to be more important than smaller farm size in farmers' decisions to use improved fertility maintenance practices that are perceived to lead to sustainable agricultural intensification. Given the decision to use some soil-fertility maintenance input, the choices of soil-fertility practice and intensification into integrated nutrient management practices were driven by different conditioning factors. Studies in Kenya also suggested that fertilizer market reforms improved marketing efficiency and led to changes in packaging practices that stimulated smallholder farmers' use of the

input (Mwaura and Woomer, 1999; Omiti *et al.*, 1999; Freeman and Omiti, 2001).

The results confirm the importance of agroecological factors in conditioning farmers' soil-fertility maintenance choices. Research and extension efforts to develop and widely disseminate integrated nutrient management strategies need to incorporate variability in the underlying biophysical conditions in technology testing and development. This implies a shift in emphasis from research-station to on-farm experimentation under the heterogeneous soil and climatic conditions where farmers make adoption decisions. Efforts to conduct complementary farmer-managed and researcher-managed trials on farmers' fields are therefore likely to yield a significant pay-off.

The finding that farmers growing coffee were more likely to intensify into integrated nutrient management strategies points to the importance of economic incentives in farmers' choice of technology. Coffee provides an interesting example, because it demonstrates how incentives to intensify into more sustainable NRM practices depended on incentives that extended beyond the farm gate. In this case, incentives were created by market and non-market arrangements arising from the vertical coordination among input distribution, product marketing and credit. This implies the need to broaden the paradigm for soil-fertility management research from its current production orientation to a focus on food subsystems that encompass markets and institutions that are likely to reduce risk and transaction costs.

The results underscore the importance of farmer knowledge and learning in farmer adoption of improved soil-fertility maintenance practices, given that many of these technologies are knowledge-based and management-intensive. Because agroecological factors play a significant role in conditioning adoption of improved soil-fertility maintenance practices, the results suggest the need for greater emphasis on the process through which farmers learn about the ecological context within which adoption decisions are made. Researchers also need to focus on the mechanisms by which farmers learn about improved soil-fertility practices.

Asymmetric flow of information between research, extension and farmers appeared to be an even greater constraint to farmer adoption of grain–legume soil-fertility-enhancing practices than availability of technologies that are consistent with farmers' socio-economic circumstances. This calls for a major rethinking of formal mechanisms for disseminating information on improved NRM practices to farmers. Alternative cost-effective mechanisms for disseminating knowledge and information on improved soil-fertility practices to farmers need to be explored. This need is urgent, given the serious funding constraints facing many extension systems in Africa. Freeman (2001) argues that, with the private sector assuming greater responsibility for input supplies under liberalized markets in Africa, there are likely to be substantial pay-offs from strengthening the role of private traders in providing informal extension services for farmers.

Given that farmers' soil-fertility maintenance choices are conditioned by different factors, research to test and develop soil-fertility maintenance practices increasingly needs to define technology options in partnership with farmers. Researchers need to improve their understanding of the questions farmers are asking and the sorts of experiments they are conducting. Greater emphasis must be given to facilitating farmer experimentation. In this regard, more attention has to be given to learning tools, including simulation modelling and participatory testing of technologies in the development of best-bet options for farmers. Simulation models can speed up technology testing by helping researchers to better evaluate a wider range of soil-fertility management options, paying particular attention to variability in biophysical conditions, as well as risk and farmers' priorities. Active farmers' involvement in the research process also ensures that research results remain relevant for the target group. Best-bet technologies, however, need to be feasible and profitable, in addition to their technical performance, for them to be attractive to farmers. Bio-economic models offer good prospects for improving technology targeting and the adoption of best-bet soil-fertility maintenance options. These modelling tools integrate the biophysical processes that track basic NRM variables and household decision-making, including behavioural and resource constraints.

Note

[1]   We thank Chris Barrett, Frank Place, Abdillahi Aboud, Said Silim and two anonymous reviewers for useful comments on earlier versions of this chapter. We are grateful to Wachira Kaguongo for research support. The Rockefeller Foundation and International Crops Research Institute for the Semi-Arid Tropics (ICRISAT) provided funding for this study under the Collaboration on Agricultural Resource Modeling and Applications in Semi-Arid Kenya (CARMASAK) project, a collaborative research effort between the Kenya Agricultural Research Institute and ICRISAT. The views expressed in this chapter are, however, those of the authors.

12 Agroforestry for Soil-fertility Replenishment: Evidence on Adoption Processes in Kenya and Zambia

Frank Place, Steven Franzel, Joris DeWolf, Ralph Rommelse, Freddie Kwesiga, Amadou Niang and Bashir Jama
International Centre for Research in Agroforestry (ICRAF), PO Box 30677, Nairobi, Kenya

Sub-Saharan Africa (SSA) is the only region in the world where the number of poor people has been increasing in the past decades. The World Bank (2000a) estimated the number of poor (less than US$1 day^{-1} of income) to have risen from 242 million in 1990 to 291 million in 1998, nearly half of the population. Unlike other regions of the world, most of the poor in SSA continue to reside in rural areas, owing to poor growth in industrial and service-sector jobs.

A key component of the rural poverty complex is the declining per capita food-production trend over the past 30 years (Badiane and Delgado, 1995; World Bank, 1999). While opportunities for agricultural land expansion are worsening, yield increases in relation to the rest of the world have been dismal. One of the reasons for poor yield performance is the woefully low use of mineral fertilizers (FAO, 1996; Larson and Frisvold, 1996; Gladwin *et al.*, 1997a; Mwangi, 1997). Another reason is the estimated high rate of soil erosion due to insufficient conservation measures and consequent loss of fertility (Bojo, 1994; Bishop, 1995). Together with poor cycling of nutrients, negative farm nutrient balances have been found to be common (Stoorvogel and Smaling, 1990; Soule and Shepherd, 2000).

A host of factors have contributed to these unfortunate trends. First, a rapid decline in average farm sizes has necessitated substantial changes in traditional farming methods.[1] However, government policies did not help to facilitate such changes. Roads and extension services have been severely constrained in SSA, so that new information is difficult to disseminate. Research and extension systems emphasized production of the staple cereal crop and, in some cases, the major cash crop of the country, paying insufficient attention to some potentially more profitable farm enterprises. Finally, agricultural input and credit subsidies were eliminated. In response, rural households have developed a highly varied and complex set of livelihood strategies, of which farming is but one component. But the fact that poverty persists in rural Africa indicates that development opportunities for the poorest households are inadequate. Identifying and promoting feasible options for these households is high on the policy agenda in Africa.

Scope of Chapter

In the 1980s, the International Centre for Research in Agroforestry (ICRAF) began collaboration with national research partners in eastern and southern Africa. Throughout the region, farmers mentioned low soil fertility as an important constraint. In response, ICRAF and its partners launched research into identifying agroforestry options for improved soil fertility. Many agroforestry systems were tested: (i) improved fallows (the enrichment of natural fallows with trees); (ii) relay cropping (planting trees into a standing crop; the trees are cut and die before the next crop is planted); (iii) biomass transfer (applying leaves as green manure); (iv) mixed intercropping (planting trees into a standing crop; the trees remain in the plot indefinitely and are cut back when crops are planted, to minimize competition); and (v) alley farming (crops are grown between rows of trees and the trees' leafy biomass is applied to the crop).

Among these, poor technical performance in researcher-managed trials led to the discontinuation of research and development of selected technologies at specific sites. The more promising systems have been tested in farmer-managed conditions. The alley-farming system was tested over several years with farmers in western Kenya, but did not spread, due to lacklustre performance. For the relay-cropping and mixed-intercropping systems being tested in Malawi, there are insufficient data, owing to their relatively recent dissemination. Thus, this chapter will focus on three primary case-studies:

- improved fallows in eastern Zambia;
- improved fallows in western Kenya;
- biomass transfer in western Kenya.

For each of these systems, though the numbers of farmers using them are in the thousands and growing, it must be emphasized that their use has been for 6 years at the most. Thus, this study of farmer assessment, use and adoption is at a very early stage of the adoption process.

The rest of the chapter is as follows. We start with a brief description of eastern Zambia and western Kenya. This is followed by a description of the two agroforestry systems highlighted in the chapter. A section then briefly describes the methods used to monitor and evaluate the impact of the systems. The results are then presented. One section deals with the profitability, acceptability and feasibility implications of each agroforestry system. A second section examines factors associated with household-level use/adoption. A third results' section examines key factors affecting the scaling up of adoption in much wider areas. The final section contains a summary of the major points and implications.

Description of the Study Sites

Western Kenya

The research in western Kenya is focused largely on medium- to high-potential highland areas. Rainfall is good, ranging from 1200 to 1800 mm year^{-1}, with two cropping seasons annually. The short rainy season is much less reliable in terms of total rainfall and length of growing season. The topography is undulating, with moderate slopes. Soils are of generally good physical structure but are nutrient-depleted. In many parts of the region, phosphorus is the major limiting nutrient, but nitrogen and potassium limitations are also prevalent.

High population densities prevail, ranging from 500 to 1200 km^{-2} in Kakamega, Siaya and Vihiga Districts. Poverty rates are very high, exceeding 50% of the population. Farm sizes are small, generally 1 ha or less. The acute land pressure has led to significant rural-to-urban migration and as many as 30% of households are headed by females. Landholdings consist mainly of a single parcel of land and land tenure is relatively secure. Most land is either inherited or purchased and around half of households hold title to the land.

Farmers generally keep relatively few cattle. There are a variety of crops grown in western Kenya, but maize and beans are the dominant enterprise. Other crops include sorghum, cassava, kale and cabbage, with some areas able to accommodate coffee and tea. But productivity remains relatively low, with minimal mineral-fertilizer use. Two recent studies have found that over 50% of farmers fallow some of their land for at least one season (Ohlsson *et al.*, 1998; DeWolf *et al.*, 2000). Farmers in western Kenya plant a significant number of trees and one can find high densities of a limited range of trees planted on a given farm.

Eastern Zambia

Eastern Zambia is distinctively drier than the East African highlands, with rainfall ranging normally between 800 and 1000 mm and occurring in a single cropping season. Eastern Zambia is situated on a plateau with mainly gently sloping land, dissected by low-lying areas that are moist during the long dry season. The clay soils of eastern Zambia are generally good for farming and often maintain adequate levels of phosphorus. The sandier soils, also common, are less favourable for farming and are subject to compaction during the dry season and nutrient leaching during the rainy season.

Eastern Zambia has one of the lowest population densities in eastern and southern Africa, at 10–50 persons km^{-2}. Farm sizes are mainly in the range of 3–6 ha and natural fallowing is still common, though increasingly limited in duration (e.g. 2–3 years). Farms consist of a single and substantial upland field and, in many cases, a small second field located in a low-lying area. Land is normally acquired through inheritance, and households have secure, long-term rights to land.

Cropping patterns are dominated by maize; cash crops are few in number. As late as 1990, over 80% of cultivated area in Eastern Province was estimated to be under maize (Celis *et al.*, 1991), though this was found to be lower in 1999 (Peterson, 1999). Prior to liberalization, farmers had been high users of fertilizers and hybrids (Kwesiga *et al.*, 1999), but these have fallen into disuse following removal of input price subsidies

and credit facilities. Livestock are very important in this relatively dry climate. Cattle are used as a store of wealth, to buffer against risk and to provide draught power and manure. Livestock is free-grazed and this can pose problems for the growth and survival of tree seedlings. With ample Miombo woodland, farmers have not gained a tree-planting culture.

Description of the Key Agroforestry Systems

The improved-fallow and biomass-transfer systems are described below. Much more information on these systems' biophysical aspects can be found in Cooper *et al.* (1996) and Sanchez (1996).

Improved fallows

In Kenya, the majority of farmers plant improved fallow trees into an existing crop, while in Zambia most farmers establish them in an uncultivated field. The dominant crop for which fallows are used is maize in Zambia and maize/bean in Kenya. In western Kenya, farmers direct-seed or broadcast at high density one or more of several species, with *Crotalaria grahamiana* and *Tephrosia vogelii* being the most popular. In eastern Zambia, sesbania, the preferred species, is established in a raised-bed nursery and then transplanted to the target field. Tephrosia is direct-seeded and, for both species, the density of choice among farmers in Zambia is about 10,000 ha^{-1}. In western Kenya the trees are planted around the time of the second weeding of a maize field and then occupy the field alone during the following rainy season (mainly the short season). In eastern Zambia, the trees grow in a pure stand for two seasons (or along with the crop for part of one season and for a full second season). In both countries, the tree fallows are cut and the leaves incorporated into the soil during land preparation.

The tree-fallow species contribute high levels of nutrients, both through incorporation of leafy biomass and from underground root biomass. Leguminous trees, such as those used, have the added benefit of fixing atmospheric nitrogen. An improved tree fallow can produce up to 150 kg ha^{-1} of N and recycle 50–60 kg ha^{-1} of K and about 5–10 kg ha^{-1} of P during the fallow phase. In addition to nutrients, improved fallows contribute organic matter and soil carbon, improve soil structure and infiltration and can significantly reduce weeds. In addition to these soil benefits, the trees produced by the fallows can also provide important by-products, such as stakes, firewood or, in the case of tephrosia, insecticide prepared from the leaves.

Biomass transfer

Biomass-transfer systems in Kenya involve the growing of trees or shrubs along boundaries or contours on farms (or the collection of the same from off-farm niches, such as roadsides) and applying the leaves on the field at planting time and sometimes later in the season. In western Kenya, *Tithonia diversifolia* became the preferred species used by farmers. This has been tested on maize, kales, French beans and tomatoes. Given the small farm sizes in Kenya, farmers generally utilize the green manure on smaller plots, often preferring higher value vegetables.

Tithonia does not fix atmospheric nitrogen but is a good scavenger of nutrients and thus its leaves have high concentrations of N, P and K. It benefits the soils and crops in much the same way as fallows, except that the lack of the tree in the field will limit the effectiveness of biomass-transfer systems on soil structure, infiltration and weed control. Some farmers claim that tithonia helps to prolong harvest periods of vegetables and to improve plant resilience during dry spells. Tithonia hedges do not provide significant secondary products.

Overview of Methods Used

Data used to assess the feasibility, profitability and adoptability of soil-fertility technologies come from three types of trials (Franzel *et al.*, 2001): (i) researcher-designed–researcher-managed; (ii) researcher-designed–farmer-managed; and (iii) farmer-designed–farmer-managed. Researcher-managed work is used to assess biological responses to specific agroforestry treatments in comparison with other alternatives. Data from researcher-designed–farmer-managed trials are used to evaluate biological performance under farmer management across a range of ecological and socio-economic conditions and are relied upon to examine labour and profitability aspects of agroforestry systems. Finally, farmer-designed trials are used to assess the acceptability and feasibility of agroforestry systems. Our examination of the use, expansion or adoption of agroforestry systems includes farmer experimenters, as well as farmers who have spontaneously adopted without any interaction from researchers.

ICRAF and partners employ qualitative (e.g. ranking of outcomes) and quantitative (e.g. financial returns) methods to measure farmer use and assessments of agroforestry systems. Early testers of the systems are monitored over time to examine dynamic processes, such as costs and returns from different land-use practices, adaptations of technologies and changes in the extent of the use of technologies. Larger samples of farmers within and adjacent to early dissemination points are surveyed to study the types of households that decide to use the systems. In both sites, there has been considerable attention devoted to the measurement and classification of households according to wealth, in order to allow for the testing of whether the technologies are being utilized by the poor.

Feasibility, Acceptability and Profitability

Farmer assessments covering a range of questions and applying a range of tools are used to analyse the feasibility and acceptability of the agroforestry systems. For profitability, all costs (including family labour) and benefits are recorded in the appropriate season and discounted at a rate of 20% per year. All prices and costs (including labour) are from local market surveys and are constant within a site.[2] Land costs are not included directly (because the size of land is held constant across comparative land-use practices), but the opportunity cost of foregone production is factored into the analysis. We calculate net present values for different land-use options and are able to directly compare them by holding the time period and land size constant. We also compare the same systems in terms of returns to a labour day, where the local wage rates in both Kenya and Zambia are near US$1.

Biomass transfer in western Kenya – feasibility and acceptability

Obonyo (2000) conducted a survey of 69 farmers in Vihiga District who had received the information on biomass transfer from extension agents (called extension farmers). She also interviewed 53 farmers from Vihiga who initially tested biomass transfer in collaboration with the ICRAF team (called research farmers).

Labour for collecting, transporting and incorporating biomass, which all occur at peak labour times, was the biggest problem noted in the case-studies. At the inception of farmer experimentation, collecting, transporting and incorporating 1 kg of biomass from off-farm took an average of 4 min among research farmers. After some years of experience with the system, farmers began to grow tithonia on farm and reduced this time to 2 min. Cutting and applying tithonia is not physically demanding and is done by men and women.

About one-fifth of the extension farmers had, by the time of Obonyo's (2000) study, planted tithonia on their own farm. They had shifted from using biomass mainly on maize/beans to using it mainly on kales and other vegetables. They valued increased yields as well as increased crop quality and sometimes prolonged harvest periods, each of which can enhance the market value of the product. The average size of field on

which biomass transfer was used increased from 196 m² to 252 m² over five seasons for extension farmers and from 79 m² to 344 m² for research farmers.

Biomass transfer in western Kenya – profitability

Financial returns have been analysed for maize, kale and tomatoes under various treatments, including biomass transfer. For maize, calculations were made based on researcher-managed trials. The application of tithonia biomass at 0.91 or 1.82 tons of dry matter ha⁻¹ (during the first season) increased yields and profits substantially. However, the biggest increases occurred when tithonia was integrated with phosphorus fertilizer. The returns to land and labour were highest when 1.82 tons ha⁻¹ of biomass (dried equivalent) were applied along with 50 kg of phosphorus ha⁻¹ (e.g. the returns to labour were four times those of the unfertilized continuous-maize treatment).

Table 12.1 provides evidence on the economic returns to biomass transfer on kale and tomatoes. The biomass transfer system is more profitable on these higher-valued crops as compared with maize. Due to high costs of labour and pesticides, vegetable production is not profitable in the absence of soil-fertility amendments. The addition of tithonia alone (row 4 under each crop) was not profitable for kale production but was profitable for tomatoes. This most probably reflects the fact that the phosphorus status of soils varies somewhat in the region. As was the case with maize, the largest impacts occur when some phosphorus is added. For both crops, the most profitable systems used tithonia combined with a low dose of phosphorus.

Improved fallows in western Kenya – feasibility and acceptability

In western Kenya, two surveys were conducted with farmers to assess management and innovation in the use of improved

Table 12.1. Economic analysis of biomass transfer on kale and tomatoes in western Kenya (farmer-managed trial).

Tithonia fresh weight (tons ha^{-1})	N input from tithonia (kg ha^{-1})	P input from rock phosphate (kg ha^{-1})	Costs for labour (US$ ha^{-1})	Costs for capital (US$ ha^{-1})	Return to land (US$ ha^{-1})	Return to labour (US$ day^{-1})
Kale						
0	0	0	571	286	−857	−0.47
0	0	33	571	339	116	1.12
0	0	65	571	393	311	1.44
10	49	0	628	286	−801	−0.26
10	49	33	628	339	985	2.39
10	49	65	628	393	820	2.14
Tomatoes						
0	0	0	929	500	−1012	−0.08
0	0	32.5	929	554	−728	0.20
0	0	65	929	607	752	1.68
10	49	0	985	500	201	1.12
10	49	32.5	985	554	1854	2.68
10	49	65	985	607	1677	2.51

fallow and its feasibility and acceptability with farmers of different characteristics. The first survey in 1998 involved 140 farmers (DeWolf *et al.*, 2000) and a second in 1999 involved 67 farmers (Pisanelli *et al.*, 2000). From the larger survey, it was found that, from a technical point of view, farmers had little trouble in establishing their fallows. Most (70%) did so in an existing crop to save on land preparation and weeding, although, in 28% of these cases, farmers reported a negative effect on that season's maize crop. The other significant problem reported was a caterpillar attack on crotalaria, which occurred mainly during the 1998 El Niño period. None the less, 79% of farmers reported that subsequent crop yields were positively affected by the fallows, through soil-fertility improvement and weed reduction (notably striga).

Labour and land constraints were also investigated. About one-third of farmers said that land preparation after an improved fallow was more difficult than after a natural fallow (more felt otherwise). Only one farmer discontinued the use of the technology for this reason. The study by Pisanelli *et al.* (2000) found that 55% of the fallows were cut by women, 35% by men and 10% by mixed groups, so improved fallows do not

appear to be less acceptable to women for physical or cultural reasons. An improved fallow of 1000 m^2 requires 1.6 days to establish, 1 day to cut and 1 day in additional land preparation as compared with continuous cropping. However, over a four-season rotation, the fallow system uses only 83% of the labour of the continuous-cropping system. Though there was concern about the small farm sizes in the area, the average improved tree-fallow size in 17 study villages increased from 134 m^2 to 247 m^2 between 1997 and 1999 (DeWolf *et al.*, 2000).

Improved fallows in western Kenya – profitability

Table 12.2 presents an analysis of two farmer-managed trials in western Kenya. The first trial was for four seasons and the second for three seasons.[3] The crop following the fallow was maize or maize/bean. In the first trial, the natural fallow system was found to be unproductive and not financially attractive compared with all other systems. The tephrosia fallow without phosphorus inputs was the most economically attractive by the criteria of both returns to land and returns to labour. The

Table 12.2. Economic analysis of improved fallows on maize and beans for three seasons in western Kenya (farmer-managed trial).

Land-use system	P rate (kg)	Average total yield: maize (kg)	Average total yield: beans (kg)	Total costs (US$)	Return to land (US$)	Return to labour (US$ day^{-1})
Trial 1 (total *n* = 34)						
Continuous cropping	0	4390	969	585	405	1.74
	250	5025	1191	1047	108	1.14
Natural fallow	0	2626	519	442	148	1.36
	250	3573	681	904	−131	0.63
Crotalaria fallow	0	3964	855	484	397	1.87
	50	5191	1035	588	528	2.13
Tephrosia fallow	0	5122	962	495	588	2.31
	50	5440	867	588	534	2.14
Trial 2 (total *n* = 61)						
Continuous cropping	0	4160	0	388	242	1.53
	50	4505	0	481	189	1.40
Crotalaria fallow	0	4498	0	313	351	2.04
	50	4414	0	404	249	1.71

crotalaria system, favoured by most farmers, gave poor results in the first season and thus was superior to the continuous-cropping practice only in returns to labour. For this system, the addition of phosphorus increased returns substantially. A second trial involving more farmers (about 30) found that the crotalaria fallow system without any additional fertilizer was far superior to that of the continuous cropping system. The returns to land and labour were 45 and 33% higher, respectively.

Improved fallows in eastern Zambia – feasibility and acceptability

A survey of 108 farmers who first planted improved fallows in 1994/95 was made in 1999 (Franzel *et al.*, 2000) to assess their experiences in managing the technology. Establishing nurseries was not a major problem because of the relative availability of low-lying moist areas. Sesbania had higher survival rates than tephrosia, since the latter is direct-seeded. Both species were found to perform much better on the higher clay soils than on the sandy soils (75% versus 41% survival rates, after 1 year (Franzel *et al.*, 1999)). Over time, farmers have managed to increase the land area devoted

to fallows from an average of 0.04 ha to 0.07 ha between first and third plantings. Peterson (1999) found larger fallows among a smaller, more dispersed sample.

Neither tree planting nor cutting seemed to be a problem and the improved-fallow system as a whole required 11% less labour than a continuous unfertilized-maize alternative. If farmers were to plant 0.3 ha each year,[4] this would increase labour by only less than a day for direct-seeded tephrosia or by 6.3 days for the transplanted sesbania system. Cutting of the fallows generally took less time than for planting, was not difficult for women and took place during a slack labour period. In Peterson's study (1999), the major reasons for not trying improved fallows were related to lack of awareness or germ-plasm, but not to farm size or labour. Similarly, household resources were not found to be significant factors in explaining who tested or expanded use of improved fallows in the four ICRAF pilot villages (Franzel *et al.*, 1999).

Improved fallows in eastern Zambia – profitability

Twelve farmers who agreed to compare continuous unfertilized maize, continuous fertilized maize and maize following a 2-year fallow, were intensively monitored for inputs and outputs. Over a 5-year period (for a full fallow cycle), the improved-fallow system delivered almost twice as much total yield as the unfertilized maize, though cropped for two fewer seasons. Table 12.3 shows the mean net returns to land and labour for the three systems. The net present value from the improved-fallow system was US$203 ha^{-1} as compared with only US$5 for continuous cropping without fertilizer. The fertilized-maize system was far superior in total maize production, but the improved-fallow system performed better in terms of discounted returns to land. The fallow system was slightly below fertilized maize in terms of mean return to labour, but was higher on nine of 12 farms.

Results from Studies of Farmer Use, Adoption and Expansion

This section presents results from statistical analyses of the use and expansion of agro-forestry and other soil-fertility replenish-ment practices. The analyses range from simple cross-tabulations to logit and multi-nomial logit regressions. We focus on the influence of household factors, rather than broader community-level factors, as the latter are not possible to ascertain at this early stage in the dissemination process. Binary

Table 12.3. Profitability of 2-year *Sesbania sesban* improved fallow compared with continuously cropped maize in eastern Zambia (farmer-managed trials, *n* = 12).

Option	Returns to land: net present value (US$ ha^{-1})	Returns to labour: net returns per workday (US$ day^{-1})
Continuous unfertilized maize (5 years)	5	0.42
Continuous fertilized maize (5 years)	160	1.02
Sesbania sesban fallow (2 years fallow, 3 years maize)	203	0.93

logit models are estimated separately for a variety of soil-fertility practices in both Zambia and Kenya. The larger number of observations in Kenya further allow us to specify a multinomial logit regression, which considers the influence of household factors on the use of alternative combina-tions of soil-fertility practices. In both cases, the explanatory variables come from surveys completed prior to or at an early stage of farmer interaction with the agroforestry systems.

Eastern Zambia – improved fallows

In Zambia, two separate exercises were done to look at early use/adoption of improved fallows. First, monitoring by ICRAF/partner scientists was done in four pilot villages (*n* = 218). Secondly, a study was made of the early testers (located in a wide number of villages) of fallows to understand which types of farmers were continuing the use of fallows and which had discontinued (*n* = 101). In the pilot villages, analysis has focused on the effects of two variables: gender of household head and household wealth. There was little dif-ference in use between men and women, where the percentages using were 32 and 24, respectively. Wealth was ascertained by wealth-ranking exercises by villagers who placed one another into four categories (well-off, fair, poor and very poor). The use of fallows is higher among wealthier households, who appear to be leading the testing and adaptation process. While 53% of the well-off farmers were using improved fallows, this percentage drops to 16% for the very poor households.

A study of 101 early (1994) testers of the fallows was undertaken in 1999. Informa-tion was collected about improved fallowing practices over the years, the use of other soil-fertility practices and household and farm characteristics. Farmers were par-titioned into those who have continued the use of the fallows (planted at least once after 1995) and those who have discontinued (not planted following the initial 2-year trial period). Similarly, we distinguished between users and non-users of chemical fertilizer and animal manure. These dichotomous soil-fertility practices were then related to several household factors, using logit regression models, and the results are given in Table 12.4. The models presented are in a reduced form in the sense that many choice variables of farmers that might be related to the use of soil-fertility practices (e.g. number of cattle, off-farm income) are omitted in order to examine the full effect of exogenous household factors.[5]

Table 12.4 shows that none of the household variables were significantly related to continued use of the improved-fallow system. Two higher-scale variables are significant. The strongest is whether the household resided in one of the four pilot villages. The positive effect reflects the greater attention and technical advice given by extension and researchers. Such effects are expected at early stages of dissemination, but would be expected to be overtaken by variables reflective of farmer demand over the course of time. The only other variable that is significant in the improved-fallow regression is one of the location variables. Continued use rates were significantly higher in the base (and omitted) camp location. We found that rates of use of chemical fertilizer and animal

Table 12.4. Logit regression results for soil-fertility replenishment options in eastern Zambia (*P* value of Wald ratios in parentheses).

	Dependent variable		
Independent variable	Continued use of improved fallows	Chemical fertilizer	Animal manure
Constant	−1.3394	1.9773	−0.3417
	(0.2683)	(0.0844)	(0.7592)
Age	0.0279	−0.0262	0.0101
	(0.2507)	(0.2659)	(0.6611)
Completed primary education	−0.1989	−0.4903	0.3626
	(0.7623)	(0.4552)	(0.5755)
Secondary education	0.0068	−0.2631	0.0382
	(0.9920)	(−0.7260)	(0.9561)
Male-headed, single or polygamous household	−0.2437	0.8409	−0.7624
	(0.7235)	(0.3265)	(0.3045)
Female-headed, married, single or widowed household	0.4101	−0.6199	−0.3026
	(0.6133)	(0.3796)	(0.6704)
Farm size	−0.0935	0.0416	0.1436
	(0.2261)	(0.6208)	(0.0796)
Number of household members over 13 years	0.1406	0.1694	0.1085
	(0.2909)	(0.2758)	(0.4838)
Pilot-project village	3.3225	n/a	n/a
	(0.0032)		
Camp 1 dummy	1.3949	−0.2119	−0.7738
	(0.1408)	(0.7963)	(0.3000)
Camp 2 dummy	0.2738	−1.1187	−3.8124
	(0.6962)	(0.1179)	(0.0001)
Camp 3 dummy	−0.5546	−1.5226	−2.4405
	(0.4820)	(0.0612)	(0.0042)
Camp 4 dummy	−1.5272	−3.6013	−2.1430
	(0.0546)	(0.0001)	(0.0027)
Number of observations	101	101	101
% of users of technology	54.5%	55.4%	40.6%
% correctly predicted by model	76.2%	76.2%	77.2%

n/a, not applicable.

manure were similarly higher in this camp (see following section).

Eastern Zambia – comparison among soil-fertility options

The only significant result among the fertilizer and manure regressions was that farm size was positively associated with the use of animal manure. Hence, manuring was slightly less feasible or desirable for smaller farms (e.g. through having smaller herds). Otherwise, land, labour and education factors were not related to the use of the three soil-fertility technologies. Similarly, household type was not related in any of the cases – they appeared to be neutral with respect to gender. However, it is important to highlight the significance of the community-level variables in explaining the use of soil-fertility options. This indicates that access to or incentives for such practices vary widely across space and these appear to be sufficiently strong to have similar impacts on all types of households within communities. Analyses at the community level are critically needed to shed light on this.

Western Kenya – improved fallows

Two studies were undertaken to examine the factors associated with the early use/uptake of improved fallows in western Kenya. The first was a detailed examination of early testers ($n = 99$) to identify factors associated with the continuation and discontinuation of the technology (Pisanelli *et al.*, 2000). A second analysis involved a regression of significant use (used the improved-fallow system for at least two of three seasons) opposed to non-use (never used) or discontinuation (used in first season only) among farmers in the pilot areas of Vihiga and Siaya Districts.[6] In all, 1131 households were included in the analysis.

Among the early adopters, Pisanelli *et al.* (2000) used regression analyses to study continued use of the system (logit

model) as well as the size and proportion of land area devoted to fallows (ordinary least squares (OLS) regressions). They found that households more likely to continue use of improved fallows had a male decision-maker, were from Siaya District (i.e. Luo as opposed to Luhya) and were more likely to have off-farm employment. Regarding area planted, variables that were strongly associated included use of chemical fertilizer (positive), female head of household (negative) and area normally under traditional fallow before the use of improved fallows (positive). Finally, the percentage of area planted with improved fallows is associated with only one household variable, farm size, which had a negative effect on percentage area. None of the results are particularly surprising, though the complementary relationship between organic and inorganic fertility measures is interesting.

Table 12.5 shows the results of the logit regression for improved fallows, as well as other soil-fertility techniques (which are discussed below). The most important variables related to the use of improved fallows were size of farm, number of adult household members, residence in a primary pilot village and residence in Siaya, each being positive in sign. Farm size appears to be the most important of the household variables, especially as the land–adult-member ratio was also found to be positively related in an alternative regression. This result, however, needs to be qualified, because farm sizes vary in a relatively narrow range and 89% of households had farms of less than 3 acres. As expected, residence in one of the primary pilot villages (comprising ten of the 17 villages studied) is positively and significantly related to the use of improved fallow. Controlling for this effect, it is useful to note the strong effect of the Luo community in Siaya District. We believe this to be related both to demand considerations – that is, more interest and cohesion among these farmers – and to supply considerations – enhanced interest on the part of extension, non-governmental organizations (NGOs) and project technicians in response to farmer enthusiasm.

Table 12.5. Logit regression analysis of household factors affecting the use of several soil-fertility replenishment options from 17 villages in Vihiga and Siaya Districts, western Kenya (*P* value of Wald ratios in parentheses).

Independent variable	Dependent variable				
	Biomass transfer	Improved fallows	Chemical fertilizer	Animal manure	Compost
Constant	−2.3876	−3.0774	−3.3823	−0.0555	−0.1618
	(0.0000)	(0.0000)	(0.0000)	(0.8575)	(0.5634)
Age of household head	−0.0104	−0.0048	−0.0111	−0.0005	−0.0210
	(0.1101)	(0.4067)	(0.0372)	(0.9086)	(0.0000)
Lower primary education	0.4445	−0.1665	0.1984	0.3382	−0.0568
	(0.1065)	(0.5019)	(0.4307)	(0.0600)	(0.0000)
Upper primary education	0.4747	0.1227	0.6334	0.0716	−0.3768
	(0.0754)	(0.6050)	(0.0077)	(0.6924)	(0.0286)
Secondary education	0.5417	0.1046	0.7877	−0.0040	−0.3682
	(0.0934)	(0.7150)	(0.0036)	(0.9861)	(0.0761)
Male-headed − single or polygamous	−0.1676	0.0799	−0.0692	−0.5240	0.0829
	(0.5201)	(0.7462)	(0.7491)	(0.0052)	(0.6348)
Female-headed − widowed	−0.2187	0.1125	−0.1289	−0.2370	−0.2016
	(0.3577)	(0.6059)	(0.5326)	(0.1500)	(0.1980)
Female-headed − husband away	−0.1425	0.1005	−0.9199	−0.2904	0.5487
	(0.7106)	(0.7305)	(0.0014)	(0.1937)	(0.0051)
Wealth index	0.0027	0.0579	0.5904	0.3368	0.3147
	(0.9693)	(0.3462)	(0.0000)	(0.0000)	(0.0000)
Owned farm land	0.0772	0.2136	0.0414	0.0700	0.0878
	(0.0792)	(0.0000)	(0.2557)	(0.0903)	(0.0095)
Number of household members	0.0613	0.0474	0.0062	0.1010	0.0244
	(0.0144)	(0.0459)	(0.7806)	(0.0000)	(0.1564)
Pilot village with relatively high interaction with project	0.5994	1.1308	n/a	n/a	n/a
	(0.0009)	(0.0000)			
Luo	1.3094	1.2005	1.5595	−0.7975	0.0717
	(0.0000)	(0.0000)	(0.0000)	(0.0000)	(0.5288)
Number of observations	747	1131	1620	1623	1621
% of users of technology	30.6%	23.8%	20.5%	71.0%	40.6%
% correctly predicted by model	70.4%	77.5%	80.8%	72.5%	63.8%

Western Kenya – biomass transfer

Two sets of studies were conducted to investigate the household characteristics associated with the use of biomass transfer. In Vihiga, chi-square tests and *t* tests were made to examine the relationship between continued use of biomass transfer and 15 household characteristics for 65 farmers who were exposed to the technology by extension agents and had tried it at least once (Obonyo, 2000). Secondly, a logit regression was made to analyse the effect of similar household factors for 747 farmers located in the pilot-study villages of Siaya and Vihiga. In the Vihiga extension sample,

the major finding was the strong effect of gender. While 43% of households where males were the primary decision-maker were classified as continuing use of the technology, only two of the 14 (14%) female counterparts were still using the system. Frequency of contact with extension agents was the only other significant relationship (age, education and reliance on non-farm activities were not related).

The regression analysis (Table 12.5) follows closely the results on feasibility. It indicates that farmers using biomass transfer are more likely to have a larger number of family members. This is congruent with the farmer recognition of labour as the major constraint to biomass transfer. The system was also more likely to be practised on larger farms, though this is somewhat less statistically significant (8% level). It was also found that more-educated farmers were most likely to use biomass transfer. This may be explained by several factors, but the most likely may be that the more educated are more likely to have adopted cash crops (significant in separate regressions), which are increasingly becoming the favourite target of the biomass. Unlike the results from the Vihiga extension sample, the use of biomass transfer in the pilot villages was not related to the household type of decision-making (this holds even when land, labour and wealth are omitted from the model). The results for the pilot-project and Siaya-District residence variables are the same as for improved fallows and the reasons for these are believed to be similar (see above for detailed explanation).

Kenya – comparison among soil-fertility options

Table 12.5 shows, in addition to regressions for improved-fallow and biomass-transfer use, the use of chemical fertilizer, animal manure and composting by farmers. Some interesting similarities and contrasts emerge. The likelihood of practising most of the options increases as land and labour increases. The exception to this rule is the use of chemical fertilizer that seems scale-neutral. Land and labour endowments, however, are not necessarily reflective of wealth.[7] Results on the wealth variable show that the agroforestry systems are being used to a larger extent than other practices by the poorer households. In terms of gender, again, the agroforestry systems are similar in their apparent neutrality. In contrast, households where females are heads while husbands are away are much less likely to use chemical fertilizer but more likely to use composting than male-headed monogamous households. Education is an important criterion in the use of biomass transfer and chemical fertilizer, where the impact is positive, and in the use of composting,

where, surprisingly, the impact is negative. Lastly, older household heads appear to be less likely to use compost and chemical fertilizer. Looking at the set of options, there appear to be some that satisfy each of the different characteristics and demands of households, with the possible exception that none of the inorganic techniques seemed highly attractive to those operating the smallest farms.[8]

Experiences and Issues in Scaling Up Adoption

The early successes with improved fallows in Zambia led to wider dissemination in 1996 (Kwesiga et al., 1999), and in Kenya wider dissemination of improved fallows and biomass transfer was initiated in 1997. By 2000, several thousands of farmers were using the technologies in each location.

The agroforestry-based soil-fertility replenishment systems are information-intensive in the sense that they are not simple adjustments to current or past farming practices. They require much more learning and interaction. Indeed, the analyses in this chapter provide ample support for this. In both Zambia and Kenya, use rates were significantly higher in pilot villages, marked by the greater farmer access to technical support than in nearby villages. Further, the use of some of the technologies, including biomass transfer, was positively related to education levels. Because pilot development projects cannot be replicated in many other areas and extension systems are often weak, the question of how to support farmer learning and sharing of information at wide scales is a difficult one.

There are, however, some positive signs. A large number of NGO partners and community groups have incorporated these agroforestry systems into their programmes. In both western Kenya and eastern Zambia, researchers have found extension agents to be enthusiastic partners. They are certainly motivated somewhat by the increased attention and modest incentives, such as bicycle-repair allowances, they are offered. But they are also motivated by the real possibilities to provide their farmer clients with improved practices that are feasible for a large number of households. This experience would suggest that extension systems may be able to play a vital role in the scaling up of these information-intensive technologies.

The improved-fallow technologies require large quantities of germ-plasm. The quality of the germ-plasm is less of an issue because of the nature of the final tree product desired and, more importantly, because, with such a large quantity needed, farmers would not be willing to pay for increased quality.[9] In theory, farmers can provide for their needs by collecting seed or even saving relatively few trees, since all the species are relatively good seeders. But, in practice, this is not easy to achieve. In Kenya, DeWolf *et al.* (2000) found that 131 of 150 improved-fallow users did not have enough seed to plant the size of fallow desired. In eastern Zambia, there were reports of farmers travelling over 20 km with ox carts to find sesbania seedlings (Kwesiga *et al.*, 1999). The involvement of the formal private sector (e.g. retailers) is non-existent and the relatively low value of fallow seed would suggest a limited role. There is more potential for local informal initiatives in germ-plasm production and distribution. The increased farmer-to-farmer trade in seed and establishment of local seed orchards managed by groups or individuals are positive steps towards increasing supplies of tree-fallow seed.

The biomass-transfer system, using tithonia, poses few germ-plasm problems in the areas where it is found. Tithonia proliferates rapidly and, in areas where it had been introduced, it multiplies quickly. Farmers may also easily establish the plant in preferred niches by transplanting tithonia cuttings.

The results of this study showed that the technical and financial performance of specific agroforestry systems varies across spatial location. Further, the feasibility and acceptability of these systems varied over different household types. Therefore, in order to be able to satisfy the needs of a large number of farmers, it is necessary to develop a range of species and management options suitable to these different conditions.

Participatory research is important in achieving this, as is a commitment by researchers to work in a range of biophysical and socio-economic conditions.

Conclusions

This study has reviewed the experiences of farmers practising improved-fallow and biomass-transfer agroforestry systems in Kenya and Zambia. These options have been available to farmers for only a small number of years and therefore the analyses pertain to the very early stages of adoption processes. At these early stages, we have found that improved-fallow and biomass-transfer systems are feasible and acceptable to farmers, at least at the modest levels with which they are initially being used. Economic analyses have also found the systems to be profitable to farmers in terms of returns to land and labour. Lastly, the systems are being used by a wide range of farmers. Unlike other soil-fertility options, there is evidence that improved fallows and biomass transfer are being used by large numbers of women farmers. In places like western Kenya, where the number of female-headed households is large, this is very important. Also, the percentages of poor households using these agroforestry systems exceeds their use rates of most other soil-fertility options.

Thus, there are positive signs that the agroforestry options may be useful for disadvantaged groups in rural Africa, as well as for other farming households. But there remain difficulties in reaching the very poor or near-landless with agricultural technologies. Whether agroforestry systems can indeed catalyse or contribute to processes to alleviate poverty and, if so, whether these systems can be effectively disseminated to resource-poor farmers are critical remaining areas for research and development.

Notes

[1] Evidenced by strong positive correlations between farm size and age of household head (see

Migot-Adholla *et al.* (1990) for Kenya and Blarel (1994) for Rwanda).

[2]   Details on prices, quantities and analytical methods can be found in Rommelse (2000).

[3]   For both trials, the pattern was for a fallow–crop combination in season 1, a fallow in season 2 and then crops in the seasons that followed.

[4]   This would be the level required for a 'full adopter', that is, a farmer who fully practises a fallow rotation within her/his farm.

[5]   These choice variables are important to analyse in generating predictive models of likely adopters and also for *ex ante* impact assessment.

[6]   In some cases, households could not yet be classified as users or non-users. These households were eliminated from the regression.

[7]   Wealth-ranking exercises with villagers found that farm size was not an important criterion in differentiating households by wealth. Variables used to construct the wealth index included livestock holdings, hiring of labour and purchasing of fertilizer.

[8]   These results are confirmed by a multinomial logit regression that focused on combinations of methods. The role of farm size, household labour and education were all important in explaining the concurrent use of two or more different methods.

[9]   In contrast, when buying a relatively small number of fruit-tree seedlings, quality is the overriding issue.

13 Evaluating Adoption of New Crop–Livestock–Soil-management Technologies using Georeferenced Village-level Data: the Case of Cowpea in the Dry Savannahs of West Africa

P. Kristjanson,[1] I. Okike,[2] Shirley A. Tarawali,[2,3] R. Kruska,[1] V.M. Manyong[3] and B.B. Singh[3]

[1]*International Livestock Research Institute (ILRI), PO Box 30709, Nairobi, Kenya;* [2]*International Livestock Research Institute (ILRI), Oyo Road, PMB 5320, Ibadan, Nigeria;* [3]*International Institute for Tropical Agriculture (IITA), Ibadan, Nigeria, c/o IITA, L.W. Lambourn & Co., Carolyn House, 26 Dingwall Road, Croydon CR9 3EE, UK*

Genetically improved crop varieties and improved, low-input soil and livestock management techniques are recognized as key ways of improving and sustaining productivity among small-scale mixed crop–livestock farmers in Africa (McIntire *et al.*, 1992; Smith *et al.*, 1997). Unfortunately, there are few technologies available, particularly in the drier areas, that achieve this goal without involving purchased inputs and increased labour outlays beyond the reach of most smallholder households: hence the depressingly low adoption rates for new agricultural technologies seen throughout Africa (Sanders *et al.*, 1996) and the importance of examining adoption behaviour by farmers (Feder *et al.*, 1985; Feder and Umali, 1993).

An encouraging new technology that increases productivity and improves the natural resource base emerging in Nigeria is genetically improved cowpea (*Vigna unguiculata* (L.) Walp). Cowpea is a versatile

legume crop that is an important source of food for both people and livestock in Africa. Cowpea grain, valued for its high nutritive value and short cooking time, is a major source of protein in the daily diets of the many rural and urban poor. Its spinach-like leaves are eaten as a vegetable, as are its immature pods and seeds (Singh and Tarawali, 1997). Cowpea leaves and stems provide a highly digestible, high-protein, high-mineral fodder for livestock, and the plant's ability to fix atmospheric nitrogen helps maintain soil fertility. It is a drought-tolerant crop well adapted to drier areas that are often not suitable for most other crops (Singh *et al.*, 1995). Its ability to stimulate suicidal germination of the seeds of *Striga hermonthica*, a parasitic plant that causes huge cereal losses throughout Africa, also makes it a desirable risk-reducing crop choice for farmers (Quin, 1997). There are roughly 12.5 million ha of cowpea worldwide (8 million of which are attributed to Africa), but these figures are questionable, as cowpea is considered a minor crop and reliable area and production estimates are not available for most countries. Nigeria is thought to have around half the Africa total, or 4 million ha planted in cowpea, and Niger an additional 3 million ha (Singh and Tarawali, 1997).

Scientists at the International Livestock Research Institute (ILRI) and the International Institute of Tropical Agriculture (IITA) have been working together with numerous West African national agricultural research system researchers to develop improved dual-purpose (DP) cowpea varieties with pest and disease resistance and better quality (i.e. higher digestibility and nutritional attributes) and quantity of fodder. The new varieties have significantly higher cowpea grain and fodder yields through incorporation of genes for resistance to various pests, tolerance to drought and shade and enhanced nitrogen fixation, with efficient use of soil fertility, into local landraces (IITA, 1996). Unlike earlier crop-breeding efforts, the components of fodder quantity and quality have been included in the breeding and selection efforts.

The new varieties, along with simple techniques aimed at improved and sustainable soil-fertility management (e.g. minimum input levels and optimal timing, crop geometry) are now at the initial stages of adoption in the dry savannahs of northern Nigeria. Understanding the factors influencing the adoption process will be crucial for catalysing dissemination efforts. In order to be able to measure the actual impact of the technologies generated through this collaborative research effort in 5 or 10 years' time, now is also the critical time to gather the necessary baseline information that will allow such an analysis.

Community-level Adoption Studies

Studies of the factors influencing adoption of agricultural technologies are typically undertaken at the farm level and focus on household resource endowments, characteristics of the household head, location of the household, the nature and extent of information provided before adoption and the characteristics of the technology (for a good review of this adoption of agricultural innovations literature, see Feder and Umali, 1993). Empirical household-level studies of the determinants of adoption usually find that variables such as level of education, farm size, income and land tenure have a significant impact on adoption intensity. For each study showing such farm-household characteristics to be significant factors, however, often another study can be found that finds the same variables insignificant determinants of adoption. This may be in part because many of the factors affecting adoption (e.g. market access, population density, frequency of visits by village extension officer) are not always unique to a household but apply to the community as a whole. This suggests that it may not always be necessary to undertake costly and time-consuming formal household surveys, but that more informal (also cheaper and faster) group-survey techniques undertaken at a village level may be useful.

There are few such community-level adoption studies available. Walters et al. (1999) examine the characteristics that influence the adoption of tree planting and soil conservation in eight villages in the Philippines. They conclude that adoption cannot be explained solely by household or village characteristics, but that historical trends (e.g. migration, sociopolitical organization, etc.) often explain a great deal of the variation in adoption across communities. Pender and Scherr (1999) take a community-level approach when they address institutional and organizational development issues surrounding the adoption of natural resource management techniques in Honduras. They find that local organizations have mixed impacts on farmers' decisions to adopt resource-conservation measures.

At the community level, the use of geo-referencing – a technique to improve the ability to relate locations of study units to one another and to other phenomena in the landscape, such as towns or roads – is particularly useful. This is because aggregations of phenomena (e.g. population density) and distances to key physical or human-made structures can be important in creating incentives for behaviour (see Pender *et al.* (1999) and Wood *et al.* (1999) for discussions on the use of spatial analysis for better targeting development strategies). Primary data collection at the community level can be easily linked to other spatial variables likely to differ across communities.

We were unable to find any adoption studies combining georeferenced and community-level data. However, Staal *et al.* (1999) combine a household-level survey with geographical information system (GIS) analysis of smallholder dairy production systems in Kenya to explore the influence of several spatially related factors, such as market access, on the probability of adoption of dairy intensification technologies.

The present study uses georeferenced community-level data to study the adoption of improved cowpea in northern Nigeria. One objective of this study is to find out which factors at the community or village level are significant determinants of adoption of improved DP cowpea varieties and management techniques. A Tobit model is used to examine the factors affecting the intensity of adoption. The implications for those attempting to catalyse dissemination of the new varieties and related management techiques are explored. A second objective is to estimate the amount of cowpea currently being grown in the two Nigerian states included in the study and to extrapolate, using GIS techniques, to a wider area suggested by the study findings (i.e. places with similar agricultural potential, population density and market access across West Africa). This predicted 'recommendation domain' is applicable to the new DP varieties now being released and, together with predicted adoption rates arising from this study, will be used in a comprehensive impact assessment of the collaborative research that has resulted in this new technology.

Study Area, Sample and Survey Methods

The study area, made up of the Kano and Jigawa States of Nigeria, is located in the semi-arid agroecological zone between latitudes 10°31′N and 13°00′N. Mean annual rainfall ranges from 500 mm at its northern fringes to 1200 mm on the southern boundary. This rainfall is unimodal and provides farmers with a growing period of 100–180 days year^{-1}. In this area, a vast majority of farmers own livestock and cultivate crops. Sorghum, millet, cowpea and groundnut, grown as cereal–legume intercrops in various combinations, are the dominant rain-fed crops grown and cattle, sheep and goats are the most important livestock reared. During the dry season, vegetable and wheat production is also found along inland valleys using residual moisture and in public and private irrigation schemes. During the dry season, grazing is limited and crop residues, in particular those of the leguminous crops, cowpea and groundnut, are essential supplements to maintain livestock. When the crop-residue grazing is done on farm, the plots benefit directly from manure and urine. If the supplementation occurs at the homestead, some of the manure, often combined with other household waste, is later returned to the farm. Such farming systems that exploit the complementarities and synergies of crop–livestock interactions have a long history in this zone. Thus, targeting improved DP cowpea varieties for enhanced food and feed production towards this zone has potentially high and widespread impacts both on the environment and on people's welfare.

Earlier community-impact workshops (based on group discussions) with cowpea farmers in Bichi and Minjibir villages in Kano State (Kristjanson *et al.*, 1999, 2001) elicited information as to the perceived benefits from DP cowpea. These benefits are realized at the plot, farm-household and village/community levels, and include economic, environmental and social benefits.

From the information gained from farmers in these workshops, the following hypothesis was used to select locations for the village-level survey:

The varieties of cowpea grown and their importance to farming systems and livelihoods depend mainly on three socio-economic factors – human population density, livestock population density and access to a wholesale market (for obtaining farm inputs and for sale of produce).

The human population density (number of persons km^2) GIS layer used comes from Deichmann (1994). The spatial market-access variable used in this study was based on a 'market tension' concept developed by Brunner *et al.* (1995), and essentially accounts for travel time to the nearest wholesale market. Market tension decreases with distance from the market and decreases faster off-road than on-road and faster along dirt roads than along paved roads. Thus it corresponds to economic distance, defined in terms of transport costs, rather than straight-line distance. The market tension indicator ranged from 1 to 10, where 10 is essentially easy year-round access to a wholesale market and 1 corresponds to locations with long travel times to a wholesale market due to both distance and the condition of the roads. Both human population density and market-tension measures were derived for 1990.

Livestock population density proved to be problematic, since it fluctuates considerably during the year in northern Nigeria. In general, during the planting season, there is a higher concentration of livestock in low-population areas, where up to 25% of the land may be under fallow and livestock are allowed to graze. By harvest time, these fallow plots have become degraded and the animals are moved to more intensively farmed areas, where crop residues abound. The scale of these movements and the seasonal reversal in livestock concentration are so large that they raise the issue of whether it makes sense to use livestock population concentration as a stratification criterion for a study that is expected to span both planting and harvesting seasons (Bourn *et al.*, 1994). Thus it seemed to make more sense not to use livestock population density as a stratifying variable (but to include it as an explanatory variable in the Tobit model).

Pender *et al.* (1999) describe agricultural potential, market access and human population density as factors largely determining farmers' comparative advantage and explore how different combinations of these factors influence possible development pathways. The farmer-impact workshops suggested that these factors might also influence adoption rates of improved DP cowpea (Kristjanson *et al.*, 1999). Since agricultural potential does not vary substantially across our area of study, we took a similar approach but excluded this factor, and classified the situation in our region into four socio-economic domains, considering 'high' and 'low' levels of the market-access and human population-density dimensions (also following the approach taken by Manyong *et al.* (1996) and Okike (1999a)):

- LPLM – low human population density (≤ 150 people km^{-2}) and low market access (market tension ≤ 5, or lack of year-round road access to a wholesale market);
- LPHM – low human population density and high market access (market tension > 5, or year-round road access to a wholesale market);
- HPLM – high human population density (> 150 people km^{-2}) and low market access;
- HPHM – high human population density and high market access.

GIS tools were used to overlay geo-referenced spatial data on human population density and market accessibility and to map out each of these four zones (Government of Nigeria, 1992; Brunner *et al.*, 1995). For each of the four zones, 20 sample points were randomly generated using a computer program that provided their coordinates. Thus a total of 80 points were marked on the map, and the nearest villages to these sample points were located through the aid of a global positioning system (GPS) instrument.

The next phase of this survey involved extensive travelling – covering over 10,000 km within the study area over a period of 6 weeks – to identify the villages on the ground, verify their coordinates, validate their categorization as determined by the GIS mapping exercise and conduct the interviews. Some of the locations could not be reached because of extremely poor access, due to the particularly heavy rains that occurred during the period of the survey, or because of security concerns. Where this was the case, the closest village to the original was used as a replacement or another village from the original strata was chosen and its coordinates taken using a GPS instrument. As a result, 80 villages were surveyed but the four socio-economic domains were no longer represented in equal numbers in the final sample. Figure 13.1 shows the socio-economic domains, the length of growing period, roads and the location and relative size of towns in the Kano and Jigawa States. The study villages are represented by circles, with the smallest circles depicting zero adoption of improved DP cowpea and the largest circle corresponding to 18–38% of village crop land planted to the new varieties.

Interview approach

Generally, the rate of adoption of new technologies among farmers in many developing countries has been below expectation, in many cases hardly measuring up to the research efforts involved in developing these technologies or improving existing ones. Some authors attribute this result to the fact that traditional research approaches neglect the 'human element' in farming systems (Norman and Baker, 1986; Walker *et al.*, 1995). It has been recognized that farmers' decisions depend on and are influenced by their knowledge and perception of technology, rather than the researcher's knowledge of technology (Gladwin *et al.*, 1984; Adesina and Zinnah, 1993). This notion has led researchers to move increasingly towards more participatory and interactive approaches of information gathering from stakeholders.

An important goal in taking an approach that is more participatory in nature than formal structured surveys is the incorporation of farmers' perspectives into the research design. Since this is very much in line

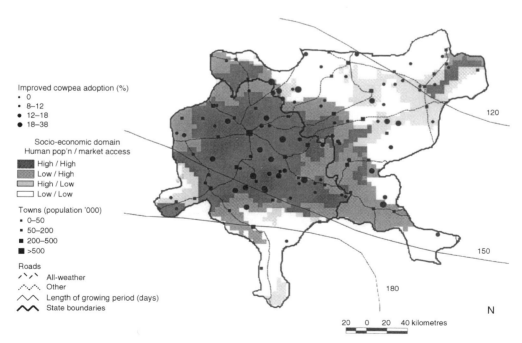

Fig. 13.1. Socio-economic domains and improved DP cowpea adoption in study area.

with the ILRI/IITA cowpea research-team objectives at this early stage in the adoption process, focus group and key-informant interviews were adopted as the major interviewing techniques for this study. The focus groups typically included the village extension agent, the chief and a number of farmers. The answers to each question were reached by consensus. Lack of consensus or other unresolved issues subsequently formed the main points of the subsequent key-informant interviews.

Tobit Model Specification

To evaluate farmers' adoption decisions on improved DP cowpea, a Tobit model (Tobin, 1958) was used. Factors affecting the intensity of adoption were estimated by examining the influence on areas planted with improved seed. Since not all villages have areas planted to improved seed, this variable has a censored distribution (i.e. the percentage of total cropped land devoted to improved DP cowpea equals zero for villages where there has been no uptake of the new varieties). This suggests that ordinary least-squares regression is not appropriate and that Tobit estimation should be used (Tobin, 1958).

Variables and Hypotheses

The dependent variable is expressed as the percentage of total village cropped area planted to improved varieties of DP cowpea. The explanatory variables related to adoption at the community level are assumed to be a function of five sets of factors: (i) the socio-economic domain (LPLM, LPHM, HPLM, HPHM); (ii) the importance and relative density of livestock owned by the villagers; (iii) the relative importance of cowpea as a crop compared with all other crops grown within the community; (iv) the frequency of visits by the village extension officer; and (v) the relative market price of grains from improved varieties compared with traditional

varieties. Specific definitions of each explanatory variable and the corresponding hypotheses made with respect to the influence of each of them are as follows.

1. Socio-economic domain. This variable, defined above, represents the interaction between human population density and market infrastructure (including roads). Higher population densities tend to be related to smaller farms cropped more continuously. This may in turn provide incentives for using technologies that maintain soil fertility and structure, including legumes such as improved cowpea. Higher population densities can also lead to more markets and roads (since the per capita costs of building roads are lower and the benefits higher in such circumstances) and, conversely, markets and roads attract migrants. Hence these two factors are difficult to separate, but it is useful to consider how different combinations of these factors influence possible adoption patterns. The socio-economic-domain variable also represents a proxy for different livestock systems, since the lower population density and remote areas are more likely to be associated with traditional, i.e. more mobile, pastoralism, whereas in more accessible areas, closer to markets, more intensive, mixed crop–livestock systems are becoming increasingly important (McIntire *et al.*, 1992; Williams *et al.*, 1999; Okike *et al.*, 2000). Because of the important crop–livestock interactions going on in more market-oriented systems, it is expected that higher-population-density villages with better market access will have a higher percentage of cropped area devoted to the new DP cowpea varieties.

2. Importance and relative density of livestock. Higher crop–livestock interactions associated with intensification also suggest a positive sign for the coefficient of the density of livestock, measured as number of tropical livestock units (TLU) per square kilometre (RIM, 1992). Villages ranking livestock as 'very important' are expected to have more area under the new varieties.

3. Relative importance of cowpea. Adoption of improved cowpea is expected to be more likely in the villages that considered

cowpea to be 'very important' as opposed to 'important'.

4. Frequency of visits by village extension officer (days per month). The more frequent the visits of village extension agents to the villages are, the more likely it is that the farmers get exposed to and adopt new technologies. This variable is therefore expected to have a positive sign.

5. Price dummy. This variable is defined as the following: if the first or second highest-priced cowpea grains are improved varieties, then it equals 1, otherwise it equals 0. Price of cowpea grains is expected to have a positive influence on the acreage devoted to new varieties.

Results and Discussion

The village-level survey turned out to be extremely useful for examining differences in broad patterns of land use and the role of cowpea and cowpea fodder across communities. A brief overview of the findings is given here. For more information, see Okike (1999b).

Land-use patterns

The results of the survey show that a typical cell (i.e. the area covered by the village extension agent) has six villages, populated by around 1600 farm families. Village size ranges from 183 farm families in LPLM domains to 1755 in the HPHM domains. Going by the ratios of extension workers to farm families (ranging from 1362 to 1745 households per extension worker (Table 13.1)), the targeted number of producers for each extension worker in an HPHM domain is within easy reach. In contrast, extension agents in LPLM domains need to travel to seven to ten settlements to reach all the farming households assigned to them.

Farmers considered millet to be the most important crop for rain-fed agriculture, followed by sorghum and then cowpea. Farming is also carried out during the dry season, using irrigation or residual moisture in inland valleys. Seventy per cent of the

Table 13.1. Description of Kano and Jigawa States (from Kano State Agricultural and Rural Development Authority (KNARDA, 1999) and Jigawa State Agricultural and Rural Development Authority (JARDA, 1999)).

	Kano	Jigawa
No. of farm families	840,895	457,510
Area (km²)	20,877	23,396
No. of village extension agents (% female)	482 (17)	336 (12)
Ratio of farm families : extension agent	1745 : 1	1362 : 1

farmers in HPHM domains practised dry-season farming, while only 32% did so in LPLM domains. Irrespective of farm location, the overall picture was that, during the dry season, the following crops were considered as important: wheat, tomatoes, peppers, spinach/lettuce, onions, carrots, sugarcane and cowpea. The results also highlight the decreasing role of fallows for maintaining soil fertility. Across the entire study area, 60% of the farmers cropped their land continuously without fallow, 35% cropped continuously for a period of 5–20 years before a break and the remaining 5% cropped for less than 5 years before fallowing. Those with fallow typically rest the land for only 1 year after continuous cropping. Villages in LPLM domains were more likely to fallow (40%) than were those in HPHM areas (20%). An emerging trend is for farmers to regard sole cropping of cassava as 'resting the land'.

Role of cowpea and cowpea fodder

The most important reason given for the popularity of a particular cowpea variety in any location was high grain yield. With cowpea being such an important source of income for many, the market value of the different types plays an important role in farmers' choice processes. The second most important consideration cited for choosing a particular cowpea variety was the perceived adaptability of the variety to the local environment. The third most important factor cited was fodder yield, as producers are concerned with availability of fodder for their own livestock. Although many villages had no households selling cowpea fodder as yet, they none the less indicated a desire to be able to do so.

In the last 5–10 years, changes have occurred in the cultivation of cowpea. An increase in the area planted to cowpea, without any change in cowpea types that are planted, was reported by 48% of the communities. Nineteen per cent reported an increase in both area planted and use of new varieties. In villages where access to a wholesale market was relatively good, population pressure seems to influence uptake of improved DP varieties. Only 8% of crop land was sown to improved DP cowpea within the low-population-density domain, whereas 15% of the total area cropped was planted to improved DP varieties in the high-population domain (Table 13.2). Thus the patterns seen by comparing socio-economic domains seem to support the hypothesis that a certain degree of land pressure due to increasing human population density may be necessary before farmers search for improved crop varieties (Boserup, 1981; Ruttan and Hayami, 1991).

In villages where human population density is high and increasing, with little corresponding market-infrastructure development and a land constraint, farmers reported that they were increasing cowpea production through use of improved varieties and higher levels of inputs (i.e. through yield increases alone). These HPLM communities on average planted 5% of their crop land to improved DP cowpea. With good wholesale-market access (plus high population and land pressures, i.e. HPHM), however, respondents reported increases in both area under cowpea (through crop substitution) and the uptake of improved cowpea varieties. Thus the survey results suggest that market access elevates the attractiveness of improved cowpea varieties as an option, and which crops are being substituted for improved DP cowpea warrants further exploration at the household level.

The results of the community-group interviews suggest that every farm family owns livestock in addition to their cropping activities. On average, across the surveyed villages, 75% of farming households use cowpea fodder both to feed their livestock and to receive income from the sale of

Table 13.2. Description of villages sampled (from village-level survey ($n = 80$)).

Characteristics	Low human population density (< 150 persons km^{-2})		High human population density (> 150 persons km^{-2})	
	Poor market access	Good market access	Poor market access	Good market access
Av. family farm size (ha)	3.75	3.67	3.11	2.71
Av. area per family under DP cowpeas (ha per family)	0.78	0.94	0.45	0.62
Av. area under cowpea (ha)	255	324	947	1,896
Total area under DP cowpea (ha)	6,073	4,608	3,118	9,383
Av. area under DP cowpea (ha)	145	219	445	938
Total farm land (ha)	28,275	17,551	20,375	46,650
Farm land under DP cowpea (%)	21	26	15	20
Farm land under improved DP cowpea (%)	6.5	8.1	5.0	14.8
Percentage of villages saying livestock is very important	47.5	21.3	5.0	7.5
Percentage of villages saying cowpea is very important	37.5	17.5	7.5	7.5
Percentage of villages that ranked white DP cowpea as the most important cowpea type	33.8	20.0	2.5	8.8

fodder; 23% use all their cowpea fodder to feed their livestock; while only 2.5% leave the fodder behind in the field after harvest. It is noteworthy that none of the farmers sold all of their cowpea fodder, even in the HPHM villages.

One of the reasons sales of fodder appear low may be the extent of cowpea-fodder exchanges that are occurring. Fodder is exchanged for manure or animal caretaking (typically in arrangements with mobile pastoralists), is used as a payment for land preparation to farmers owning bulls for animal traction and is exchanged for transportation of products to market or between the farm and homestead.

Extrapolation to estimate the dual-purpose cowpea area in West Africa

Since information on the area devoted to DP cowpea varieties is not available or is based on pure speculation, an important contribution of this study, through the synthesis of GIS and the village-level survey data, allows such an estimate to be made.

As shown in Table 13.2, although total area under DP cowpea is highest in HPHM, the average area per family under DP cowpea is highest where human population density is low and market access is good (LPHM) and lowest in HPLM. The percentage of farmland planted to DP cowpea ranges from 15 to 26 by domain and is higher in the low-population-density domains, reinforcing

the relative importance of DP varieties in more remote areas. In Fig. 13.1, the relative percentage of farmland planted to improved DP cowpea is shown for the 80 surveyed villages, with the small circles corresponding to zero adoption and the largest circles representing 18–38% of farmland under improved DP cowpea. There are few villages with significant adoption that are not located near an all-weather road or a town with more than 50,000 residents.

Extrapolating to all of West Africa (by applying the same percentages for each domain) implies an area of approximately 3.4 million ha devoted to DP cowpea (Table 13.3). Given that the 'easiest' transition for farmers to make will be from traditional DP to improved DP cowpea (although there may well be some shifts from grain types or from other crops into improved DP cowpea), this estimate provides baseline information for future *ex post* assessments that attempt to measure actual benefits from this new technology. Coupled with adoption information from the household survey that follows this study, it will also provide the basis for an *ex ante* impact assessment.

Village-level factors influencing adoption

The results of the Tobit regression analysis are presented in Table 13.4. The log-likelihood and the likelihood ratio (LR) χ^2 statistics indicate a good fit. Forty-two of the villages had no improved DP cowpea

Table 13.3. Extrapolation to West Africa (from Kruska, ILRI GIS calculations, based on RIM (1992) database).

Characteristics	Low human population density (< 150 persons km^{-2})		High human population density (> 150 persons km^{-2})	
	Poor market access	Good market access	Poor market access	Good market access
Total cropped area (ha): W. Africa (LGP 90–209 days)	12,968,088	1,319,040	1,074,106	859,375
Farm land under DP cowpea (%) for 80 villages surveyed	21	26	15	20
Estimate of DP cowpea area: W. Africa (LGP 90–209 days)	2,723,298	342,950	161,116	171,875

LGP, length of growing period.

Table 13.4.   Results of the Tobit regression analysis (Y = per cent of village cropped area planted to improved DP cowpea).

Variable	Coefficient (SE)	t Statistic
Socio-economic domain	0.105 (0.049)	2.128
Frequency of visit by VEA (days month^{-1})	−0.0035 (0.002)	−1.674
Importance of livestock (v. imp; imp.)	0.117 (0.044)	2.653
Livestock density (TLU km^{-2})	0.003 (0.0014)	2.160
Importance of cowpea (v. imp; imp.)	−0.036 (0.022)	−1.607
Price dummy (1 if improved varieties have first or second highest price, 0 otherwise)	0.203 (0.037)	5.434
Constant	−0.326 (0.136)	−2.387
42 left-censored observations (% improved DP land = 0)		
38 uncensored observations		
LR χ^2 (6) = 42.82		
Prob. > χ^2 = 0.0000		
Pseudo R^2 = 1.4585		
Log likelihood = −46.73		

SE, standard error; VEA, village extension agent.

varieties on their farms (censored at 0%), while the remaining 38 had varying percentages of crop land sown to the new varieties. In the model, no cut-off points were specified for these farms. The explanatory variables that had a significant and positive influence on area devoted to improved DP cowpea were socio-economic domain, importance and density of livestock and price of grain. HPHM villages had a higher intensity of adoption than the others, supporting the hypothesis that important 'drivers of change' are high population pressure coupled with good market access. Intensity of adoption was significantly and positively influenced by both the perceived importance of livestock and by the number of livestock owned (TLU density) within the village. Not surprisingly, the price of the improved cowpea grain relative to traditional varieties had a highly significant influence on the percentage area planted to improved DP cowpea. Cowpea was considered 'very important' or 'important' in almost all the locations and the lack of variability may account for this variable being non-significant.

The fact that intensity of adoption is higher in the more densely populated, better-market-access domains, despite the fact that DP varieties and livestock are more important in the other domains, highlights the opportunities for ongoing dissemination efforts. Expanding the availability of information and improved seeds to these more remote areas is the challenge.

The frequency of visits by an extension agent was negative and significant at a 10% level. In an earlier model estimated using a logit model (where villages were classified as either adopters or non-adopters), the extension variable was again negative, as well as significant at the 1% level. This suggests that the more often the extension agents visit, the less likelihood that new cowpea varieties are adopted. Implications from this finding are somewhat worrying, and possible reasons merit some discussion.

Extension agents are currently expected to be the major channel for disseminating new knowledge and technologies to farmers. They are also supposed to influence research priorities based on feedback from farmers. Our results suggest that this is not happening and this is probably not peculiar to improved cowpea varieties, as it is a similar finding in related studies (e.g. Okike, 1999a). Under Nigeria's agricultural extension system, each extension agent is expected to carry messages to farmers concerning crops, livestock, fisheries and forestry, regardless of their educational background. It is doubtful that these 'generalists' sufficiently understand all the diverse material they are expected to extend to farmers. Our results suggest that the extension problem goes beyond the issue of frequency of visits, since a very high frequency of visits, tested by including a squared frequency of the extension-visits variable, still retained a negative and significant sign.

Since traditional dissemination pathways do not appear to be working, national and international agricultural researchers need to either strengthen these institutions or explore other pathways for dissemination of their research results. Another possible interpretation is that new technologies must be attractive enough to stimulate horizontal farmer-to-farmer diffusion (Inaizumi *et al.*, 1999). The poor roads and communication infrastructure will tend to hamper this dissemination means, however. The follow-up study to this one, a household-level survey, will explore in more depth through what channels farmers are getting information and new technologies and the extent to which the private sector is (or is not) reaching farmers that the public sector is apparently not able to reach.

In terms of how well we can predict factors affecting adoption of improved DP varieties from the results of this study, a few cautions are in order. First, improved DP varieties are very new, and it is too soon to say much about the process of adoption (e.g. some of the farmers using them may still be at the experimental stage and it would be incorrect to call them 'adopters'). Secondly, there are numerous questions relating to adoption that cannot be answered by a community-level study, particularly those relating to within-community variation in adoption patterns.

Conclusions

This study has shown that the combination of spatial data defining socio-economic domains, using factors such as population density and market access, and village-level survey data can be useful in explaining adoption behaviour. This is good to know, since the time and resources needed to carry out detailed household-level surveys are not always available. Surveys such as this one allow extrapolation to wider areas,

allowing broader impact assessments and are appropriate for a wide range of technologies. None the less, there are issues that spatially referenced community-level adoption surveys cannot address. These include detailed information on who exactly is adopting the new technologies, how exactly they are using them (and if disadoption is occurring) and how they are getting access to information and the technologies. However, the ideal of following many households that have had access to a new technology for many years is not always feasible. If the goal is a comprehensive impact assessment, for example, which requires an understanding of adoption patterns, ideally you also need to collect information from all levels – plot, household and community – as well as broader spatial data describing the overall recommendation domain for the technology in question.

Both area planted to cowpea and the number of varieties being cultivated are increasing in northern Nigeria. An important finding of the survey for researchers and policy-makers is that approximately 20% of the total cropped area is being devoted to DP cowpea, as existing statistics on the area sown to cowpea are extremely shaky and are not broken down by type of cowpea. Using GIS, these results were extrapolated to estimate that there may currently be roughly 3.4 million ha of DP cowpea across West Africa within similar agroecological zones. This information is useful for several reasons, including an *ex ante* impact assessment under way of the potential returns to DP cowpea research and as a baseline for a future *ex post* assessment of the impact of this technology.

The perception within the majority of villages is that the local or traditional cowpea varieties are still the farmers' first choice when deciding what varieties to plant. Key respondents explained this choice as being related to the perceived need to spray improved varieties with insecticide. The fact that some producers have been sold adulterated seed was also mentioned. However, in the high-population areas with good market access, there are villages that rank the improved varieties first. This raises the issue of whether access to information and improved seeds are much higher for producers in these villages and through what pathways dissemination is occurring. It also suggests that, while the highest impact may come from targeting new varieties and techniques to the HPHM areas, there is in fact a lot of unexploited potential for adoption, particularly for DP varieties, in the other areas where livestock and cowpea are rated higher in importance for a household's livelihood strategies.

The results of this analysis support the findings of other studies in SSA relating to integrated crop–livestock systems (e.g. Manyong *et al.*, 1996; Vosti and Reardon, 1997; Okike, 1999a; Pender *et al.*, 1999; Staal *et al.*, 1999; Okike *et al.*, 2000), namely that growth in human population and improved market access are particularly strong drivers of change and influence the uptake of new technologies. It is interesting, and not necessarily intuitive for a technology such as cowpea that improves the natural resource base and is an important source of household food and fodder, that market access and prices should have such a strong influence. The policy implication of this is that good roads and market infrastructure matter – a lot.

An important message for researchers from this study is that agronomic research should not lose sight of the importance of product quality while pursuing yield improvement. Better product quality is reflected in high market prices (e.g. for higher-quality or more desirable products), which were demonstrated in this analysis to have a significant impact on the probability of adoption. Also, the recent shift in focus from grain yields to both grain and fodder yields and quality is equally well placed. The survey suggests that, for improved

cowpea varieties to become and remain popular in all locations, they must first be high-grain-yielding; secondly, pest resistant; and, thirdly, yield enough fodder to support the crop–livestock enterprises in which each and every farm family in the study area is involved.

The adoption of improved DP cowpea varieties appears to be a 'win–win' situation with respect to improvements in natural resource management in these intensive, integrated crop–livestock systems, particularly soil fertility. Those more market-oriented producers more interested in selling the cowpea grain leave the residues to rot in the fields, adding nutrients to the soil. Those that prefer the DP types and feed the residues to their animals return a significant amount of nutrients to the soil via the manure. The challenge remains, however, to 'scale up' the encouraging trend and results regarding uptake of improved DP cowpea occurring in two states in northern Nigeria to reach the hundreds of thousands of potential farming households that have not yet heard about or do not have access to these improved varieties.

Acknowledgements

This research was supported by both ILRI and IITA, and benefited greatly from the logistical and technical support provided by IITA-Kano. We would like to thank Jimmy Smith and the System-wide Livestock Programme for financial support. Thanks to Boru Douthwaite for some insightful comments on an earlier draft. We would also like to acknowledge the assistance of Mr Shehu Yahaya, who conducted the group interviews and key-informant interviews in the local language.

14 Contradictions in Agricultural Intensification and Improved Natural Resource Management: Issues in the Fianarantsoa Forest Corridor of Madagascar[1]

Mark S. Freudenberger[1] and Karen S. Freudenberger[2]

[1]Landscape and Development Interventions (LDI) Program, c/o Chemonics International, 1133 20th Street, NW, Washington, DC 20036, USA; [2]FCE Railway Rehabilitation Project, c/o Chemonics International, 1133 20th Street, NW, Washington, DC 20036, USA

The international development profession is periodically buffeted by the infusion of new conceptual approaches that have a broad impact on both policies and project interventions at the local level. In Madagascar, as in many other countries, the government and donor organizations are turning to an ecoregional approach to natural resource management. Ecoregional planning is an outgrowth of the integrated conservation and development strategies (ICDPs) that prevailed in the 1980s and 1990s (Larson et al., 1998). It represents a major shift from the previous focus on preserving species diversity to a more encompassing view of maintaining habitat diversity, evolutionary phenomena and adaptations of species to different environmental conditions around the world. As articulated by the World Wildlife Fund (one of the leaders of the new conservation paradigm), ecoregions are defined as:

> relatively large units of land or water containing a geographically distinct assemblage of natural communities sharing a large majority of their species, dynamics, and environmental conditions. Ecoregions function effectively as conservation units because their boundaries roughly coincide with the area over which key ecological processes most strongly interact.
>
> (Olson and Dinerstein, 1998)

Because the approach favours the protection of larger areas, ecoregional conservation and development strategies often seek to connect existing protected areas with biological corridors. As one of the key proponents of the approach notes:

> in order to stop the destruction of native biodiversity, major changes must be made in land allocations and management practices. Systems of interlinked wilderness areas and other large nature reserves, surrounded by multiple use buffer zones managed in an ecologically intelligent manner, offer the best hope for protecting sensitive species and intact ecosystems.
>
> (Noss, n.d.)

Corridors connecting one protected area to another enable the flow of species across larger distances, thereby contributing to species survival over evolutionary time.

In Madagascar, policies have shifted away from the focus on protected-area management to a broader regional and spatial analysis and a set of corresponding interventions at multiple scales from the village to the national level. The second phase of the 15-year Madagascar National Environmental Action Plan strongly supports a spatial perspective. The US Agency for International Development (USAID) has both actively promoted this approach and revised its own environmental programme in response to these concerns.

The ecoregional approach, in so far as it looks at vast territories and focuses on more than excluding populations from circumscribed protected areas, implies a close attention to the human element of conservation. In reality, it is in most cases impossible to exclude populations from such vast areas. Instead, it is necessary to solicit their collaboration in conserving natural resources that would otherwise be subject to human threat. In this context, issues such as population, migration and large-scale agricultural intensification necessarily take on a high profile in ecoregional planning.

This chapter reviews the way one USAID-financed project – the Landscape Development Interventions (LDI) Program – has addressed these issues in a programme that seeks to reduce human pressure on a forest corridor that connects two major national parks in the Fianarantsoa region of Madagascar. Initially, LDI focused on agricultural intensification as the primary strategy for reducing pressure on the natural forest. As the project evolved, however, several contradictions implicit in this approach began to surface. Among these were: (i) the possibility that intensification and increasing incomes for some farmers might actually be contributing to greater rates of deforestation; and (ii) the realization that structural factors related to the deterioration of transport systems were likely to dwarf the positive impacts of the project's extension activities. In the absence of reliable transport, farmers were likely to disintensify their agricultural production and replace

sustainable agricultural production systems with unsustainable practices. This would, in turn, result in a corresponding increase in the rate of deforestation as people sought to expand agricultural holdings.

This chapter focuses on these complexities, with specific attention to the relationships between transport, agricultural intensification and sustainable natural resource management as these three factors are currently playing out in the LDI conservation and development programme. It focuses especially on the impact the Fianarantsoa–Côte Est (FCE) railway has on both agricultural intensification and on conservation in the Fianarantsoa Region as it crosses the threatened forest corridor on its way from the highland to the coast.

The hypothesis that the train line has a positive impact on the forest is suspect for many in the conservation community, which is more accustomed to viewing transport systems as threats to nature than as potential saviours. Indeed, road and rail transport systems do often facilitate immigration into otherwise inaccessible areas and promote exploitation of resources that might otherwise be spared from human pressures. These criticisms may ignore another set of equally important questions, however, including the critical role of transport in permitting agricultural diversification and intensification and the role that those processes play, in turn, on reducing pressures on the forest.

The Application of the Ecoregional Approach in the Malagasy Highlands

The forest corridor

In the Fianarantsoa Region, LDI's ecoregional approach is focused primarily on preserving the highland forest corridor between Ranomafana and Andringitra National Parks (Fig. 14.1). The corridor in the Fianarantsoa Region (which is actually part of a longer corridor that stretches over much of the Malagasy highland) is a 450 km long band of forest. It is the last vestige of what was once a vast forest that covered the north–south escarpment between Madagascar's east coast and the highland plateau. Now the moist-forest ecoregion has been reduced to a narrow band, ranging from about 4 to 15 km wide. Rainfall ranges from a high of 2500 mm year^{-1} on the eastern escarpment to about 1500 mm year^{-1} on the high plateau to the west.

One particularly key part of the corridor from a conservation perspective (and therefore the area where LDI has invested the greatest effort) is the forest that connects Ranomafana National Park with Andringitra National Park approximately 160 km to the south. This 1,005,395 ha of low montane and high montane forests represented about 10% of the province's total land mass at the time of the national forest inventory of 1994. These parks and the corridor between them represent especially interesting ecosystems, because of the great altitude differences in a relatively small area. Endemic species of flora and fauna abound both in the world-renowned protected areas and in the corridor that connects them.

This forest keenly interests both conservationists, who are concerned with the maintenance of Madagascar's extraordinarily rich biodiversity, and local populations, for whom the forest serves multiple functions, including the protection of watersheds and the purveyor of goods of economic value (wood, crayfish, frogs, etc.). Yet the corridor is becoming increasingly fragmented and there is a widespread concern that the forest may disappear entirely as farmers expand their fields and clear-cut the hillsides for *tavy* (slash-and-burn) agriculture. This fragmentation is particularly severe in the low-altitude (500–800 m) humid tropical forest (S. Goodman, 2000, personal communication). As land is cleared and thinned by agriculture and forest-product extraction, there is a reduction in habitats essential for the reproduction of flora and fauna, leading eventually to declines in biodiversity. Indeed, there is evidence that loss of the forest corridor may lead to certain species' extinctions, since some animals (such as the charismatic bamboo lemurs, *Hapalemur aureus* and *Hapalemur simus*) have no other known habitat on earth.

In addition to these biodiversity concerns, which attract the attention of the world community, the disappearance of the remaining forest would have an immediate impact on neighbouring populations, who use a multitude of forest products in their daily lives; they collect wood and vines for construction and gather honey, crayfish,

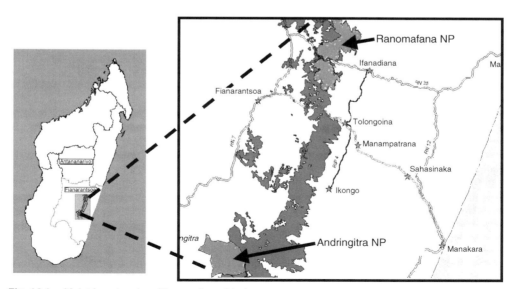

Fig. 14.1.   Moist forest region, Fianarantsoa, Madagascar.

firewood, resins and medicinal plants from the forest. But the effects will not be limited to those who live in and near the forest. Deforestation may also affect vast expanses of productive valleys and rice-fields below if the hydrological balance of the region's watersheds (most of which originate in the corridor) is disrupted. While there have been few rigorous hydrological studies of these questions, farmers fear that deforestation will lead to flooding during the rainy season (because water no longer infiltrates the denuded hillsides above) and water shortages during the dry season. Some farmers already suspect that forest clearing is leading to diminished rice harvests and they fear that, if the forest is cleared, they will no longer be able to produce two rice crops a year in their lowland fields (Freudenberger, 1999).

Pressures on the forest corridor

While maps of the corridor make it look as though this is an intact and uninhabited band of forest, in fact studies of the region have found ample evidence of fragmentation (Hagen, 1999; A. Dehgan, 1999, personal communication). Before 1990 there were very few people who actually lived in these primary forests. Most activity was limited to collecting wood and other products, but in a way that did not significantly alter the structure of the forest. This situation is rapidly changing, however. As recently as 10 years ago, villagers report that there were no more than a few dozen families living in the forest; now there are as many as ten to 20 families from each village adjacent to the corridor who have moved into the forest and begun cultivation (Freudenberger, 1999; Freudenberger et al., 1999). This is true on both the east and west sides of the forest band. The Tanala ethnic group inhabits the east side of the forest, while the west is the traditional property of the Betsileo. Both groups are now sending pioneers to occupy lands that fall within their traditional ethnic territories but are well inside the forest. Villagers estimate, for example, that as much as 25% of the forest corridor under Tanala control has already

been claimed by private interests. At this time, the majority of these property 'rights' have not yet been activated and the land is not yet cleared, but there are clear signs showing intent and indicating ownership (such as banana trees planted in the far interior of the forest) (Freudenberger et al., 1999).

At the outset of the LDI Program, intensive field research was conducted on the factors motivating Betsileo and Tanala families to occupy the corridor and the impact of their activities on the environment (Freudenberger, 1998, 1999; Freudenberger et al., 1999). These studies found that, while the patterns of occupation of the two groups are somewhat different, the effect is the same: scattered homesteads sprinkled throughout the forest. The Tanala, who favour upland farming, tend to choose farm sites that are high in the forest, where their upland fields will have good sun exposure. The Betsileo, who are specialists in irrigated and terraced rice, look for water catchments and depressions, where they cultivate irrigated rice, only later expanding to the surrounding hillsides. In looking for the most propitious sites for their preferred type of agriculture, both groups tend to penetrate far into the forest, rather than limiting their exploitation to the more easily accessible parcels on the periphery and adjacent to existing communities. These initial homesteads then become magnets attracting new colonists and accelerating further fragmentation.

The now familiar factors driving this occupation are also similar on the east and west side of the corridor, though there are some relatively minor differences. The principal driving forces in both cases are rapid population growth and declining soil fertility, with pursuant reductions in agricultural production per capita or per family. While population statistics are not highly disaggregated, statistics from the region and interviews with farm families suggest that growth rates in both the Betsileo and Tanala areas adjacent to the forest exceed 3%. As many as half the families in the case-study villages had more than ten children. This results, of course, in serious land fragmentation, as holdings are subdivided with each successive generation. This problem is even more acute on the Tanala side, where both male and female progeny inherit land and families are left with a multitude of often widely dispersed and very small parcels (Equipe Ralaivao, 2000).

Agricultural production typically consists of a field of irrigated rice (usually a very small parcel of 0.25–0.5 ha for all but the wealthiest families and one or more upland fields that rarely total more than 0.5 ha). These upland fields are planted with a rotation of rice and manioc, followed by at least 3 years of fallow. In the traditional system, after only five to ten such cycles, the land is considered barren by the farmer and is left in long-term fallow, often for 50 years or longer. Yields on both the irrigated rice-fields and the upland fields are notoriously low. A 1 ha lowland rice-field typically produces 0.75–1.5 tons of rice, while a 1 ha upland field produces only 300–400 kg (Equipe Ralaivao, 2000). There are few efforts to enhance soil fertility (apart from multiyear fallows) in the traditional system. In most of these communities, only the wealthiest 20% own cattle and, while those few farmers who apply manure do get significantly better yields, this is not an option for the vast majority of farmers.

Most farm families (especially on the Betsileo side of the corridor, where water shortages limit irrigated rice production to one harvest a year) produce only a fraction of their food needs (typically enough to cover 4–6 months of consumption). For the rest of the year, they undertake a variety of off-farm or salaried jobs to eke out their survival.

One of the longer-term survival strategies of these populations is to expand their landholdings by moving into the forest and clearing a new parcel. On the Betsileo side of the corridor, where food security is more precarious, this strategy is primarily used by the wealthiest 20% of the families, who are food-secure but are trying to assure the future of their offspring by expanding family landholdings (Freudenberger *et al.*, 1999). (As discussed below, this raises one of the fundamental contradictions

in LDI's agricultural-intensification activities by challenging the assumption that intensification, by raising revenues and increasing food security, will reduce pressures on the forest.) The remaining 80% essentially find themselves too poor to think very far into the future. They need activities that will immediately feed their families from day to day, and are thus too preoccupied with day-to-day survival strategies during the hungry season to be able to invest in forest clearing (Freudenberger *et al.*, 1999). On the Tanala side, farmers (at least those who own irrigated rice-fields) can usually get two harvests a year and are thus largely food-self-sufficient; as a result, poverty poses less of a constraint to acquiring forest lands (Freudenberger, 1999) and farmers from all economic classes engage in forest clearing.

Because rapid population growth and declining agricultural production are not restricted to the areas immediately adjacent to the forest corridor, on both sides immigration into the corridor is occurring both from the villages who are the traditional landowners of the corridor (and have been saving the forest as a sort of land reserve for many generations) and from areas further away. This finding argues strongly for the ecoregional approach and, as we shall see shortly, the need to scale up agricultural-intensification activities in communities that may be quite distant from the forest. The area of Masoabe, on the Betsileo side, for example, is one of the most densely populated rural areas of the Fianar region and is a major source of immigrants into the corridor. On the Tanala side, many of the immigrants come from the plains closer to the coast, where there are no longer existing reserves of fertile land, and fragmentation and soil infertility have made farming a precarious undertaking for much of the population.

When farmers move into the forest, they reproduce the same (unsustainable) production techniques that they use on their more established fields. While there is a perception that newly cleared fields produce higher yields, farmers report that even these fields often produce yields that are disappointingly low (Freudenberger *et al.*, 1999).

Other farmers complain of severe declines in *tavy* yields after the first couple of years, and the yield of all crops is reduced by bird and wild-boar damage. After 2–3 years of cultivation, soil fertility begins to decline, once again requiring fallow and the need to clear more land. Hence, intensification efforts are relevant not only for the sending communities but equally (and perhaps especially) for the new homesteads in the forest, which, it must be noted, are in most cases extremely difficult to reach and therefore hard to serve with traditional extension approaches.

In short, the Tanala on the eastern and the Betsileo on the western side of the corridor face pressures that, while perhaps different in their details, are fundamentally similar: too many people trying to make their livelihoods on too little and too unproductive land. And so the forest corridor is threatened, caught between the demographic and economic pressures of the Betsileo on the high plateau to the west and the Tanala who occupy the lower hillsides to the east . . . all of whom are motivated primarily by the basic human need to assure their and their children's food security.

Agricultural Intensification in an Ecoregional Strategy

As noted above, LDI Fianarantsoa's ecoregional strategy consists primarily of reducing the pressures on the 160 km long forest corridor that extends from the northern borders of Ranomafana National Park to the southern tip of the Pic d'Ivohibe (south of Andringitra National Park). Initial project strategies consisted primarily of attempts to reduce anthropogenic pressures by promoting agricultural intensification and rural income diversification through conservation enterprises, including ecotourism. (The project also works closely with a USAID-funded health and family-planning programme to extend services into these areas.) The strategy was to restore more village land into production by rehabilitating the large areas of hillside farmland that are now so degraded that they are no longer planted

and to increase yields on currently farmed upland and irrigated fields. As such, the theory held, farmers would be able to get more from what they already own rather than having to clear new land in the forest.

Given the agricultural production system described above, it was quickly evident that intensification strategies would have to address irrigated fields, upland agriculture and degraded lands that have been withdrawn from production. There are four principal reasons for this. First, a significant number of poorer families have no, or almost no, irrigated land and would thus be excluded if the project did not have an upland component. Secondly, the cultural value Malagasy place on their irrigated rice-fields makes many reluctant to experiment on these fields, and initial evidence suggested that adoption rates might well be low and slow for the irrigated-rice package. Thirdly, the irrigated-rice package requires a sufficiently sophisticated set of requirements and inputs (particularly regarding the control of water) for many farmers not to have the resources needed to adopt the package. And, fourthly, given the high rate of population growth, it is unlikely that, even under the most optimistic projections of adoption rates, the impact of any two interventions alone would be sufficient to significantly alleviate pressure on the forest. In short, given the high rates of population growth (which will result in a doubling of the population in the next 20 years even if nascent family-planning programmes begin to have an impact), all production systems will have to become both higher-yielding and more sustainable in order to have any reasonable chance of significantly alleviating pressures on the remaining natural resources.

In terms of potential yield increases, the most promising results have been experienced in irrigated-rice production (Uphoff, 1999). The amount of land that can be put into irrigated-rice perimeters is a small fraction of most village territories, however. The far greater land area is devoted to upland crops, which are, in the vast majority of cases, devoted to the highly unsustainable production of annual crops, such as rice, manioc and beans. The project is encouraging farmers to replace these annual crops with perennial tree crops that will protect the soils against erosion, maintain soil fertility and produce continuously without the need for a fallow that increases the amount of land that must be put into production. Unfortunately, as we shall discuss in the next section, this critical element of the project (because it is the only viable option for the vast majority of village lands) has run up against serious constraints that severely hamper adoption.

The packages proposed by the project thus included the following components.

● *Système de riziculture intensive* (SRI) and *système de riziculture amélioré* (SRA) techniques, which include improvements to soil fertility through heavy composting, introduction of new seed varieties and cultural practices such as water control. Most emphasis is placed on SRA, because the introduction of the proposed techniques requires less labour and fewer capital investments.
● Production of royal carp in fish-ponds associated with rice production.
● Rehabilitation of degraded hillsides (*tanety*) with agroforestry (especially fruit-tree production) and vetiver grass plantings.
● Diversification of the household economy with conservation-based enterprises, such as beekeeping, production of essential oils and citrus production.

LDI Fianarantsoa began its interventions by working with some 700 farmers in communities immediately adjacent to the forest corridor. Two years later, interventions had expanded to some 2200 farmers belonging to 164 rural associations. The project offered agricultural-intensification techniques to these farmer associations in return for their pledge to try to reduce slash-and-burn agriculture and other destructive extractive practices. Those agreeing to these conditions received agricultural training and information, supply inputs and micro-credit.

In most villages, increases in rice production on small experimental plots have averaged about 20% and in some cases have risen to as much as 70%. Rates of adoption of improved potato production, beekeeping and fish culture were quite high during the first year of activities: among the 700 farmers associated directly with LDI Fianarantsoa since the beginning of the programme, 23% have adopted fish-culture practices and constructed 143 fish-ponds. During the first year alone, farmers purchased 15,000 royal carp fingerlings at market price. Approximately 22% of all farmers now grow potatoes as an off-season crop. Within the first 6 months of the project, over 40% of the farmers built and stocked over 300 beehives. Fifty-one per cent of all farmers now build compost piles.

The rehabilitation of *tanety* hillsides has commenced in 70% of all villages with the extension of practices such as planting vetiver grasses on contours, planting biomass banks of leguminous shrubs for compost and as a source of pollen for bees, or tree planting for individual and community wood lots.

Demand is growing rapidly as non-participating farmers observe the successes of their neighbours. While LDI has conducted participatory field evaluations to determine what social category is participating most actively in the programme activities, no general trends have yet emerged (LDI, 2000). LDI Fianarantsoa works in most of the villages directly adjacent to the forest corridor (some 169 villages), but project staff remain concerned that, even with these positive rates of adoption, the impact is still very small on the broader landscape. One problem is that the project's activities have focused on the communities immediately proximate to the corridor, while studies now suggest that much of the migration pressure emanates from highly populated communities far from the corridor. LDI possesses neither the financial resources nor the institutional capacity to start programmes in these distant places.

Agricultural intensification and the FCE railway

LDI's initial interventions were very much focused on finding effective ways to extend techniques for agricultural intensification and necessary inputs for participating farmers in the corridor region. A combination of increasing knowledge of the regional economy (developed through a series of rapid rural appraisal (RRA) case-studies) and a calamitous but highly revealing natural disaster that visited the region, in the form of two back-to-back cyclones in the early part of 2000, brought the project face to face with another issue that has profound implications for efforts to save the corridor through agricultural intensification: the extreme fragility of rural transport systems in the region. It became increasingly clear that farmers' decisions regarding land use in general (and, in particular, their decisions regarding the adoption of *tanety* rehabilitation packages or, alternatively, the expansion of *tavy* into the forest) are greatly influenced by their access to transport. One of the vital

transport arteries in the region is the FCE railway (Fig. 14.2).

The FCE railway is a rickety, dilapidated, narrow-gauge railway line that runs from Fianarantsoa in the highlands, crosses the forest corridor about 10 km south of Ranomafana National Park, and then continues to the port city of Manakara on Madagascar's east coast. It was constructed between 1926 and 1936 as part of the French colonial policy to promote export crops (notably coffee) and to transport rice out of the highlands. The line tests the limits of railway engineering as it descends from an altitude of 1100 m in Fianarantsoa to sea level in Manakara over a distance of 163 km. It passes through some of Madagascar's most spectacular scenery, particularly in the highlands, and is considered a treasure by adventure-seeking tourists.

Unfortunately, the same characteristics that give it a certain romantic charm (its ancient locomotives and rails and dramatic scenery) also increase both the general costs of maintenance and its vulnerability to the cyclones that sweep across the region at least once or twice a decade. In February and

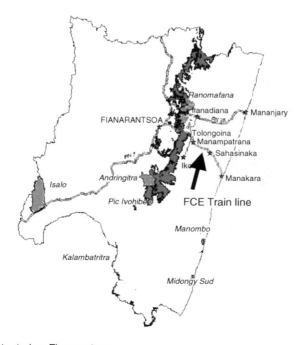

Fig. 14.2. Transport arteries, Fianarantsoa.

March 2000, Fianarantsoa was hit by two massive storm systems: Cyclone Eline and Tropical Depression Gloria. The impact on the FCE was immediate and terrible: the line was blocked by 280 landslides that dumped over 150,000 m³ of earth on the track. Four major wash-outs left 100 m stretches of rail suspended over thin air. For nearly 2 months there was no transport of bananas, coffee or fruit from the region. In one highland station, a research team found 54 tons of rotting bananas waiting for a train that never came. A rapid reconnaissance of the region told a story of severe hardship for people living up and down the line: local rice crops were badly damaged; the price of rice in village shops (transported to many rice-deficit villages even in normal times on the train) immediately rose by 30–50%; farmers were unable to transport and sell their fruit and therefore had no revenues to buy rice or other foodstuffs.

In addition to the disaster-relief implications, however, the railway line closing also highlighted a more fundamental concern. Given the extreme fragility of this line (which is threatened as much by mismanagement and the progressive decay of both rails and rolling stock as it is by more dramatic events such as cyclones), what would be the impact on the corridor and LDI's intensification activities if the line were to close permanently? A series of quantitative and qualitative studies designed to answer just such questions was already in the planning; the cyclones provided the opportunity to sharpen these questions but also forced the project to reconsider whether it should expand its transport interventions as part of the larger agricultural intensification/conservation strategies.

The train serves a population of about 800,000 people and is the only means of transport for some 100,000 people who live in areas that have virtually no road service. As such, it plays a crucial role both in exporting products from the region (especially commercial crops) and importing rice into an area that produces only a fraction of its rice requirements. Approximately 3000 tons of coffee, 6000 tons of bananas and other fruits, including lychees, oranges and avocados, are shipped by train to markets in Fianarantsoa and Manakara each year. In general, coffee is collected from a radius of about 25 km from the train line (especially to the south, since there is no alternative road service), while bananas (which must be transported to the railhead by porters) come from a distance of approximately 10 km.

Transport and sustainable production systems

Case-studies of communities served by the train (for example, on the eastern flank of the forest corridor) found that the economy of these Tanala villages has developed around commercial agriculture (Deeg and Freudenberger, 2000). Typically, in these villages, a farm family cultivates a small parcel of irrigated rice in the valley while planting tree crops on the surrounding slopes. In short, these families are already practising the relatively sustainable production system based on lowland rice and hillside tree crops that LDI would like to see practised across the region. Studies in these villages revealed that pressures to expand agricultural lands are significantly reduced where farmers engage in permaculture, producing primarily tree crops. These methods do not solve the problem of increasing population and the subdivision of lands through inheritance with each succeeding generation. They do, however, at least preserve the productivity of existing fields. Many of the lands on which trees are planted have been producing for 50 years or more; if such land had been planted in annual crops, it would have long since been taken out of production due to decreased soil fertility, requiring the owner to clear new and fertile fields.

These studies also showed clearly, however, the extent to which this relatively sustainable tree-based production system depends entirely on a functioning transport system to get the crops to market. Like the families in LDI's other intervention areas, most of the families in our studies produce only a small fraction (less than a quarter) of their rice needs on the small irrigated parcels they farm on the valley floor. Unlike the other farmers, however, they are able to purchase the rice needed to complete the family ration, using proceeds from the sale of their bananas, coffee and other tree crops. Even the families who are landless, or nearly so, gain revenues from these activities. They earn money, which is used to buy rice, by working on the coffee plantations of landed families or transporting bananas from the field to the train line. If there is no transport system in the region and therefore no way to get crops to market, the entire population (whether landed or not) will have to find other means of procuring their subsistence food needs.

Interviews with farmers in the region (Deeg and Freudenberger, 2000) suggest that, should transport systems (and especially the train, which is critical for the shipment of low-value, high-volume crops such as bananas) no longer operate, they will adopt a common strategy to protect their livelihoods. Farmers who have land will immediately cut down their tree crops and replace them with upland rice or manioc. As these fields, inevitably, become infertile, they will search out new lands in the corridor. For farmers without land, there is no buffer. To feed their families they will need to acquire fields; most plan to move into the corridor in search of this land.

In conjunction with these qualitative findings, a cost–benefit analysis (Railovy et al., 2000) attempted to quantify the pressure on the forest with and without an operating railway system. The study in no way minimizes that existing threat to the forest. Indeed, it predicts that, no matter what happens, at least 110,300 ha are likely to be cleared in the next 20 years. However, should the train cease to operate and should farmers currently practising permaculture switch to annual crops in an effort to produce their own food requirements (the outcome predicted by local residents), the rate of deforestation will be dramatically higher. The Railovy et al. (2000) study predicts that, in such a case, at least 207,700 ha of forest will be cleared over the same 20-year period.

Implications for agricultural intensification

The implications of these findings for LDI's agricultural intensification efforts are sobering. First of all, it is clear that the impact of the train on sustainable production systems and the corridor dwarfs by far the potential impact of the programme's agricultural extension efforts. If Railovy et al.'s (2000) conclusions are correct, maintaining the train in operation will spare about 95,000 ha of forest that would otherwise be cleared. The potential impact of agricultural extension activities in the region is a tiny fraction of that figure, even if adoption rates continue to be high. Currently the LDI project works with 1300 farm families, a number that may double by the end of the project. Even the most conservative estimates relating to families living in the communes directly served by the train line suggest that the FCE has an immediate impact on the production decisions (specifically, whether to farm sustainable fruit-trees or unsustainable hillside rice) of at least 15,000–20,000 families.

Secondly, this story underlines the absolute necessity of functioning transport systems in the regions where the project is promoting agricultural intensification on tanety fields. With the exception of breadfruit (which can be used as a subsistence food crop, but which only grows at altitudes up to about 400 m), all the other tree crops being promoted by the project depend on transport to regional, national and, in some cases (coffee), international markets. Evidence from the region shows that, if such transport is available, farmers are willing to forgo rice production on upland fields and to buy rice with fruit revenues (Deeg and Freudenberger, 2000). Equally clearly, however, if there are no such transport options, they will insist on growing crops for home consumption . . . and at the moment those crops are produced in ways that deplete soil fertility, render the land unproductive and promote expansion into the forest.

For the LDI Fianarantsoa programme, the continued operation of the FCE railway is thus critically important to efforts to save the forest corridor. From an ecoregional perspective, infrastructures like the railway and feeder roads provide the foundations for a market-oriented agricultural system, which in this case turns out to be more ecologically sustainable than the subsistence alternatives.

In the light of these analyses, the LDI project took a much more active role in maintaining and expanding the transport networks serving the region. The project was deeply involved in cyclone recovery efforts to restore the service on the FCE, and has recently been instrumental in attracting additional funds for the rehabilitation of the railway system, as well as several roads that are key to the transport of commercial crops from the areas where the project promotes agricultural intensification activities. The project continues to put significant resources into promoting agricultural intensification, but also recognizes that, if there are not adequate transport systems, most of its extension efforts will be futile.

Further contradictions and complexities

While the LDI Programme remains convinced that, in the long run, agricultural intensification is essential to promoting a more rational use of natural resources in the region, the project staff is perplexed by an evident contradiction in the approach: it is clear that on the Betsileo side of the corridor, it is the wealthiest 20% of the population that is engaging in forest clearing. In effect, the biggest constraint to *tavy*, at least on the west side of the corridor, is poverty. If agricultural intensification activities are successful in making families more food-secure, recent history suggests that at least some of these families will invest their surpluses in clearing forest lands (Freudenberger *et al.*, 1999). However, since most people would agree that maintaining people in poverty is an unacceptable approach to conservation, we are obliged to consider other strategies.

This brings us back to an earlier point raised in this chapter: while intensification may be a necessary strategy over the long term, it is certainly not sufficient to save the corridor. It is critical that these activities be complemented by equally rigorous efforts to control access and types of exploitation of forest lands. It will also require policies restricting those types of natural resource exploitation that threaten the biodiversity and watershed functions of the forest and the serious enforcement of those policies. Local populations can play a key role in developing and enforcing policies that favour the collective benefit of saving watersheds relative to the private benefits of clearing individual parcels. Successful agricultural intensification interventions are a critical prerequisite to enabling the enforcement of such policies, since, short of military-style occupation, there is little likelihood that exclusion or even restriction of activities will be successful if people do not have viable alternatives for assuring their livelihoods.

Realistically, however, we must also recognize that such participatory planning approaches are notoriously hard to implement, particularly in the remote, dispersed and often distrustful communities adjacent to the corridor. Madagascar's Government Forestry Service has not proved to be a particularly helpful partner in managing the forest corridor, wracked as it is by internal conflicts, competing political interests and a personnel that is as scarce on the ground as it is demotivated. Promoting and implementing a common vision of effective forest management among various stakeholders is perhaps the area where LDI has been the least successful to date.

Finally, it is hard to be optimistic in the face of population growth rates that will result in a doubling of the number of people trying to make a living from these lands in only the next 15–20 years . . . even if family planning programmes are successfully introduced and widely adopted. Indeed, these future farmers of Madagascar – and potential practitioners of slash-and-burn agriculture – are already born. These statistics allow us little time for idle despair, however. Instead they argue even more forcefully for agricultural intensification, and especially for identifying the major structural factors, such as transport, that 'warp' the whole economy in ways that are more (we hope) or less favourable to the adoption of sustainable agricultural practices.

Conclusions

If the LDI project and other regional actors are to successfully slow the rate of occupation of the corridor, it must find ways to reduce the multiple pressures that are motivating people to increase their land area and expand on to forest lands. The principal factors behind the current expansion appear to be: (i) demographic pressures and people's concern that their children will inherit parcels too small to support their food needs; and (ii) the progressive declines in soil fertility that render even newly cleared fields largely unproductive after only five or six production cycles. Strategies to reduce these pressures should thus logically include actions: (i) to introduce and/or strengthen the provision of family planning services; (ii) to improve the productivity of fields currently under production; and (iii) to rehabilitate degraded upland fields that are now more or less unusable due to soil infertility.

Agricultural intensification has an absolutely critical role to play in reducing pressures on a forest corridor that is both an international treasure, in terms of its biodiversity, and a regional treasure, in terms particularly of the role it plays in maintaining watersheds. For a project like LDI, however, there is a danger that, in getting too caught up in the admittedly challenging questions of how to increase adoption rates at the household level, the project risks failing to deal with the unexpected – and in our case counter-intuitive – consequences of 'successful' adoption, such as intensification actually in some cases accelerating deforestation, as profits are used to expand agricultural holdings. And, similarly, managers may fail to see that the most effective programmes may not always be to increase adoption rates among new converts, but rather to ensure that the conditions needed to assure the continued practice of existing sustainable agricultural systems are maintained. In the case of the Fianarantsoa corridor, this has meant fighting (often in the face of doubting conservationists) for the future of a very rickety train line whose survival will do more for sustainable agriculture in the region than anything the project's knowledgeable and highly committed extension agents can hope to accomplish.

Note

[1] The views expressed in this chapter are those of the authors and do not engage USAID or Chemonics International.

15 Synergies between Natural Resource Management Practices and Fertilizer Technologies: Lessons from Mali

Valerie Kelly,[1] Mamadou Lamine Sylla,[2] Marcel Galiba[3] and David Weight[4]

[1]Department of Agricultural Economics, Michigan State University, Agriculture Hall, East Lansing, MI 48824-1039, USA; [2]Office de la Haute Vallée du Niger, BP 178, Bamako, Mali; [3]Sasakawa Global 2000, BP E3541, Bamako, Mali; [4] Institute of International Agriculture, Michigan State University, East Lansing, MI 48824, USA

Successful agricultural development has resulted in substantial alleviation of poverty and food security in Asia and Latin America since the 1960s. In sub-Saharan Africa (SSA), however, productivity levels remain stagnant. Low soil fertility is increasingly recognized as one of the primary biophysical constraints blocking agricultural development in SSA (Mokwunye and Vlek, 1986; Penning de Vries and Djiteye, 1991; Pieri, 1992; van der Pol, 1992; Sanchez et al., 1997b, citing numerous others). This low soil fertility can be attributed to soil degradation due to soil mining (associated with long-term low-input agriculture), tillage, accelerated erosion and poor maintenance of soil organic matter and soil organic carbon.

Fertilizers are considered by many to be critical inputs for restoring soil fertility and increasing crop yields in SSA (Mokwunye and Vlek, 1986; Buol and Stokes, 1997; Quinones et al., 1997), but farmers have not adopted fertilizer in a sustainable manner. Unfavourable and volatile input/output price ratios, poor crop response, poorly developed markets and inappropriate macroeconomic and sectoral policies have all been blamed for low fertilizer-adoption rates (Shepherd, 1989; Runge-Metzger, 1995; Larson and Frisvold, 1996; Yanggen et al., 1998). Given the seemingly insurmountable problems associated with fertilizer promotion in SSA, some (Reijntjes et al., 1992; Pretty, 1995a; Jiggens et al., 1996) have argued that the solution to soil degradation and agricultural productivity problems lies in the promotion of better natural resource management (NRM) practices (e.g. agroforestry, anti-erosion measures, better use of organic amendments). Yet here too adoption has been slow. Poor researcher understanding of farmer priorities, land tenure, high labour demands and small or deferred yield and income benefits are among the constraints noted (Hijkoop et al., 1991; Bationo et al., 1996; Napier, 1996; Neef et al., 1996).

Recent research and writings support the use of fertilizers in combination with organic inputs as part of intensification strategies to drive sustainable productivity growth and end this long cycle of agricultural and economic stagnation (Bationo and Mokwunye, 1991a,b; Pieri, 1992; Swift,

1996; Bekunda *et al.*, 1997; Quinones *et al.*, 1997; Reardon, 1997b; Wallace, 1997; Breman and Sissoko, 1998; Yanggen *et al.*, 1998; Weight and Kelly, 1999). There is a consistent perspective in these works that neither input strategy, on its own, is capable of achieving production goals and food security.

The two case-studies examined in this chapter provide evidence that some Malian farmers are now adopting these combined strategies and realizing important benefits as a result. The case-study subjects are the NRM programme of the Office de la Haute Vallée du Niger (OHVN) and the Sasakawa Global (SG) 2000 programme (Fig. 15.1). The OHVN is a regional development programme providing extension and support services in a cotton/coarse-grain cropping system, where farmers are already accustomed to using purchased inputs, such as fertilizers and pesticides. The SG programme is introducing fertilizers and pesticides into a predominantly subsistence coarse-grain (millet/sorghum) cropping system that includes some production of groundnuts and cowpeas.

Mali: a Conducive Environment for Agricultural Intensification

Agricultural productivity growth based on the adoption of improved technologies does not take place in a vacuum; farmers adopting new technologies are influenced by the physical, political and socio-economic environment in which they live. During the last two decades, Mali implemented numerous reforms that improved the economic and political environment. Among the most important were the privatization of some state enterprises, extensive reform of others and a reduction in trade barriers. Private businesses and associations now compete in areas formerly reserved for the state. The liberty of association and expression also improved as a result of the democratization movement begun in 1991. The 1994 devaluation of the Communaute financière d'Afrique (CFA) franc (CFAF) provided a major stimulus to the cotton, rice, livestock and horticulture sectors. Although Mali remains one of the poorest countries in the world, with a gross domestic product (GDP) per capita of US$250 in 1998 (equal to US$720, using purchasing-power parity estimates), the economy has recently been growing more rapidly than the population (3.8% vs. 2.8% from 1994 to 1998).

Despite general economic progress, agricultural productivity (measured as yields per hectare) is not growing at desired rates (e.g. cotton yields declining during the 1990s, millet and sorghum yields stagnant). There is ample evidence from agronomic research that Malian farmers can raise yields by adopting improved cultivars, fertilizers and a variety of NRM practices (Henao *et al.*, 1992). Unfortunately, the vast majority of Malian farmers are not yet using these improved techniques, so recent increases

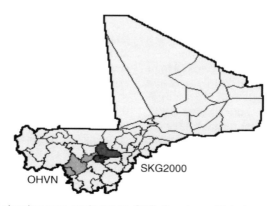

Fig. 15.1.   Map of Mali showing case-study zones. SKG, Sasakawa Global.

in aggregate crop production have been realized primarily through expanding production to land previously considered inappropriate for agricultural purposes or reducing fallows. Consequently, soil degradation (through nutrient mining, erosion and failure to maintain soil organic matter) is a concern throughout the country, although estimates of the magnitude of the problem differ.[1]

The Case-study Programmes

OHVN's NRM programme

The NRM programme began in the late 1980s but did not take off until the early 1990s, when donor support increased. The programme goal is to train communities in improved NRM and crop production techniques so that they will realize increased levels of food security and monetary income, while ensuring continued access to adequate supplies of water, wood and pasture for animals.

The OHVN extends across four agroecological zones (Sahelian, North and South Sudanian, and Sudano-Guinean), with rainfall ranging from a low of 400–900 mm in the Sahelian zone to 1000–1200 mm in the Sudano-Guinean zone. The soils (predominantly ferric luvisols) are characterized by high erosion and degradation, with deforestation a contributing factor. The consequence – low yields and incomes – stimulated high rates of out-migration from the zone during the 1980s and early 1990s.

The NRM programme has focused on the eastern and southern parts of the OHVN, where rainfall exceeds 800 mm year^{-1} and there is a history of cultivating cash crops, such as cotton. A key tenet of the OHVN programme is that strong economic incentives are needed to stimulate NRM adoption.[2] Thus far, that incentive has been the opportunity to increase household income through cotton production on improved land.

The NRM programme uses a participatory approach, only intervening in communities that are openly receptive to making changes and willing to invest human and/or financial resources in these changes (e.g. by forming a village association or obtaining literacy training for association members, so that records can be kept and credit applications can be prepared, etc.). Literacy and numeracy training was provided by the national literacy programme and assistance with organizing village associations came largely from a Cooperative League of the USA (CLUSA) project.

Once a village is selected to participate in the NRM programme, OHVN agents train a technical team composed of five to ten villagers (selected by their peers) who have completed literacy training and are willing to devote 1 day per week to learning NRM techniques, training others in the village, organizing community-level NRM activities and keeping records of both individual and community NRM activities. The team members receive no salary or special benefits from OHVN, but most are remunerated (usually in kind) by their communities. Most team members are relatively young – an interesting change to observe in a society where leadership roles have traditionally been held by village elders.[3] After the training, extension agents provide backup support for the team only when requested. The goal is to promote village-run extension services. At present, 20 villages have attained this status.

The Sasakawa Global 2000 programme

The SG programme in Mali has three themes: (i) restoring and improving soil fertility through improved fallows, using nitrogen-fixing legumes and natural phosphates; (ii) intensification of cereal production, using improved seeds, pesticides, fertilizer and cultural practices; and (iii) formation of savings and loans associations for financing input acquisition. SG/Mali is part of an Africa-wide programme that is generally associated with the promotion of green-revolution technologies through the use of farmer-managed demonstration and control plots (Easterbrook, 1997; Quinones et al., 1997).

SG/Mali began in 1996, primarily in areas not served by major development projects such as OHVN. SG zones are characterized by Sudanian climates, but border on Sahelian zones in the north and Guinean zones in the south, with typical rainfall in programme areas ranging from 400 to 800 mm since 1996. Soils are similar to the soils in the OHVN. Commercial crops and markets (e.g. groundnut, tobacco, cotton) are not well developed in areas of SG intervention; rain-fed millet/sorghum systems predominate and some maize is produced in higher-rainfall areas.

Mali is one of SG's first efforts to introduce seed/fertilizer technologies into the drier, riskier, production systems of West Africa, where the yield potential for the most common coarse grains is low.[4] The SG case-study concerns only the 1998 millet programme in the Segou Region. In 1996 and 1997, SG promoted improved fallows in Segou (dolichos and phosphate rock from local sources). Farmers used the improved fallows but, when SG stopped purchasing dolichos seed from farmers in 1998, demand for the technology fell abruptly and the other millet programme became more popular.

Farmers' Response to the Programmes and Their Perceptions of Them

Because information on the OHVN and the SG 2000 programmes come from unrelated studies, there is a lack of strict comparability in the information presented. Some information is from official OHVN or SG 2000 service statistics, some from a February 2000 rapid appraisal conducted by two of the authors in seven OHVN villages and some from a 1999 Institut de Sahel (INSAH) survey of SG participant farmers in which two of the authors were involved. Only after the survey and rapid appraisal were completed did the idea of documenting the two case-studies arise.

Response to the OHVN NRM programme

Table 15.1 shows 1996–1999 adoption levels in physical measures (metres, hectares, number) for 22 themes. Anti-erosion techniques (rock lines, gully plugs, contour ploughing, vegetative bands) and improved organic matter (manure and compost pits) were among the more popular themes. Growth in livestock stabling (the pinnacle of manure management) is slow, due to the high costs (construction of holding pens and high labour demands), but it is highly appreciated.

Table 15.2 shows that since the early 1990s, 60% of villages and 47% of farms have adopted one or more of the NRM themes. A major accomplishment has been the restoration of abandoned or severely degraded fields – estimated to represent approximately 17% of total cultivated area in 1999. Approximately 3900 farms have been able to fix their cultivated area for at least 3 years (i.e. no clearing of new land) while maintaining or increasing production.

Figure 15.2 shows graphs of cotton, millet, maize and sorghum yield trends for individual farmers who used anti-erosion practices in combination with improved organic matter for 5 or more years. The average yields obtained by these farmers during the 1990s exceed the average yields reported in OHVN statistics for every crop; the differences were greatest for maize and cotton – the two crops that benefit most directly from applications of fertilizer and organic amendments that are used more intensively once land is stabilized through the adoption of anti-erosion practices. The overall pattern for these farmers is one of steadily increasing yields, with the exception of cotton, which exhibits interannual fluctuations. Volatility in cotton yields is thought to be the result of changes in fertilizer and pesticide use following input price increases associated with the 1994 devaluation.

In the absence of adequate data to evaluate the costs and benefits of adopting these practices, we made a rough estimate of the probable differences between the present value of production for a 9-year millet/cotton rotation cultivated with and without NRM practices. Data for the without scenario were hypothetical, assuming that yield trends would follow patterns of stationary millet yields. Yield data for the NRM scenario came from a participating farmer using rock lines, vegetative bands, ploughing perpendicular to the slope, manure, fixation of plots and end-of-season ploughing. In an effort to make this a conservative estimate of returns, we used output prices (70 CFAF kg^{-1} for millet and 130 for cotton) that were lower than average prices reported during the last

Table 15.1. Physical indicators of NRM adoption (from OHVN, December 1999, report and other OHVN survey data).

	Adoption (units)				
NRM themes	Prior to 1997	New in 1997/98	New in 1998/99	New in 1999/2000	Cumulative sum
Rock lines (m)	79,400	6,485	10,076	6,581	102,543
Branch barriers (m)	18,500	780	2,011	2,333	23,624
Small dykes (m)	38,900	1,492	775	457	41,624
Vegetative bands (m^2)	8,998	1,341	4,000	5,653	19,969
Living fences (m)	127,022	12,000	11,831	12,893	163,746
Permanent field markers (ha)	1,098	599	846	799	3,322
Protected areas (ha)	450	450	615	750	2,265
Diversionary gullies (*n*)	1,417	625	1,171	100	3,313
Fire breaks (m)	5,250	1,406	615	500	7,771
Controlled land clearing (ha)	140	300	–	–	440
Village-managed forests (*n*)	1,620	35	–	–	1,655
Wells (*n*)	120	13	13	9	155
Improved water-holding capacity of ponds (*n*)	68	2	1	2	73
Improved low-land irrigation (ha)	20	–	–	–	20
Village tree nurseries (*n*)	57	15	5	28	105
Plants from tree nurseries (*n*)	178,800	13,318	14,640	45,576	252,334
Village wood lots	447	23	19	18	507
Improved cooking stoves (*n*)	2,340	745	312	631	4,028
Manure pits (*n*)	2,268	265	338	–	2,871
Stables for collecting manure (*n*)	13,608	140	135	–	13,883
Permanently stabled livestock (*n*)	146	8	–	–	154
Compost pit (*n*)	1,303	399	490	50	2,242

m, metres, *n*, number.

Table 15.2. OHVN adoption: number of villages, farms, recovered hectares and settled farms (from OHVN, 1999, survey data).

Sector	Villages	Farms	Recovered area (ha)	Settled farms
Kangaba	53	1529	3027	296
Bancoumana	57	2335	3221	234
Ouélessébougou	97	3628	7604	1054
Dangassa	33	534	434	180
Fouani	110	3295	7264	600
Kati	70	1787	1303	241
Faladié	35	951	2274	275
Koulikoro	73	1358	2075	449
Sirakorola	79	2220	7656	552
Per cent coverage	60	47	17	–

5 years. NRM practices averaged 346 kg ha⁻¹ more millet and 486 kg ha⁻¹ more cotton each year. Using a 10% discount rate, the estimated present value of the increased production over 9 years was 277,380 CFAF ha⁻¹ for the rotation starting with millet and 119,670 CFAF ha⁻¹ for the one starting with cotton.[5] As most NRM costs are for labour, we converted the monetary returns to labour-day equivalents at the current agricultural wage rate of 750 CFAF day⁻¹. This gives 370 labour days for the rotation beginning with millet and 160 labour days for the other. As OHVN extension personnel claim that the labour that went into the NRM practices during the 9 years would not have approached 370 days ha⁻¹ and was unlikely to have exceed 160 days, this rudimentary attempt to examine financial returns to NRM practices suggests that returns were probably positive for this farmer.[6]

Support for the hypothesis of positive financial returns to NRM adoption also comes from farmers who discussed the impacts of NRM adoption with the rapid-appraisal team. Specific examples of NRM impacts reported are summarized below.[7]

- Yields of all crops are increasing for farmers adopting NRM intensification methods.

- The village youth are staying at home to farm rather than migrating. This was very evident in all villages visited; youth were present at all meetings, they play important roles in the management of farmer associations and they were very active participants in rapid-appraisal discussions; the sharp reduction in out-migration was viewed as a very positive result by all.

- Farmers are investing heavily in agricultural equipment, traction animals and livestock. When asked what they were doing with their increased incomes, the most common response was investment in equipment and/or livestock.

- Farmers are diversifying, with many new forays into dry-season crops and tree crops. Increased production of horticultural products during the dry season (green beans for Europe, onions/tomatoes and bananas for Bamako) is one of the reasons for the reduction in out-migration; marketing remains a problem, but the farmers' associations appear to have a level of management skills permitting them to deal with the setbacks and move ahead. Tree-crop production (particularly teak for the production of construction poles)

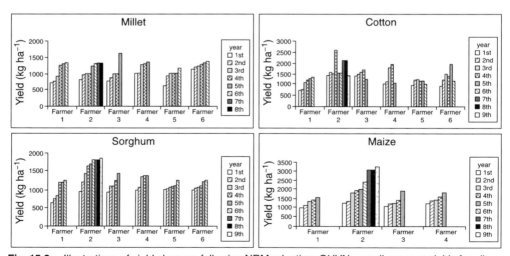

Fig. 15.2. Illustrations of yield changes following NRM adoption. OHVN overall average yields for all farmers during this period were: millet 921 kg ha⁻¹; sorghum 998 kg ha⁻¹; cotton 1056 kg ha⁻¹; maize 1137 kg ha⁻¹.

through the development of village and private wood lots is expanding slowly, but examples seen had not yet generated income.

- Farmers are unanimous that life is better now than 10 years ago: they eat better (more food and better variety); they dress better; they travel more easily (mobylettes have replaced bicycles in many cases); schools and health services are more accessible; and they are better educated (literacy programmes and CLUSA management training).
- Farmers are optimistic and enthusiastic about the future.

The team noted a marked difference between the optimism and enthusiasm for farming expressed by the farmers in villages where NRM programmes had been under way for more than 5 years and the one village visited where the programme had not yet started. These differences cannot, however, be attributed to the lack of an NRM programme, because there was also another important difference: cotton production was just getting started in the village. It will be interesting to see if this village joins the NRM programme and eventually realizes

the same types of productivity and income impacts as the others, even though it is located in a more agroecologically difficult environment.

Expanding the NRM programme beyond the initial adopters, who tend to be the better-off farmers in the higher-rainfall areas, remains a key challenge for the OHVN. As illustrated by aggregated OHVN yield trends in Table 15.3, the positive impacts of NRM adoption described above are not yet widespread enough to be reflected in zone-level yield trends, which are stagnant or declining. Identifying the key ingredients contributing to current successes will play an important role in developing a strategy for scaling up the programme. Some of those key ingredients appear to be:

- Identification of a broad list of technologies capable of increasing yields.
- Potential for increased cash income from expansion of cotton production.
- Community approach to implementation.
- Focus on youth.
- Initial focus on villages/farmers most likely to benefit from NRM.

Table 15.3. Area, production and yield data for the OHVN: 1991/92–1998/99 (from OHVN, 1999:16).

	1991	1992	1993	1994	1995	1996	1997	1998	Trend
Cotton									
Area (ha)	10,506	12,201	8,624	11,692	14,605	23,158	30,750	35,816	+
Prod. (tons)	11,842	12,494	10,684	13,097	16,167	21,990	28,927	33,740	+
Yield (kg ha^{-1})	1,127	1,024	1,239	1,120	1,107	950	941	942	−3%
Millet									
Area (ha)	30,906	31,516	31,892	34,188	36,660	35,732	38,149	37,422	+
Prod. (tons)	30,226	23,900	26,700	31,800	32,441	36,095	38,714	35,595	+
Yield (kg ha^{-1})	978	758	837	930	885	1,010	1,015	951	Stagnant
Sorghum									
Area (ha)	46,603	48,334	48,140	51,213	56,009	59,431	66,390	72,572	+
Prod. (tons)	50,508	43,911	44,622	47,904	50,292	64,638	73,047	75,901	+
Yield (kg ha^{-1})	1,084	908	927	935	898	1,088	1,100	1,046	Stagnant
Maize									
Area (ha)	11,099	11,485	11,648	12,157	12,834	13,072	14,411	15,457	+
Prod. (tons)	13,845	13,110	13,938	11,214	12,929	14,594	16,814	20,033	+
Yield (kg ha^{-1})	1,247	1,141	1,197	922	1,007	1,116	1,167	1,296	Stagnant

Yield trends were estimated using linear regressions. Stagnant indicates no statistically significant trend at a 0.90 level of significance.

- Use of demonstration effect through model farmers and model villages.
- Incremental training (literacy, technical skills, community organization, management skills using the CLUSA model).
- Support services offered: improved feeder roads (funded through a US Agency for International Development (USAID) project); credit guarantees for a limited period following management training; input/output transport assistance; regular supervision and support for trainees; some free equipment for implementing NRM activities; and market research by OHVN to help with crop diversification.

Among the factors listed above, three stand out as being absolutely essential for the sustainable adoption of an approach to agricultural production that includes both NRM practices and improved inputs.

- A profitable cash crop with reliable markets and stable prices.
- Improved, affordable technologies that benefit both cash and food crops.
- Training programmes that equip young farmers with the literacy and management skills needed to function as effective commercial farmers, both independently and in associations.

Response to the Sasakawa Global 2000 programme in the Segou Region

Approximately 100 farmers participated in the survey of SG participants in the Segou Region. A review of the characteristics of the household heads interviewed suggests that they, like the NRM adopters in the OHVN, are among the better-educated and wealthier farmers in the zone (43% were literate, average farm size was 10 ha, average assets included 3.7 traction animals, 1.8 ploughs/cultivators and 1.2 carts). In addition, 26% had been participants in other types of demonstration programmes. Although participants appear to be among the better-off farmers, the zone is not a wealthy one. Farmers are heavily

dependent on millet – a low-value, low-productivity crop – for income. Eighty-eight per cent of farmers claimed that millet production represented more than 50% of their total income (cash and in kind); 37% of this group declared that millet represented at least 75% of total income. In addition, only 27% claimed to have other sources of income (e.g. livestock or non-farm activities) that they would be willing to use to repay input credit should their millet crop fail.

The millet programme in Segou offers three incremental levels of technology, representing increasing costs, risks and yields. This incremental approach permits farmers to move gradually from current practices to input-intensive practices, improving their skills and financial capacity to work with new technologies year by year. The three themes were:

- Level A: improved seed and mildew protection.
- Level B: package A plus light fertilization using improved compost.
- Level C: package A plus heavy fertilization.

The Level A package is an anti-mildew treatment called Apron+ and a short-cycle seed variety. Moving up to Level B, farmers currently using compost pits (a theme promoted by the government's training and visit extension programme) can improve their compost by adding locally produced phosphate rock. Although some farmers successfully used this theme in 1996 and 1997, many do not have compost pits or found the costs of the phosphate rock a constraint. In 1998, SG did not have the financial resources to adequately promote both Level B and Level C technologies, so they focused on the latter. Farmers having tried the first two levels as well as those willing to move directly to this level were invited to participate. Level C consists of improved seed, Apron+, compost, natural phosphate (25 kg 0.25 ha^{-1}) and bulk-blended 23–13–13 with S/MgO/Zn (25 kg 0.25 ha^{-1}).

During the 1999 survey, farmers were asked to identify the three most important risk factors constraining their production. In this zone, one tends to think of poor rains as the major constraint, but farmers gave slightly more importance to the problem of unreliable access to inputs (score of 132 for inputs vs. 116 for rain[8]), with bird damage and declining soil fertility coming in third and fourth positions (scores of 92 and 64). Having identified the three principal factors thought to increase production risk, farmers evaluated the extent to which SG technologies alleviated these problems. Their replies strongly contradicted the conventional wisdom that the use of external inputs (particularly expensive fertilizers) increases risk; most participants (96%) claimed that the SG technologies reduced the risk of crop loss associated with the above-mentioned problems; only 3% – all Level C farmers – viewed the technologies as risk-augmenting.

Farmers were asked to explain their perceptions of these risks. The most common reply (62%) was that the entire combination of inputs diminished the risk of getting very low millet yields. Reduction of risk due to attacks by birds, rats and termites was the next most common reply (16%); apparently the Apron+ repels insects and animals. Several farmers (8%) commented that SG technologies reduced risks associated with poor access to land because they could increase yields on existing land. Reduction of risks associated with poor rains was mentioned by 5% – primarily a reference to the shorter-cycle seed varieties. The key concern of the 3% of farmers indicating that SG technologies increased risk was the problem of not being able to reimburse credit for the Level C package.

Crop budgets estimated from survey data collected for the SG participants' test and control plots (Table 15.4) show that returns for Level A technology were much greater than for Level C. All Level A farmers realized positive returns. The value/cost ratio for the package is 10 and the average net benefit was 9615 CFAF. This net benefit is the equivalent of what one would earn by hiring out labour services for 13 days at the prevailing agricultural wage. Farmers were unanimous in their praise for this package,

given that 60% of a crop can be lost to mildew.

Average returns to the Level C package were 3858 CFAF and the value/cost (v/c) ratio only 1.5.[9] Although a common rule of thumb is that a v/c ratio must be at least 2 to stimulate demand for a technology package (even 3 or 4 in risky environments like the Sahel), 53% of the farmers thought the package was 'profitable'. The average return masks a high degree of variability. Ten farmers had losses ranging from 1000 to 9000 CFAF. Four of these ten farmers, who had losses from 1000 to 5000 CFAF, considered the package profitable. Among the farmers with positive returns, 43% found the package either unprofitable or only marginally profitable; returns for this group were in the 500–10,000 CFAF range. The five farmers with returns greater than 10,000 CFAF were unanimous that the package was profitable.

Given the small sample size for the Level C technology package, linguistic problems associated with translating words such as 'profitable' into local languages and participating farmers' limited experience with purchased inputs and agricultural

Table 15.4. Yields and benefits of SG technologies (from INSAH/MSU/SG, 1999, survey data, Segou Region).

	Level A package	Level C package
Cases	40	26
Av. yield increase	133	138
Av. value of increased production (CFAF)	10,640	11,031
Av. supplemental cost for test plot (CFAF)	1,025	7,173
Net benefit (CFAF)	9,615	3,858
Value/cost ratio	10	1.5

INSAH, Institut de Sahel; MSU, Michigan State University; SG, Sasakawa Global 2000. US$1.00 = 600 CFAF. Millet price is 80 F kg^{-1} (1998/99 mean in the study zone). The Level C package costs are 8150 CFAF; average supplemental costs are slightly less because we made adjustments for cases where farmers used extra seed or Apron+ on their control plots. Costs and returns are for 0.25 ha plots.

production for commercial purposes, we do
not want to make too much of the apparent
differences between the profitability analysis and farmers' perceptions of profitability
at this point in time. Informal discussions
with farmers provide some insights. Several
farmers appeared to have evaluated the
technology from a whole-farm perspective;
although the cost of inputs was greater than
the market value of increased production,
farmers valued the increased cereal production enough to cover input costs with
receipts from other farm activities (animal or
groundnut sales). In the long run, producing
more millet on less land frees land (and
perhaps labour) for the production of other
crops. Another insight from informal discussions was that many farmers in this drought-
prone zone (where one measure of social
standing is the number of full granaries a
household possesses) might have been more
interested in maximizing cereal production
than in cash income. These insights suggest
that SG may want to spend more time in the
future doing profitability analyses jointly
with farmers, in an effort to improve
researchers' and extension agents' understanding of farmers' evaluation methods and
criteria.

Given that SG efforts to introduce
improved techniques to Segou cereal farmers is in its infancy (third year of test plots)
and SG was unable to properly test their
intermediate Level B technology, it is too
soon to draw broad generalizations about
the overall effort. The programme design,
based on the sequential introduction of more
expensive and risky technologies in this
zone of relatively poor farmers, has substantial merit. Farmers gained experience
in the SG approach the first year, using
the very low-risk Level A package – they
were extremely satisfied with the results.
The rapid jump up to Level C technology
appears problematic, given our analysis
of financial returns, but farmers remain
enthusiastic about the higher yields
obtained. The next few years of the
programme will be critical, as SG searches
for some combination of NRM practices and
inorganic fertilizers that can be financially
sustainable over time.

Lessons Learned

A first lesson drawn from the two studies is
that both SG and OHVN/NRM managed to
get farmers engaged by focusing their initial
extension efforts on alleviating a problem
identified by farmers themselves: erosion
for the OHVN farmers and mildew for SG
farmers. Although the costs (in terms of
labour, community organization and equipment needed) for the anti-erosion work
were generally higher than those for the
Apron+ seed treatment, both solutions
were low-cost enough to build a critical
mass of early adopters who provided a
demonstration effect for the spread of the
practices to other farmers.

A second lesson is that an extension
programme with a long list of à la carte
options from which farmers can select
improvements that best meet their
resources, problems and willingness to
bear risk is more likely to maintain farmer
interest and high levels of participation
than a programme with a limited number of
options. SG, for example, has a relatively
short list of technical options and has not
yet identified a sufficiently profitable technology to introduce after getting farmers
engaged with the short-cycle seed and fungi-
cide. The very long list of practices offered
by OHVN has encouraged farmers to make
adoption and improvement a way of life
rather than a special event associated with a
short-lived project.

A third lesson is that the demonstration
effect of test and control plots can be
an extremely powerful extension tool when
used well. SG test plots are to be commended
as a particularly good way of demonstrating
to both participants and non-participants
the yield differences attributable to a technology. OHVN has no formal programme for
comparing yields of different practices and
not all OHVN/NRM practices lend themselves to this type of comparative method;
nevertheless, the yield impacts of some of
the OHVN recommendations could be better
appreciated using this type of demonstration
approach.[10]

A fourth lesson is that test and control
plots need to be used to demonstrate not

only the yield differences but also to study the differences in net returns to the technologies being compared.

A fifth lesson is that basic literacy and management training make a tremendous difference in farmers' ability to manage new technologies on both an individual and a community level. OHVN farmers who had received CLUSA training were much more in charge of their farms and their communities than OHVN and SG farmers who had not benefited from comparable training.

A sixth lesson is that the adoption of both NRM practices and external inputs is facilitated by the presence of a cash crop in the cropping system. SG introduced fertilizer to Segou farmers before promoting NRM practices. This is similar to what OHVN did before the creation of the NRM programme, but the key difference is the nature of the output markets for the crops concerned. Millet is a semi-subsistence crop with poor marketing prospects (volatile prices and weak urban demand) and cotton is a cash crop with a guaranteed market and relatively stable producer price. Under current market conditions in Mali, it remains questionable whether there can be an adequate incentive for any type of agricultural intensification in the millet/sorghum production systems of the Segou Region if there is no viable commercial crop. This problem needs to be addressed by research to identify new crops as well as by developing markets for existing crops (e.g. through processing or animal-feed industries and responding to regional rather than only national demand).

A seventh lesson is that NRM as well as external inputs tend to be adopted by better-off farmers first. One expects external inputs to go to the wealthier farmers, given the costs, but it is often thought that NRM is a better way to go for poor farmers. For most farmers, building anti-erosion barriers means having access to carts and draught animals; even when barriers are built by community effort, farmers able to provide the equipment will be among the first to benefit. Farmers wanting to use improved manure can make small advances by digging manure and compost pits, but only the farmers who can afford to own many animals are

able to move up to permanent stabling of livestock. Unless adoption goes well beyond current levels, aggregate measures of agricultural productivity are not likely to increase and aggregate measures of soil degradation are not likely to decrease.

SSA has a historic opportunity to reverse the current trends of stagnant productivity and declining soil fertility. This means that long-term fallows, which maintained soils and productivity in the past, need to be replaced by (or adapted to) appropriately integrated systems that include fertilizers or other effective input sources, no-till (or mulch tillage), cover crops, rotations and/or agroforestry practices based on sound agroecological principles that take advantage of natural restorative processes and are therefore efficient in terms of fertilizer and water requirements, as well as costs and labour. This is especially critical for smallholder farmers, who make up the vast majority of agricultural producers in SSA and who are faced with severe economic and technical constraints. Lessons summarized above provide some guidance, but expansion of these programmes to poorer farmers in more difficult environments will require that research and extension services be constantly looking for new and better ways of promoting adoption.

A major challenge remains in zones where farmers do not produce a commercial crop. Given the important role that cash crops play in stimulating adoption (see, among others, Reardon *et al.*, 1996; Govereh and Jayne, 1999; Strasberg *et al.*, 1999), we find ourselves suggesting that SG – whose mission is to deal exclusively with food crops – reconsider the zones in which it intervenes, as it would be in a better position to foster the use of improved technologies on food crops if it were working in an area that already had a cash crop.

For the OHVN, we suggest promoting the use of fertilizer on food crops for farmers already using the input on cotton and having made major anti-erosion investments. The NRM practices promoted by OHVN have improved cereal yields to some extent, but inorganic fertilizer is seldom used on cereals and yields remain below potential. This is

the type of situation where an SG programme could make a real contribution
by identifying and promoting fertilization
technologies appropriate for farmers who
do not have the resources to permanently
stable their livestock or for farmers who have
yields that are consistently above 1 ton ha^{-1}
but want to aim for yields of 1.5–2 tons.
These appear to be issues that a collaborative
SG/OHVN effort would be well placed to
resolve, given the complementary
approaches and firm evidence from the
scientific literature concerning NRM/
external input complementarities.

Notes

[1]   Nutrient-balance studies based on research
trial data have reported annual depletion rates as
high as 25 kg ha^{-1} of nitrogen and 20 kg ha^{-1} of
potassium (van der Pol, 1992), while results from a
biophysical model designed to reflect farmers'
actual production practices confirmed the presence of nutrient mining and erosion but at a far
less alarming level than that suggested by others
(Dalton, 1996).

[2]   This is an unusual approach for an NRM
programme, as NRM practices are often targeted at
semi-subsistence farmers considered too poor to
purchase improved inputs such as fertilizers and
pesticides.

[3]   The requirement of literacy may be driving
this, as younger farmers tend to be more responsive to the literacy training programme than older
ones.

[4]   In Mali, millet yields of 300–700 kg ha^{-1} can
be expected using traditional technologies and
800–1200 kg ha^{-1} using improved technologies;
in Benin, where SG also works, maize yields
of 1000 kg ha^{-1} with traditional techniques and
3000 kg ha^{-1} with SG techniques are the norm.

[5]   The starting crop makes a difference, because
weather and pests in a given year may be more
favourable to one crop than the other. The farmer
supplying the data had both millet and cotton
fields both years, so we were able to take these
differences into account.

[6]   The most labour-intensive activity is building rock lines. Rough estimates (Hijkoop *et al.*,
1991) suggest that 1 m of rock-line construction
averages 0.75 to 1 h of labour time. Although the
amount of rock line per hectare varies, typical
fields tend to require less than 50 m ha^{-1} (i.e. 50 h
of labour).

[7]   Of seven villages visited, only six had active
NRM programmes. About 100 farmers participated in the rapid-appraisal discussions, of whom
approximately 80 had NRM experience.

[8]   Scores reported are weighted frequencies:
the most important problem received a weight of
3, the second 2 and the third 1.

[9]   A shortcoming of the partial budget analysis
is that the slow-release phosphate rock is unlikely
to have produced a yield response in the year it
was applied. A multi-year analysis of the package,
capable of measuring residual effects, could raise
the profitability.

[10]   For example, many farmers are not adding
crop residues to the manure being collected from
stabled animals, although research results show
that the yield impact of adding these residues
is important. Two plots grown side by side with
and without the crop residues might be a way of
demonstrating that the extra effort is worth the
investment.

16 Soil and Water Conservation in Semi-arid Tanzania: Government Policy and Farmers' Practices[1]

N. Hatibu, E.A. Lazaro, H.F. Mahoo and F.B.R. Rwehumbiza

Soil-Water Management Research Group, Sokoine University of Agriculture, PO Box 3003, Morogoro, Tanzania

Introduction

The semi-arid zone occupies over 50% of Tanzania's total area, extending north-east to south-west across the central part of the country. The semi-aridity results from low rainfall, high evapotranspiration rates and erratic temporal and spatial distribution of rainfall (Nieuwolt, 1973). The primary problem facing farmers in the semi-arid areas is therefore insufficient soil-water availability. Consequently, the semi-arid areas exhibit low and unreliable crop and livestock productivity. For example, maize yields in central Tanzania are only 800 kg ha^{-1} as compared with the national average of 1400 kg ha^{-1} (MoAC, 1998) and the average live-weight of cattle is only 200–250 kg (Hatibu and Mtenga, 1996).

Sustainable agricultural development in semi-arid Tanzania depends heavily on effective utilization of scarce rainwater. This requires policies and methods that emphasize improving soil-moisture availability for crop and pasture, as well as improved practices to make effective use of soil and water.

The objective of this chapter is to assess national natural resource management (NRM) policies, with particular reference to soil and water management in relation to farmers' actual practices in semi-arid areas.

We review polices on land, agriculture, forestry and water and then describe farmers' practices in three case-study areas: Dodoma District, Shinyanga District and the western Pare lowlands (WPLL) of Mwanga and Same Districts. These case-studies lead to a synthesis of the factors influencing the adoption of rainwater harvesting (RWH) technologies. Throughout the chapter we demonstrate the gap between government policy and farmers' practices and priorities.

Background on Relevant Tanzanian Policy

Government policy: agriculture and livestock

The development of Tanzanian agricultural policy began with directives by the ruling party. The first of this kind was the *Siasa ni Kilimo* ('Politics is Agriculture') directive of 1972 (MoA, 1982). Soil erosion was recognized as a major problem, but the focus was put on rehabilitating highly eroded areas, such as Kondoa (Christiansson *et al.*, 1993). No mention was made of the causes of erosion and how to protect cultivated lands.

The second major directive, *Kilimo cha Kufa na Kupona* ('Agriculture as a matter of life and death'), issued in 1975, aimed at food self-sufficiency. The outstanding outcome was the rapid expansion of urban and peri-urban agriculture. Today, more than 30% of Tanzania's urban population considers agriculture to be their main source of income (Planning Commission, 1996).

In 1983 a comprehensive agricultural policy was built around crops and livestock (MoA, 1983a,b). The policy objectives made no mention of land resources management and conservation. The policy focus on land and water conservation was limited, emphasizing tree planting, protection of water sources and erosion control on steep lands.

A new agricultural and livestock policy was enacted in 1997, its goal being the improvement of the farmers' well-being. For the first time, national policy included a specific objective 'to promote integrated and sustainable use and management of natural resources such as land, soil, water and vegetation' (MoAC, 1997).

There followed six policy statements focusing on drought-resistant crops as a strategy for overcoming soil-moisture deficits. Management of rainwater for crop production was given very little mention in the policy. Strategies for concentrating, using and/or storing rainwater appear nowhere in the policy. Meanwhile, strategies for 'soil conservation and land use planning' focus on 'soil conservation' rather than 'soil and water management'. This reveals an important weakness – indeed, an inherent contradiction – in the policy. Although the policy provides excellent detailed guidance as to the management of soil and water in the rangelands, it ignores the main rain-fed cropping areas, where soil-moisture deficiencies are the main production constraint.

Government forestry policy

A forestry policy has existed in Tanzania since 1953. The central objective of the policy was to preserve forests for public interest (Legislative Council of Tanganyika, 1953). The policy focused on forestry management for the purpose of sustainability and meeting the needs of society. A new policy was approved in 1998 with the overall goal to 'enhance the contribution of the forestry sector to the sustainable development of Tanzania and the conservation and management of natural resources' (MTNR, 1998). Most NRM programmes in Tanzania have been implemented under the forestry policy.

Government water policy

Tanzania's current water policy dates back to 1991 (MWEM, 1991). The policy recognizes that large quantities of rainwater are lost without being utilized and that there should therefore be an emphasis on RWH through construction of dams and *charcos* in drought-prone regions and collection of water from roofs and storage in tanks. However, the policy has focused mainly on water supply for industry and domestic needs. Only limited Ministry of Water effort has been directed to the development and management of water for agriculture.

Approaches to policy implementations

Most of the NRM objectives in central government policies are pursued through programmes and projects implemented by non-governmental organizations (NGOs) and/or the government. Programmes in the semi-arid areas, including *Hifadhi Ardhi Dodoma* (HADO), *Hifadhi Ardhi Shinyanga* (HASHI) and *Hifadhi Mazingira Iringa* (HIMA) (Kerkhof, 1990), have focused on two major issues: soil-erosion control and soil-fertility improvement. The management of the runoff water for the purpose of increasing its productivity was not given consideration. Livestock were excluded in some areas because they were considered to be agents of erosion. These programmes have been oriented towards conservation for its own sake rather than for the purpose of increasing land productivity to benefit rural people (MTNRE and Sida, 1995:5).

Government programmes have also emphasized tree planting as the front-line soil-conservation measure. It is often not realized, or at least acknowledged, that trees compete heavily with crops and pasture for scarce soil moisture. From the rainfall-consumption point of view, water demands by trees are higher, because up to 25% of the rainfall may be intercepted and evaporated from the canopy, and evapotranspiration from trees is much higher than that from annual crops or pasture (Calder, 1994).

Tanzanian agricultural policy has had two major programmes: development and promotion of drought-tolerant crops and plant nutrition. Under the first, crops such as sorghum, millet and cassava have been recommended and supported in the semi-arid areas (MoAC, 1997). As a result, the focus of research and extension has mainly been on these crops. Meanwhile, farmers in the semi-arid areas have adopted the production of paddy rice at a very rapid pace (ICRA, 1991; Mashaka *et al.*, 1992; Meertens and Lupeja, 1996).

Perhaps the most striking feature of Tanzanian policy for the semi-arid areas, in agriculture, forestry, livestock and water, has been its consistent neglect of the water and associated soil-moisture constraints faced by farmers in these regions. As the next section describes, farmers have been very attentive to these constraints in spite of government inattention.

Farmers' Practices and Factors Influencing the Adoption of Rainwater Harvesting

Farmer strategies

Farmers in semi-arid areas are aware that both crop and livestock production can be improved substantially through concentration of scarce rainwater, as well as by provision of supplementary water during critical times. This strategy is manifest in the concept of '*Mashamba ya Mbugani*' (fields located at the bottom of the landscape). Farmers grow water-demanding crops, such as vegetables, rice and maize, in the lower part of landscape. The aim is to exploit the natural concentration of rainwater and

nutrients flowing into the valley bottoms from the surrounding higher lands.

This strategy differs from the government's strategy of focusing on drought-tolerant crops and trees. Critchley's (1999) survey of farmers' innovation in semi-arid areas of East Africa found that RWH innovations constituted 30% of all farmer innovations, while water management innovations more broadly comprised half of the total (Fig. 16.1). Forestry scored very low, at 4%, underscoring the divergence between policy focus and farmer practice.

RWH is widespread. A study conducted in the semi-arid WPLL of Tanzania showed that 70% of respondents adopted RWH (Table 16.1). RWH comprises a continuum of techniques, from traditional irrigation to *in situ* groundwater conservation, that focus on collecting rainfall runoff for cultivation. The systems being practised by farmers thus focus on the effective capture and management of rainwater. Three distinct management practices can be identified.

● *In situ* capture of rainwater where it falls and enhancement of its infiltration into the soil; the techniques normally used to achieve this are tillage, pitting and ridging.

● Collection, concentration and/or diversion of runoff into crop fields through catchment systems.

● Collection and storage of runoff for later use in crop fields.

In situ *rainwater harvesting*

In situ RWH, otherwise known as soil-water conservation (SWC), comprises a group of techniques for preventing runoff and promoting infiltration. Rain is conserved where it falls, but no additional runoff is introduced from elsewhere. *In situ* RWH using pitting, ridging and tillage is the most widely practised technique in the semi-arid areas of WPLL, practised on 32% of all plots cultivated by respondents (Table 16.2).

Micro- and macro-catchment

Micro-catchment RWH comprises a group of techniques for collecting overland flow

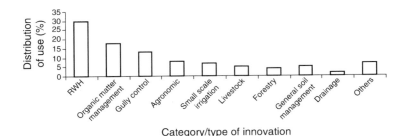

Fig. 16.1. Examples of actual practices by farmers in semi-arid areas (modified from Critchley, 1999:51).

Table 16.1. Adoption of rainwater harvesting according to topography (figures in parentheses are row percentages).

Slope	Without RWH	With RWH	Total
Uphill	25 (22)	90 (78)	115
Moderate	14 (27)	38 (73)	52
Flat and valley bottom	39 (42)	53 (58)	92
Total	78 (30)	181 (70)	259 (100)

(sheet or rill flow) and delivering it to a cropped area in order to supplement inadequate direct rainfall. This system involves a distinct division of catchment area (CA) from cultivated area (CB), but the two zones are adjacent. The transfer distance is typically in the range of only 5 m to 50 m. Both areas are normally within the landholding of an individual farmer and is therefore sometimes known as an 'internal catchment' system.

The short transfer distance ensures that the system offers relatively high runoff efficiency, possibly yielding as much as 50% of precipitation, compared with as little as 5% contribution to stream flow in a natural

catchment. The small catchment size ensures that the flow volume and speed are limited and soil erosion is therefore relatively easy to control. The main disadvantage of the system is that it involves leaving uncropped areas within the farmer's field. In evaluating the benefit, therefore, it is important to account for the opportunity cost of the uncropped areas. The system is rarely practised in the semi-arid areas of Tanzania – on only about 24% of plots cultivated by WPLL respondents (Table 16.2).

Macro-catchments generally lie outside the landholding of the farmer(s) using the runoff, so the system is sometimes known as an 'external catchment' system. The transfer distance may be in the range of 100 m to several kilometres. This distinct separation can be particularly beneficial if runoff events can be harvested at times when there is no direct rainfall in the cropped area. Runoff efficiency is normally less than for a micro-catchment system, but the large catchment area ensures that the runoff volume and flow rates are high. This sometimes gives rise to problems in managing potentially damaging peak flows, which may cause serious erosion and/or sediment deposition. Substantial channels

Table 16.2. RWH techniques used according to topography (figures in parentheses are row percentages).

Slope	*In situ* system	Rill flow	Diversion and spreading system	Diversion/ storage	Total
Uphill	28 (31)	10 (11)	16 (18)	36 (40)	90
Moderate	13 (34)	17 (45)	8 (21)	0	38
Flat and valley bottom	17 (32)	16 (30)	16 (30)	4 (8)	53
Total	58 (32)	43 (24)	40 (22)	40 (22)	181

Table 16.3. Household labour supply.

Household labour size (people)	Respondents	
	No.	%
One	12	15
Two	36	43
Three	16	19
Four	9	11
Five	4	5
More than five	6	7
Total	83	100

and runoff-control structures may be required. This usually involves collective construction and maintenance by a group of farmers, which sometimes gives rise to common-property problems over the management and distribution of water.

One finds many different macro-catchment RWH techniques in semi-arid Tanzania. Hillside systems reinforce the effects of gravity and slope through constructing cross-slope bunds and basins to intercept and store runoff, as in the *Majaluba* system of Sukumaland, or hillside conduits to redirect runoff towards cultivated lowland plots. Stream-bed systems use permeable stone dams or earth banks to spread flowing surface water across adjacent plots. Stream diversion, like spate irrigation, involves diverting water using a channel, small spillways to create a basin-to-basin cascade or a sequence of open trapezoidal or semicircular bunds.

RWH with storage systems

Macro-catchment RWH systems often yield high volumes of runoff. Simple reservoir systems, sometimes known as 'charco dams' or 'haffirs', have been used widely to store such runoff for water supply and livestock watering. In WPLL, 22% of the plots use RWH storage systems (Table 16.2). Siltation is often a problem and the labour required for sediment removal can be considerable. Household labour supply

therefore limits accessibility for the majority of respondents. Most WPLL households have two or fewer people available for farm work (Table 16.3). This helps explain why less labour-intensive *in situ* systems are the more widely practised RWH technique. *In situ* systems involve very few maintenance costs, and the cost of installation is relatively low compared with the other techniques.

Effect of cropping system on RWH and vice versa

Smallholder farmers in the semi-arid areas are mainly agropastoralists cultivating sorghum, millet, rice, groundnut and bambara nut. Four farming systems can be defined based on the intensity of cultivation of the dominant crops (rice, sorghum and millet) and livestock keeping: sorghum–livestock–millet (SLM), sorghum–livestock–rice (SLR), sorghum–rice (SR) and rice (R).

The SLM system once dominated the case-study areas of Dodoma and Shinyanga Districts. However, a more reliable water supply brought about by RWH has resulted in a shift towards the higher value of rice systems: R, SR and SLR. RWH use is high and precise under the R system and non-existent in the SLM system. Increased rice cultivation and more widespread use of RWH techniques are mutually reinforcing trends. The added expense of RWH is most easily justified for higher-value rice, which also requires more water than millet or sorghum does.

These patterns also emerge in the development of indigenous irrigation systems in the WPLL for the purpose of extending the planting season. This has created a third season outside the regular (short rains) Vuli and (long rains) Masika seasons. This season is called *champombe* or *chamazi*, which literally means growing season dependent on stream or stored water (*mazi* or *mpombe*) rather than direct rainfall. The supplementary irrigation system consists of three major components:

- the river/stream (source of water);
- a water-storage reservoir (*ndiva*);
- a water-distribution canals (*sasi*).

The system is prominent in both Mwanga and Same Districts, where a number of ponds have been constructed in various villages. In Same District, most of the irrigation supports vegetable production for cash or maize/legume intercropping. Farmers in Mgwasi village consider it impossible to harvest a crop of maize without supplementary irrigation. The *sasi* are an engineering marvel, sometimes up to 20 km long, with numerous branches. Most of the irrigated fields have stone bunds along the contour for the purpose of controlling erosion.

Positive and negative externalities of RWH

Risk of erosion

Treated catchments are prone to soil erosion because their construction and maintenance involve the removal of stones and vegetation. It is recommended that RWH programmes should emphasize the effective use of the runoff that is already occurring naturally. If it is necessary to clear an area for the purpose of enhancing runoff generation, then the catchment must be built so as to minimize erosion, for example by using stone bunds to divide the catchment into small subcatchments.

Flooding and salinization

There is often a risk of diverting too much water leading to damage of crops, fields and infrastructure. Some farmers have been reluctant to adopt RWH because the runoff amount and timing are difficult to predict. Flash floods can cause devastating damage to fields. The use of reservoirs to store a full day's rainfall can minimize this problem. The reservoir has a permanent water outlet to release water at a flow rate of minimum risk to the spreading bunds installed in the fields. The stored water drains away continuously until the reservoir is dry in a

day or two, ready to receive the next flash flood.

On the other hand, widespread use of *in situ* RWH techniques helps to reduce the amount of water reaching waterways, streams, rivers and lakes, thus reducing the risk of flooding. Vegetation cover is particularly necessary to increase the rate of soaking. Therefore, when RWH keeps vegetation growing during the dry season, it contributes to flood control. Terracing, contour ploughing and other means of *in situ* RWH can further reduce flooding.

Many salinity problems occur in arid and semi-arid regions, because evaporation far exceeds the rainfall amounts on both a seasonal and an annual basis, leaving salts on the ground surface. Over time, these salts accumulate and become harmful to plant growth. RWH can mitigate salinization problems by flushing excess salts.

Enrichment factor

Runoff always carries with it some solid particles and dissolved nutrients. The sloping areas tend to be drier and less fertile than the bottom lands or valley bottoms where water spreads and infiltrates. The 'enrichment factor' refers to the amount by which the eroded material is richer in nutrients than is the soil from which it is taken. In the semi-arid areas of Tanzania, the enrichment effect has been key to the sustainability of low-input agricultural production on the lower part of the catena, where most crop production is concentrated.

Effect of policy and policy changes

The role of policy and policy changes in regard to NRM adoption was investigated by asking groups of WPLL farmers to assess historical developments in SWC in relation to policy. Within the groups, in-depth discussions took place and a general consensus on the occurrence of the major events was reached. The farmer groups divided the history into five periods. These were pre-colonial (pre-1900), colonial (1900–1961),

post-independence (1961–1967), post-
Arusha declaration (1967–1985) and liberalization (1985 to the present).

Farmers' perceptions of the historical
development of SWC over the five periods
are summarized in Fig. 16.2, wherein
farmer's opinion regarding the expansion of
a given SWC practice is ranked on a scale of
0–4. A score of 0 means that the practice has
been entirely abandoned. For example, deep
tillage was abandoned during the liberalization period; hence it is given a 0 score. A
score of 1 during a given period means that
the farmers were of the opinion that the practice was just introduced during that period
or experienced very little expansion. A score
of 4 means that the practice expanded rapidly during that period. On the other hand, if
the practice was considered to have shrunk
during the period it was given a score of –1 to
–4, where –4 means rapid shrinking or even
disappearance. This assessment was made
using four questions:

1. Did the size of land covered by this
practice expand rapidly, stay the same or
shrink during the period?
2. Did the number of households practising or benefiting from this practice shrink or
expand?
3. Did the benefit accruing from this
practice increase or go down during the
period?
4. For the supplementary irrigation system, how well did it meet the demand?

It is evident that SWC is considered to
have expanded during the first three periods
and declined rapidly during the post-Arusha
declaration and liberalization periods. The
trends for each of the 13 SWC practices are
discussed in this subsection.

Precolonial period

Seven SWC techniques were practised during this period: supplementary irrigation,
stone bunds, trash lines, reserved forests
and bushes for traditional rituals, allocation
of grazing land and watershed protection.
The last four were the most dominant ones.
Reasons for the widespread adoption of the
practices, as given by farmers, were:

- Strong traditional rules, which were
  observed strictly.
- There were fewer people than in later
  periods, so land for agricultural production was enough to accommodate
  forest reserves and restrained use of
  rangelands and watersheds.
- Leaders were committed and
  trustworthy.
- People lived in clans that shared the
  same values and they valued their land
  highly.
- Young people were taught traditional
  values and taboos that they respected.

Colonial period

All seven SWC practices from the pre-
colonial period were maintained. Supplementary irrigation expanded further during
this era. Three more practices were introduced: live barriers, tree planting and
cut-off drains. The colonial period was the
'most successful' era in regard to SWC.
Contributing factors to this perceived
'success', as given by the farmers, included:

- Very strict laws, strictly implemented.
- Civilians (villagers) were fearful of the
  colonial government.
- Education and extension advice on
  SWC were readily available.

Post-independence period

The use of stone bunds, tree planting,
trash lines and allocation of grazing land
increased following independence. Live
barriers and cut-off drains were abandoned.
Traditional forest declined. The decline
is attributed to the attitude of politicians,
who associated SWC with colonialism. In
some cases, campaigns for independence
included promises to the effect that after
independence there would be no forceful
implementation of SWC, as was the case
during colonial time. This eroded the power
of extension workers and village leaders
in enforcing rules and regulation on SWC.
Other factors identified by farmers
included:

- The collapse of the good working relationship that existed between extension workers and village leaders.

- More politics and little action in SWC.
- The period also coincided with years of drought.

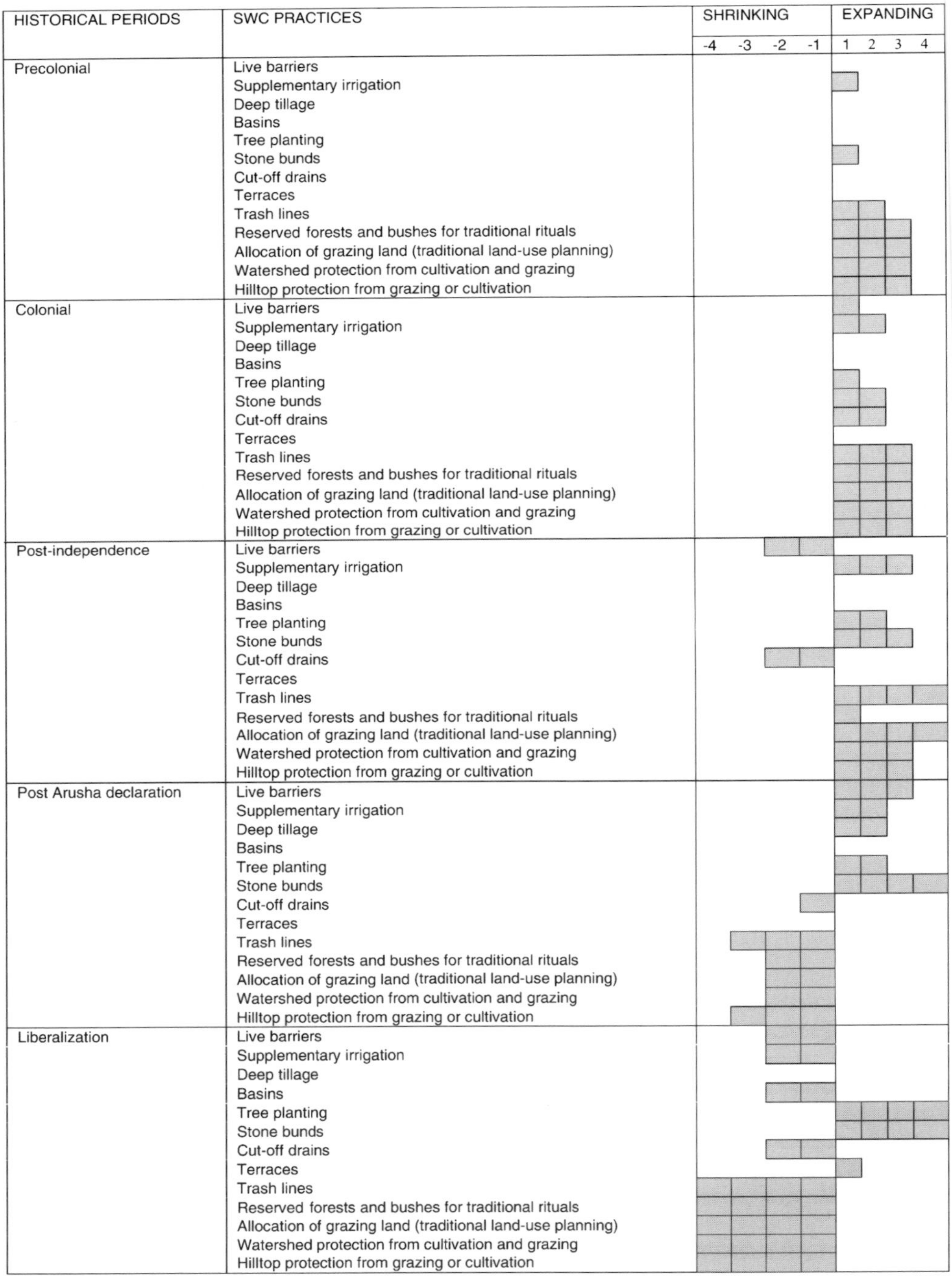

HISTORICAL PERIODS	SWC PRACTICES	SHRINKING −4 −3 −2 −1	EXPANDING 1 2 3 4
Precolonial	Live barriers		
	Supplementary irrigation		
	Deep tillage		
	Basins		
	Tree planting		
	Stone bunds		
	Cut-off drains		
	Terraces		
	Trash lines		
	Reserved forests and bushes for traditional rituals		
	Allocation of grazing land (traditional land-use planning)		
	Watershed protection from cultivation and grazing		
	Hilltop protection from grazing or cultivation		
Colonial	Live barriers		
	Supplementary irrigation		
	Deep tillage		
	Basins		
	Tree planting		
	Stone bunds		
	Cut-off drains		
	Terraces		
	Trash lines		
	Reserved forests and bushes for traditional rituals		
	Allocation of grazing land (traditional land-use planning)		
	Watershed protection from cultivation and grazing		
	Hilltop protection from grazing or cultivation		
Post-independence	Live barriers		
	Supplementary irrigation		
	Deep tillage		
	Basins		
	Tree planting		
	Stone bunds		
	Cut-off drains		
	Terraces		
	Trash lines		
	Reserved forests and bushes for traditional rituals		
	Allocation of grazing land (traditional land-use planning)		
	Watershed protection from cultivation and grazing		
	Hilltop protection from grazing or cultivation		
Post Arusha declaration	Live barriers		
	Supplementary irrigation		
	Deep tillage		
	Basins		
	Tree planting		
	Stone bunds		
	Cut-off drains		
	Terraces		
	Trash lines		
	Reserved forests and bushes for traditional rituals		
	Allocation of grazing land (traditional land-use planning)		
	Watershed protection from cultivation and grazing		
	Hilltop protection from grazing or cultivation		
Liberalization	Live barriers		
	Supplementary irrigation		
	Deep tillage		
	Basins		
	Tree planting		
	Stone bunds		
	Cut-off drains		
	Terraces		
	Trash lines		
	Reserved forests and bushes for traditional rituals		
	Allocation of grazing land (traditional land-use planning)		
	Watershed protection from cultivation and grazing		
	Hilltop protection from grazing or cultivation		

Fig. 16.2. Historical trends of SWC in the study area.

Post-Arusha declaration

There was a decline of most of the SWC practices and especially those involving reserve lands. Use of cut-off drains increased a bit, while live barriers were revived. Farmers attributed this decline to the following reasons:

- The villagization programme, through which people were uprooted from their traditional areas and concentrated in new residential and agricultural lands.
- Abolition of true local village governments, with leaders chosen by higher authorities of the central government.
- The decentralization programme, which reduced people's power.

There was introduction and expansion of deep tillage and basins. Expansion of these practices was attributed to:

- Extension of loans to villages for the purchase of tractors facilitated deep ploughing.
- Availability of extension workers and therefore their advice.
- Learning from neighbours in the new villages.

Liberalization

During this period, only three SWC practices from the colonial era survived: tree planting, stone bunds and supplementary irrigation, especially for vegetable production. Terraces were also introduced during this period. The main reasons behind the decline in the use of most SWC included:

- Reduction in the number of extension workers.
- Reduced labour force in many households.
- Rampant corruption and weak enforcement of by-laws.

The negative score given to supplementary irrigation does not mean a real decline but rather the failure of the system to meet the high demand.

A synthesis of these findings shows that, in private fields, SWC was found to thrive in a functional market economy and when there is a need for intensification of agricultural production. Important elements of the functional market economy were identified.

- Availability of social services, especially if they are commercialized.
- Easy access to consumer goods.
- Reliable markets and stable prices for agricultural outputs.
- Availability of good-quality inputs at stable prices.

With a functional market economy in place, farmers seek to produce more. Where this cannot be achieved by expanding the area cultivated, intensification becomes necessary. This creates a conducive environment for increased SWC practices. These findings agree with other findings from Africa (Tiffen *et al.*, 1994; Reardon *et al.*, 1999).

Contrary to much received wisdom in academic circles, strict customs regulations and their uncompromising enforcement were identified by farmers to be most important for promoting SWC, especially in communal areas. This was made clear by nearly all the interviewed farmers, who were more concerned with the corruption that hinders enforcement of by-laws, rather than the by-laws themselves.

Effect of external interventions on diffusion of technologies

People in rural areas relate to the macro-policies through the external interventions and assistance they receive as a result of these policies. Therefore, the perceptions on how external interventions and assistance affect adoption of SWC were also investigated.

About 61% of WPLL respondents said they have never received any assistance for SWC (Table 16.4). For the 39% who said they had received assistance, the assistance included extension, training, excursions, and credit or aid (Table 16.5). For the same group the identified sources of assistance were NGOs, extension workers, schools, neighbours and government (Table 16.6). Of

Table 16.4.   External interventions/assistance to promote SWC.

Intervention/assistance received	SWC use	Number of respondents	%
None	None of SWC methods	49	25
	One or two methods	69	35
	More than two methods	3	1
	Subtotal	121	61
Some	None of SWC methods	12	6
	One or two methods	62	31
	More than two methods	4	2
	Subtotal	78	39

Table 16.5.   Types of interventions/assistance provided for SWC.

Type of assistance	Number of times mentioned	%
Training	45	37
Extension	48	40
Excursions	14	12
Credit	6	5
Aid	7	6
Total	120	100

Table 16.6.   Identified sources of assistance given for SWC.

Source of assistance	Number of times mentioned	%
Neighbours	20	20
Extension	22	21
NGOs	38	38
Government	5	5
Schools	9	9
Others	7	7
Total	101	100

the 61 respondents who have not adopted SWC, 80% have not received any assistance on SWC, underscoring the importance of some form of assistance, typically through excursions or extensions to stimulate SWC adoption (Table 16.6). At the same time, among SWC adopters there was little evidence of differences between those who received some external assistance and those who did not. This is consistent with a farmers' innovation study (Critchley, 1999: 52) showing that most of the SWC practices result from the farmers' own initiative and needs.

Further, the source or types of external interventions have an effect on adoption levels (Table 16.7 and 16.8). For example, financial and material aid was the least effective, with 36% of those who received aid not adopting any SWC practices. The reason for this may be the reluctance by farmers to put cash into SWC, as discussed later. Training was the second least effective, with 18% of those receiving training ending up not adopting any practice. The failure to adopt was only 8% and 7% for those receiving extension and excursions, respectively (Table 16.8). Much of this 'training' was seminars held in classrooms, with very few practical activities. Excursions were more effective, as farmers tend to adopt practices better after seeing them working on other farmers' fields.

Most (88%) villagers identified their role in SWC as providing only labour (Table 16.9), while 60% of the respondents considered the supply of inputs and SWC construction materials to be the role of government (Table 16.10). A similar pattern is observed in relation to the perceived role of NGOs (Table 16.11).

Most farmers consider their responsibility in SWC to consist of:

● taking care of the individual farm and providing labour for SWC;
● cooperating with others and adopting improved farming practices.

Table 16.7. Influence of source of assistance on SWC adoption.

Source of assistance	SWC use	Number of respondents	%
Neighbour	None of SWC methods	5	25
	One or two methods	14	70
	More than two methods	1	5
	Total	20	100
Extension worker	None of SWC methods	3	14
	One or two methods	15	68
	More than two methods	4	18
	Total	22	100
Non-governmental organization	None of SWC methods	6	16
	One or two methods	29	76
	More than two methods	3	8
	Total	38	100
Government	None of SWC methods	1	20
	One or two methods	3	60
	More than two methods	1	20
	Total	5	100
School	None of SWC methods	2	25
	One or two methods	5	63
	More than two methods	1	12
	Total	8	100
Others	None of SWC methods	1	12
	One or two methods	7	88
	More than two methods	0	0
	Total	8	100

Table 16.9 shows that the overwhelming majority (88%) consider their main role in SWC to be taking care (agronomically) of the field, as well as providing labour required for implementing SWC practices. Few farmers expressed a willingness to invest cash or materials in SWC (see Table 16.12). These responses contradict sharply with those shown in Tables 16.4–16.8, which demonstrate that a lot of SWC has been undertaken with little financial assistance from outside. In fact, financial and material aid was shown to be the least effective intervention for promoting SWC. It is difficult to ascertain which is the true picture.

The contradiction can be best explained by the tendency among respondents to give an impression that they need aid whenever cash is required. Visits made to farmers' fields revealed that many had indeed made substantial investments in SWC, but these have not been revealed in the answers to the questionnaire. Farmers' investments are often very high compared with the assistance received. The question still remains as to why farmers who eventually invest considerable labour do not do it until an NGO or a project has given them assistance that is worth less than 1% of their overall investment. There are two possible explanations.

1. The opportunity value of cash in hand is much higher than that of labour, which is not easily sold.
2. Where the project is risky and the level of labour investment is very high, the farmers seek somebody else to underwrite the

Table 16.8.   Effect of type of assistance on SWC adoption.

Type of assistance	SWC Use	Number of respondents	%
Training	None of SWC methods	8	18
	One or two methods	36	80
	More than two methods	1	2
	Total	45	100
Extension	None of SWC methods	4	8
	One or two methods	40	84
	More than two methods	4	8
	Total	48	100
Excursions	None of SWC methods	1	7
	One or two methods	11	79
	More than two methods	2	14
	Total	14	100
Financial and material aid	None of SWC methods	4	36
	One or two methods	5	46
	More than two methods	2	18
	Total	11	100

Table 16.9.   Farmers' perception of owners responsibility for SWC.

Responsibility	Number of respondents	%
Providing labour and caring for the farm	122	88
Cooperating with others	3	2
Adopting improved farming practices	9	7
No responsibility	3	2
Total	137	100

Table 16.10.   Farmers' perception of government responsibility for SWC.

Responsibility	Number of respondents	%
Enactment of laws and regulations	28	23
Supply of input and materials	73	60
Plans and education	18	15
No responsibility	3	2
Total	122	100

Table 16.11.   Farmers' perception of role of NGOs and private organizations in promoting SWC.

Responsibility	Number of respondents	%
Supply inputs	21	23
Plans and education	72	77
Total	93	100

risk. The involvement of an NGO gives the farmers more confidence in choosing to invest in any activity.

Discussion and Conclusions

There has been a gap between the emphasis given in national policies and programmes and what farmers really practise in semi-arid areas. While government has, for example, focused on drought-tolerant crops and erosion control, farmers have directed their efforts to the effective management of rainwater for the production of highly water-demanding but high-value crops, such as rice and vegetables. Previous scientific studies have shown that conservation of rainwater is more important in semi-arid areas (Stocking, 1988). The case-studies

Table 16.12. Identified farmer investment priorities for Tshs 50,000/– own money.

Investment priority	Hedaru		Mgwasi		Total	
	Number of times mentioned	%	Number of times mentioned	%	Number of times mentioned	%
Agriculture improvement	57	26	36	38	93	30
House	70	32	17	18	87	28
Livestock	26	12	23	25	49	16
Farm tools	39	18	7	8	46	15
SWC	8	4	5	5	13	4
Bicycle	8	4	4	4	12	4
Fertilizer	9	4	2	2	11	3
Total	217	100	94	100	311	100

Table 16.13. Comparison of the socio-economic periods in relation to implications for SWC.

Period	Pre-1967	Post Arusha Declaration	Liberalization
Rules and regulations and their enforcement	Strict customs, rules and by-laws Non-compromising enforcement	Liberal rules and by-laws Neglect of local institutions Lax enforcement	Liberal rules and by-laws Neglect of local institutions Lax enforcement
Direct economic benefits to the individual	People needed cash to purchase nearly all services and to pay poll tax High benefits perceived due to existence of markets Individual property highly valued	Individual benefits and wealth were discouraged Lower need for cash as poll tax was abolished and services were 'free' Poor marketing system	Individual benefits and wealth are encouraged Poll tax and payment for services have been reintroduced Markets for crops have been liberalized and function better
Implications for SWC practices	Relatively high rates of SWC adoption and implementation	Relatively little SWC adoption and implementation	High success in approaches emphasizing water availability for production of high-value cash crops, such as rice and vegetables

provide evidence that farmers also find this to be the case. Farmers in semi-arid areas have been searching for ways to enhance the productivity of rainwater. As a consequence, farmers have adopted farming systems that provide these benefits, often contrary to policy statements, as shown by the concentration of water for the production of paddy rice, while government policy is advocating drought-tolerant crops.

Factors influencing adoption are categorized into several broad categories: technology, biophysical, farmer characteristics, extension assistance, market incentives and rules. Based on technology characteristics, the *in situ* system consisting of deep tillage and ridging was the most widely adopted technology. Farmers regard deep tillage as having high potential for supplying water to crops. The second best technology under this criterion is the collection of water from rills and/or sheet flow. The most important biophysical characteristic was the topography/location of the farms. There is more adoption of RWH in areas further uphill than in areas further downhill. This is partly because most of the runoff is taken off in the higher areas and there is not enough to meet the demand in the areas below. Household labour available for farm work is an important farmer characteristic influencing adoption. Farmers feel that labour requirements for *in situ* systems are relatively lower than for other techniques. However, availability of labour exchange and draught power also contributes to adoption of SWC techniques.

Farmers' SWC practices also differ significantly across five major socio-economic periods through which Tanzania has passed, due to variation in resource-use rules and regulations and their enforcement and in economic benefits to the individual farmer (Table 16.13).

SWC adoption was much better in the pre-1967 period. This does not mean to say that the force used during the colonial period to enforce SWC should be reintroduced, but rather that fair but strict customs, rules and by-laws appear necessary. These factors deteriorated during the post-Arusha declaration period and, in the current liberalization period, only one (economic benefits to the individual) of the factors has improved. Consequently, farmers are now implementing SWC in a highly selective manner, focusing on those approaches that improve water availability for high-value crops, such as rice and vegetables. For example, in the Shinyanga case-study, liberalization has led to a shift from cotton production to an accelerated practice of RWH for paddy-rice production.

In conclusion, we find that:

1. Sustainable adoption of SWC practices requires policies and strategies that ensure fair but strict customs, rules and by-laws on SWC and direct tangible benefits to the individual.
2. Farmers in the semi-arid areas have shown that they prefer SWC technologies that emphasize the management and conservation of the scarce rainwater. There is a gap between farmers' practices and government policy objectives, strategies and programmes.
3. Technology, biophysical and farmer characteristics are all important in influencing the adoption of RWH technology. The most important technology characteristics are potential to provide water and simplicity. The key biophysical characteristics are the topography and availability of a water source. Family labour availability and access to farm power are important farmer characteristics influencing adoption.
4. There is an urgent need to reorient SWC strategies pursued by government in semi-arid areas so as to focus less on drought-tolerant crops and tree planting and more on practices such as RWH that are clearly preferred by farmers.

Note

[1]   This chapter is an output from projects funded by the UK Department for International Development (DFID) for the benefit of developing countries. The views expressed are not necessarily those of DFID.

17 Initiatives to Encourage Farmer Adoption of Soil-fertility Technologies for Maize-based Cropping Systems in Southern Africa[1]

Mulugetta Mekuria and Stephen R. Waddington

CIMMYT-Zimbabwe, PO Box MP163, Mount Pleasant, Harare, Zimbabwe

Southern Africa combines old soils with resource-poor smallholder farmers. Low soil fertility is one of the major factors contributing to the low productivity and non-sustainability of the existing production systems in southern and eastern Africa (see van Reuler and Prins, 1993; Blackie, 1994; Blake, 1995; Kumwenda et al., 1996, 1997; Sanchez et al., 1997a,b; Waddington and Heisey, 1997). This is a prime reason for increased poverty and household food insecurity in the region. Soil infertility is also a major source of inefficiency in the returns to inputs and management committed to smallholder farms, including N fertilizer (Mushayi et al., 1999) and labour. Accordingly, ways to reduce and manage soil infertility have received major attention from agricultural research and development agencies and donors in recent years.

Mixed maize + legume + livestock systems dominate in the subhumid and some semi-arid zones of the region. Because soil infertility is a widespread constraint in these dominant farming systems, many research and development institutions have attempted to address this issue. In the 1990s in particular, intense work was undertaken to develop appropriate technology recommendations and approaches to achieve the wider diffusion of soil-fertility management interventions and enhance their adoption. Different research projects dealing with mineral nutrient management, organic + inorganic combinations, legumes and other soil-fertility technologies have developed viable sets of practices. To be acceptable to farmers, soil-fertility technologies have to integrate well into the existing farming system and offer something new. This integration involves not only compatibility with current practices and inputs but also, where possible, technologies that smallholder farmers judge to have several uses for them. Farmers are looking for ways to combine technology inputs, employing them in ways that minimize the requirements for additional cash, labour and land.

Developing and transferring these technologies and approaches is an enormous challenge. Given the limited financial and human resources available, organizing them within regional networks enables research, extension staff and farmers to jointly identify researchable problems and develop collaborative research programmes, whose benefits can be shared by network partici-pants. More than 15 agricultural research networks are operational in southern Africa. Since 1994, the International Maize and Wheat Improvement Centre (CIMMYT) has been working with national partners on a range of soil-fertility natural resource man-agement issues for maize-based smallholder farming systems in southern Africa. Most of this work has been conducted within two networks, coordinated at CIMMYT-Zimbabwe. These are the Soil Fertility Management and Policy Network for Maize-based Farming Systems in Malawi, Zimbabwe and Zambia (Soil Fert Net) and the Maize and Wheat Improvement Research Network for the Southern African Develop-ment Community (SADC) (MWIRNET). While some significant farmer adoption of soil-fertility technologies developed through these networks is starting to be recorded, in most cases we are at earlier stages of developing farmer awareness of technology options and learning about and addressing constraints to adoption. However, many smallholder farmers are very interested in some of the technol-ogies and their potential adoption is substantial.

In this chapter, we present the experi-ences of our regional networking efforts to promote soil-fertility technologies for maize-based systems of smallholder agri-culture in southern Africa, with special reference to Zimbabwe and Malawi. These technologies are the result of many major research initiatives undertaken by Network members, following problem diagnosis, process and adaptive research on farm and incorporating farmer assessments of tech-nologies. Some of the technologies found promising by farmers have recently been promoted widely. In the following sections we outline the more promising technologies, examine their adoption potential, describe methods used in their promotion with farmers and the support required to help adoption, and provide empirical evidence of adoption from selected case-studies in Zimbabwe and Malawi.

The Promotion and Adoption of Best-bet Soil-fertility Technologies

Since it began late in 1994, one of the main aims of Soil Fert Net has been to develop a range of organic and inorganic soil-fertility technology options for smallholders. The technologies have resulted from wide-spread participatory research and testing with the farmers on their farms in Malawi and Zimbabwe.[2]

In more detail, the criteria used in the selection of best-bet technologies have included:

- Longer-term contribution to raising soil fertility.
- Ability to raise crop yields and generate profit in the short term (1–2 years).
- Appropriate for many farmers across important agroecologies.
- Compatibility with other components of the farming system.
- Small additional cash and/or addi-tional labour requirements.
- Only a small reduction in maize yields or substitution by production of other crop.
- Where possible, little competition for arable land.

Technologies meeting most of these cri-teria should be adoptable by farmers. Most of the technologies provide some short-term soil-fertility and crop-productivity benefit and several end uses, which makes them attractive to farmers. They are compatible with farmer circumstances and effective within farmer resource constraints (cash, labour and land). Thus, these technologies offer farmers the 'best-bets' for improved productivity, sustainability, useful products and income.

In mid-1999, Soil Fert Net members held a workshop in Malawi to take stock of how far we have come with the development of best-bet soil-fertility technologies (Giller, 1999; Soil Fert Net, 1999). We identified 12 technologies as ready for promotion: seven for Malawi and five in Zimbabwe (Table 17.1). In addition, members listed six technologies (also in Table 17.1) that require some additional verification and testing.

Adoption of best-bet technologies

Table 17.1 also gives our estimated potential adoption of the best-bet technologies in Malawi and Zimbabwe. Up to the year 2000, most of the technologies have been adopted by few farmers, except for those that are existing farmer practices but are being more widely promoted into new agroecologies or improved upon, notably pigeon pea + maize intercrops in southern Malawi and groundnut + maize rotations in northern Zimbabwe. However, the potential of the best bets for adoption and impact is massive (Table 17.1).

Many of the technologies available for farmers to manage soil fertility involve the use of legumes. Despite the potential of legumes to put N and organic matter into cropping systems, there are performance, management and acceptability difficulties (and opportunities) with them on smallholder farms, which have been widely documented recently (e.g. Giller *et al.*, 1994, 1998, 2000; Kumwenda *et al.*, 1996; Hikwa and Waddington, 1998; Snapp *et al.*, 1998).

The supply and maintenance of appropriate legume seed remains a big issue. Reasons for low growth and yield of legumes on smallholder farms include limiting P supply, low pH, poor stand establishment and the high labour cost for weeding.

For many farmers there is often a conflict between the short-term requirement to meet today's food supply and building up the long-term fertility of the soil to meet tomorrow's food needs. Farmers discount the value of a benefit that will only be achieved in several years' time from investments made today. Those legume systems that are best as soil improvers (such as hedgerow intercrops, green manures and improved fallows) tend to have few other uses. They usually do not provide human food or cash. Additionally they occupy, for 1 or more years, land that farmers could plant to food crops. Consequently they are less likely to be adopted by farmers (Fig. 17.1)

unless given significant support on seed supply, provision of information and perhaps even fertilizer or labour – something that is difficult to do with many farmers simultaneously. Broadly speaking, the larger the likely soil-fertility benefit from a legume technology, the larger the initial investment required in labour and land and the fewer short-term food benefits it has. Thus, when seeking soil-fertility legume technologies that farmers will use, it is important to get assessments from farmers about their interest in and ease of adoption of a technology.

Among the legumes, annual grain legumes offer a good compromise for promoting farmer adoption (by providing some grain and sometimes leaf for human food and animals or for sale), and improving soil fertility (fixing some N and having reasonable shoot and root biomass for incorporation into soil organic matter) (see Kumwenda *et al.*, 1996; Giller *et al.*, 1998; Snapp *et al.*, 1998). Self-nodulating promiscuous soybean types with a small N and C harvest index and medium- and long-duration pigeon pea, groundnut, dolichos bean and cowpea are among the most promising in Malawi, Zimbabwe and Zambia as intercrops or sole-crop rotations with maize.

Opportunities to use grain legumes in ways that also help to reduce other production problems – such as the associated control of *Striga* and other weeds from a cowpea intercrop with maize – need to be exploited, because these will improve adoption prospects. New smallholder grain-legume cash crops, such as promiscuous soybean, also provide cash for farmers to buy mineral fertilizers for maize (Giller *et al.*, 2000).

Nevertheless, it is clear that smallholder farmers will take up even complex soil-fertility natural resource technologies if given sufficient support. An excellent example is improved *Sesbania sesban* and *Tephrosia vogelii* 2-year fallows in Eastern Province in Zambia. Such practices can triple the grain yield of the following maize crops (Kwesiga and Coe, 1994) and have economically attractive returns to labour (Kwesiga, 1998). Yet these technologies should be relatively difficult for farmers to adopt, since they require the farmer to forgo

Table 17.1. Soil Fert Net's 'best-bet' soil-fertility technologies for smallholder maize-based farming systems and a preliminary assessment of their adoption potential.

Technology or cropping systems	Target agroecology	Farm type	Ease of adoption*	Adoption potential† (number of farmers)
Malawi				
Soil-fertility technology				
Area-specific fertilizer recommendations for maize	All areas by soil type and market or home use	Richer and middle-income farmers	++	900,000
Optimum combinations of organic and mineral fertilizers	Most of Malawi		++	1,000,000
'Magoye' promiscuous soybean	All mid-elevation areas	Richer cash croppers	++	300,000
Fertility-enhancing cropping system				
Groundnut in rotation with maize, and pigeon pea intercropped with other grain legumes	All mid-elevation areas	Medium to large holdings	+++	400,000
Tephrosia undersowing of maize	Mid-elevation and lakeshore areas	Medium to large holdings	++	400,000
Mucuna + maize rotations	Most of Malawi, poorer soils	Medium to larger holdings	+	200,000
Faidherbia albida trees in crop land	Adaptation range (500–1000 masl)		++	500,000
Sesbania undersowing	Mid-elevation areas	Larger holdings	+	100,000
Pigeon pea + maize intercropping	South and central Malawi	Smaller holdings	++++	1,000,000
Soil fertility × *Striga* interactions	*Striga*-affected areas		+++	150,000?

Zimbabwe

Soil-fertility technology	Recommendation domain	Area	Adoption	Estimate
Fertilizer management package for maize (conditional on rainfall) and grain legumes	All except poorest farms in driest areas	Subhumid and semi-arid areas	+++	1,000,000
Liming on acidic sandy soils	Higher-input farms	Acidic soils in subhumid areas	++	300,000
Optimum combinations of organic and mineral fertilizers		Subhumid and wetter semi-arid areas	++	600,000
Improved cattle-manure management, including anaerobic composting	Farmers with cattle	All, except driest areas, where farmers reluctant to use manure	+	250,000
Fertility-enhancing cropping system				
Pigeon pea rotations and intercropping		Subhumid areas	++	150,000
Soybean (inoculated and promiscuous) in rotation with maize	Cash-crop farmers	Subhumid areas on better soils	+++	300,000
Mucuna + maize rotations		Subhumid areas	+	100,000
Other grain–legume rotations		Subhumid and wetter semi-arid areas	+++	700,000

*+, low; ++, moderate; +++, high; ++++, extremely high.

†From key informants. Estimated by the Soil Fert Net coordinator, using information from Soil Fert Net members, with amendments from Ken Giller and Webster Sakala.

masl, metres above sea level.

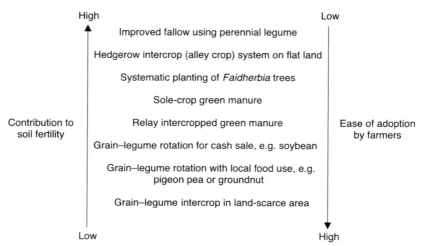

Fig. 17.1. Approximate contribution to soil fertility vs. ease of adoption for a range of legume technologies for maize-based smallholder systems in southern Africa.

a maize crop for 1–2 years and often require the nursery-rearing, planting and weeding of a non-food crop. However, the number of farmers testing the technology rose to over 4000 in 3 years to 1997 and has expanded since. This was due to a combination of careful agroecological targeting[3] and a major international and national research and development input, involving identifying and addressing constraints to farmer use, training and mobilization of the extension and non-governmental organization (NGO) community to demonstrate to farmers, provide information and inputs and encourage farmer experimentation, modification and farmer-to-farmer transfer (see Kwesiga, 1998).

Finally, it is very clear that uptake of soil-fertility technology is easier when there is good access to markets. This is well illustrated by smallholder soybean in Zimbabwe. Since about 1996, there has been a major promotion drive on soybean in smallholder areas of Zimbabwe, involving a task force from the Department of Agricultural and Technical Services (AGRITEX), the University of Zimbabwe, farmers' unions and private-sector input suppliers and grain processors (Pompi *et al.*, 1998). This has built up from 55 farmers in 1996/97 to where in 1999/2000 over 10,000 smallholders grew soybean on 4000 ha, selling 4000 tons.

Information was provided on how to grow soybean on smallholder soils. Once it can be grown well, its economics was clear, since it requires little fertilizer. Although average yields are 800 kg ha^{-1}, it is still economic, with better returns to cash and labour than with maize, and farmers are keen to grow it. Its production by smallholder farmers will help the national economy, soybean importers and smallholders themselves. The development and support of input and output markets and home utilization through the task force has been key to its success. The private sector (including seed and fertilizer companies) has been active in the supply of soybean seed to smallholder areas, and fertilizer and lime. Produce markets were assured through Olivine Industries, a major oil processor, who agreed to take a quota from smallholder areas and helped with collection of the grain through local traders. This also meant that farmers could get a better price than through the normal sales to the Grain Marketing Board. However, a recent study on economic potential concluded that rural distribution and assembly costs need to be reduced, while better communication between partners, and better local trader access to capital all needed to be addressed to continue the build-up of smallholder soybean production (Rusike *et al.*, 2000).

Promotion of best bets

Throughout 1999–2001, members of the Soil Fert Net have been promoting the best bets with farmers in many parts of Malawi and Zimbabwe. Several of the methods we have used are described below.

Promotion involves a range of partnerships with government extension services, farmer groups and NGOs in Malawi and Zimbabwe to allow widespread coverage. Many involve farmer participatory approaches to the testing and modification of the technologies. Some of the more intensely participatory initiatives involve green manures and other legumes that may be a challenge to adopt. Studies have been undertaken to prepare the groundwork for adoption of soil-fertility technologies, by getting farmer feedback on the technologies on how they see these fitting into their systems and what types of support they need.

Our case-study experiences in Zimbabwe and Malawi show that farmers have many knowledge gaps with soil-fertility technologies, but often extension workers (and many NGO staff) are not well placed to help. Most do not understand the soil-fertility technologies and the farmer participatory demonstration and experimentation methods. Training is needed to develop the capacity of partners. The concept of farmer field schools for extension and farmer training will be useful. Additionally, there is need to train soil-fertility input stockists.

Larger-scale national commodity task forces (on maize in Malawi and soybean in Zimbabwe) have proved to be highly effective ways of focusing awareness, resources and partnerships on national initiatives to address soil-fertility issues and disseminating some of the best bets. Soil Fert Net members have been the leaders in a smallholder Soybean Promotion Task Force in Zimbabwe that has involved researchers, extension, farmers' unions and crop-processing companies. This has led to the adoption of soybean by 10,000 smallholder farmers over 4 years. For several years in the late 1990s, the research and extension services within the Malawi Maize Productivity Task Force mounted thousands of on-farm demonstrations and provided thousands of brochures of area-specific fertilizer recommendations for maize and the use of a range of legumes throughout the country (Kumwenda, 1998). The task force helped the area-specific fertilizer recommendations to be accepted by the extension service in 1997 and their policy implications to be assessed with government. These more flexible recommendations are now promoted nationwide.

Additionally, Soil Fert Net members within the Maize Task Force provided the technical input on expected benefits from technology options and helped develop input support strategies for a nationwide initiative to give fertilizer and maize- and legume-seed starter packs (Mann, 1998) to all 1.8 million smallholder households in Malawi during the 1998/99 and 1999/2000 cropping seasons. Collectively, the government of Malawi, the UK's Department for International Development (DFID), European Union (EU) and World Bank provided over US$23 million for this programme in 1998/99. It has had a major impact on human nutrition and household food security in Malawi and is an excellent example of where technical scientists have influenced government and donor policy.

Because farmers and their advisers know so little about some of the technologies, information provision is vital. Soil Fert Net members are producing and distributing a range of information brochures on the best bets.

Natural resource economics and policy support

In the current socio-economic situation in southern Africa, research needs to generate and present information that will help to justify policy decisions about input support and public-good external investments into inputs such as lime, fertilizer and legume seed for smallholder farmers. Soil Fert Net is developing economic and policy information and advocacy to further assist with adoption. In 1999, Soil Fert Net set up an Economics and Policy Working Group to provide:

- A holistic framework for closer inter-action between soil-fertility experts, economists, extensionists and policy-makers on strategies for solving the soil-fertility problems of farmers in a manner that involves increased farmer participation and all stakeholders.
- Objective economic evaluation of the existing best-bet technologies for soil-fertility management.
- Priority setting and targeting of potential best bets for smallholder farmers.
- Policy research and advocacy to create an enabling policy environment to promote farmers' use of improved soil-fertility management technologies.
- Strategic and relevant partnership for scaling up the work of the Soil Fert Net.

Integrated multidisciplinary studies will help us learn more about off- and on-farm views and constraints to the adoption of soil-fertility technologies and help us better prepare for adoption.

In the following section we present initial findings of detailed case-studies, where many of the methods and approaches just described for the extension of soil-fertility technologies have been used. The first study, from Chihota, Zimbabwe, describes farmer assessments of the suita-bility of best-bet technologies and gives initial information on their adoption potential. The Murewa (Zimbabwe) and Zomba (Malawi) studies present an eco-nomic analysis of soil-fertility technologies and factors affecting adoption.

A Case-study on Promotion of Soil-fertility Technologies: Chihota Communal Area, Zimbabwe

Each of the 20 AGRITEX Agricultural Extension Workers (AEW) in Marondera District worked with one or more farmer groups of about 15 farmers each to implement 105 best-bet soil-fertility tech-nology demonstrations and experiments

throughout Chihota and Svosve Communal Areas, near Harare, in 1998/99 and in 1999/2000. The objectives of the project are as follows:

- Expose about 4000 farmers to a range of best-bet soil fertility technologies from research by mounting a set of on-farm demonstrations throughout Chihota during 1998–2001.
- Bring those farmers, extension and research closer by involving the farm-ers in the assessment of technologies and providing a facility for feedback on generated technologies to extension and research.
- Encourage farmer adoption, experi-mentation and integration of the most acceptable technology into the farming system.

The project covers the following technologies:

- Maize liming and fertilizer;
- Soybean rotation;
- Groundnut rotation;
- Bambara nut rotation;
- Velvet bean green-manure intercrop with maize;
- Velvet bean green-manure sole crop;
- Sunnhemp green-manure intercrop with maize;
- Sunnhemp green-manure sole crop.

Many of the demonstrations involved the liming of maize fields that were selected beforehand because of their low pH. Farmers have been testing these options, incorporating them into their cropping systems and providing feedback on their suitability and benefits.

An initial survey with farmer groups in Chihota in September 1998 developed farmer, climate, soil and soil-fertility man-agement taxonomies of the target farmers (Bellon et al., 1999). Extension workers were trained in the technologies and conduct of participatory demonstrations. These in turn worked with farmer groups to implement the demonstrations/experiments. The extension approach emphasizes group experimenta-tion and learning, and farmer-to-farmer transfer of information. Farmer groups are using songs and drama to distribute messages. Some 60 field-days and mid-season and end-of-season evaluations provided the farmers, extensionists and researchers with opportunities to assess the technologies in the demonstrations. Group and individual interviews were conducted during the cropping season in 1999 and 2000 to get farmers' opinions and feedback on these technologies. End-of-season replanning workshops for all stakeholders were held in mid-1999 to discuss the results of the demonstrations, paying particular attention to what the farmers thought about the various technologies, their adoption and modification (Gambara *et al.*, 2000).

Additionally, the project has attempted to reach beyond members of the experimentation groups by broadcasting programmes about soil fertility on the radio, organizing look-and-learn tours for staff and farmers to adjacent districts and villages and the production of technology fact-sheets or brochures for extension workers and farmers. These cover liming, fertilizer use and recipes for soybean. Some are in English and some in Shona. About 3000 copies of each are being produced by the AGRITEX Information Unit.

To develop a quantitative baseline of current practices against which to measure future adoption of the technologies, a baseline survey questionnaire about farming resources and soil-fertility practices for Chihota was implemented with 258 farmers during September 1999. This is being analysed and the report will be available shortly.

A preliminary economic analysis on the technologies as implemented in the 1998/99 demonstrations (Gatsi, 1999) suggested that liming and intercropping of maize with sunnhemp and soybean are the most economic of the interventions. The sole-crop green-manure packages (with maize yield forgone for 1 year) were considered the least economic.

AGRITEX staff monitored the amount of adoption of liming, maize + legume rotations and green manuring within and outside the farmer groups in Chihota. Their calculations for early 2000 are in Table 17.2. The highest adoption has been lime for maize and legumes, which has been promoted for several years among many farmers, and in grain-legume rotations. Green manures are the least attractive of the three technologies. Most farmers appreciated the benefits of applying lime, but had many questions about its availability and management. Some farmers were very interested in sunnhemp and velvet-bean green manures and had many questions about the inputs, management and benefits from these. Farmers said soybean was new to them and they needed help on how to use it.

A survey of farmer perceptions about the technologies was undertaken during the cropping season in February 2000. Ninety-nine members of the farmer groups experimenting with the soil-fertility technologies were interviewed individually (Gatsi *et al.*, 2000). Some preliminary findings from the survey are in Tables 17.3 and 17.4. Ninety-seven per cent of farmers said they would take up at least one of the soil-fertility technologies. Most farmers would take up the technology because they felt it would improve soil fertility and crop yields (Table

Table 17.2. Farmer adoption of three best-bet soil-fertility technologies in Chihota, Zimbabwe, 1999/2000 (from Gambara *et al.*, 2000).

| | Group members | | | Non-members |
Technology	Total number	Number of adopters	Per cent	(number of adopters)
Liming	572	250	43.7	531
Green manure	277	64	23	28
Legume rotations	299	142	48	184

17.3). Input unavailability was judged a serious constraint for the legume technologies (Table 17.4).

Economic Analysis of Soil-fertility Technology Adoption in Murewa and Zomba Districts of Zimbabwe and Malawi[4]

MWIRNET supported two research activities on the economics of soil-fertility technologies in Zimbabwe and Malawi. The Zimbabwe study was undertaken as collaborative work between AGRITEX and the Department of Research and Specialist Services (DR&SS) and the Malawi study was carried out by the Agricultural Policy Research Unit (APRU) of Bunda College of Agriculture. The broad objectives of these studies was to determine the economic

levels of chemical fertilizer, manure and compost application and to understand adoption constraints faced by smallholders.

The Zimbabwe study was conducted in Mangwende Communal Area in Murewa District of Mashonaland East Province, which is a high-rainfall area receiving 750–1000 mm year^{-1}. The sampling frame included poor and wealthier wards, providing a representative sample of the district. Six wards were first purposively selected to capture the variability in the wards. The three wards where the survey was conducted were selected, together with one ward where cotton was also a major crop. Two wards that were considered poor (from an initial survey) were also selected, thus ending up with six wards. Within the six wards, 11 villages were randomly selected, followed by a random selection of 213 households from the 11 villages.

Table 17.3. Reasons given by participating farmers for adopting soil-fertility technologies from group demonstrations or experiments in Chihota, Zimbabwe, 2000 (from Gatsi *et al.*, 2000).

	Technology being tested				
Reason for adopting	Lime and fertilizer	Green manure	Legume rotation	Lime and green manure	Lime and rotation
Improves yields	54%	25%	40%	100%	42%
Suppresses weeds	–	–	–	–	–
Corrects pH	14%	–	–	–	–
Not expensive	–	–	–	–	–
Improves soil fertility	11%	100%	30%	–	75%
Good crop growth	18%	–	10%	–	–
Increases fertilizer use efficiency	11%	–	–	–	–
Less labour	–	–	–	–	–
Multipurpose use of legumes	–	–	40%	–	42%
Controls diseases	–	–	–	–	–
Others	4%	–	20%	–	8%
Number of farmers	28	4	10	1	12

Table 17.4. Farmer perceptions about resource requirements for adopting soil-fertility technologies in Chihota, Zimbabwe, 2000 (from Gatsi *et al.*, 2000).

	Liming		Green manure		Legume rotation	
	Yes	No	Yes	No	Yes	No
Are the inputs available to you?	63%	39%	43%	57%	31%	69%
Is land a constraint?	3%	97%	29%	71%	4%	96%
Will you need to hire labour?	23%	77%	43%	57%	39%	61%
Number of farmers	60		7		29	

Various studies have been conducted on the adoption of chemical fertilizers by the smallholder subsector in Zimbabwe. In separate studies, Chipika (1988) and Kupfuma (1993) concluded that communal farmers use chemical fertilizer on staple crops, especially maize, and not other crops. They concluded that communal farmers use more fertilizer in high-crop-yield-potential areas than in low-potential areas. Dlamini (1993) highlighted the relationship between fertilizer consumption and variables such as extension-worker contacts and farm sales and concluded that extension advice has a significant positive impact on the adoption of basal fertilizer use and that this was also the case with farm sales. Murwira *et al.* (1998) evaluated the agronomic effectiveness of low rates of cattle manure and combinations of nitrogen in Murewa (Mangwende)

and other areas to determine the profitability of using fertilizer. The study showed that the returns per unit of fertilizer N were higher in Murewa (Z$43.47) than in any other site. This revealed that it is economic to apply fertilizer in most sites (especially Murewa). However, the long-term profitability of using fertilizer at most sites is not clear, since the study was done for one season. It was noted that cattle manure applied in small amounts immobilized N, hence depressing the yields at some sites.

Profitability of soil-management technologies in Murewa

Table 17.5 shows the profitability of the various soil management technologies. The 1998/99 cropping season, however, was

Table 17.5. Results of the gross-margin analysis for the soil-fertility management technologies in Murewa, Zimbabwe.

	Low*	Medium*	High*	Fert + Manure[†]	Manure[‡]
Benefits					
Yield ha^{-1}	500.00	700.00	1053.00	1200.00	200.00
Price of grain kg^{-1}	4.20	4.20	4.20	4.20	4.20
Gross income	2100.00	2940.00	4422.60	5040.00	840.00
Variable costs ha^{-1}					
Seed	178.00	178.00	178.00	178.00	178.00
Compound D	374.12	567.10	967.10	544.63	0.00
Ammonium nitrate	338.20	338.20	1292.28	669.28	0.00
Total	890.32	1083.30	2437.38	1391.91	178.00
Ha ha^{-1}					
Ploughing	700.00	700.00	700.00	700.00	700.00
Planting	12.87	12.87	12.87	12.87	12.87
Weeding	227.22	227.22	227.22	227.22	227.22
Fertilizing	92.96	92.96	92.96	92.96	0.00
Manure application	0.00	0.00	0.00	70.00	70.00
Harvesting	594.00	594.00	594.00	594.00	250.00
Total labour costs	1627.05	1627.05	1627.05	1697.05	1260.09
Total variable costs	2517.37	2710.35	4064.43	3088.96	1438.09
Gross margin ha^{-1}	−417.37	229.65	358.17	1951.04	−598.09
Gross margin (excl. labour)	1209.68	1856.70	1985.22	3648.09	662.00
Gross margin per labour day	−11.28	6.21	9.68	52.73	−23.92
Gross margin per $ invested	−0.17	0.08	0.09	0.63	−0.42

*Low, 0–150 kg ha^{-1}; medium, 151–300 kg ha^{-1}; high, over 300 kg ha^{-1}.
[†]151–300 kg fertilizer + ten Scotch carts' (animal-drawn metal carts used to carry inputs and produce in rural Zimbabwe) manure.
[‡]Five Scotch carts of manure.

very wet, reducing crop yield. A general trend seems to be noticeable, with higher yields coming from a combination of fertilizer and manure and also high amounts of Compound D and ammonium nitrate (AN) (yields of 1200 and 1053 kg ha^{-1}, respectively). Application of manure alone resulted in depressed yields (200 kg ha^{-1}). The application of small amounts of chemical fertilizer, which is the case with most smallholder farmers, also resulted in low yields (500 kg ha^{-1}). The medium and higher rates of fertilizer and the combination of fertilizer and manure seem to be profitable with gross margin (GM) ha^{-1} of Z$229.65, Z$358.17 and Z$1952.04, respectively. However, the combination of fertilizer and manure gave higher returns per labour day (Z$52.73) compared with all the other options, which is higher than the opportunity cost of labour (Z$20 day^{-1}) in Murewa District and the commercial-farm wage rate (Z$43 day^{-1}). Results of the sensitivity analysis suggest that, with a 20% increase in the price of grain, *ceteris paribus*, the GM ha^{-1} for the medium levels, the high levels and the combination of fertilizer and manure will increase to Z$789, Z$1200 and Z$2911, respectively. By removing imperfections in the output markets, the returns to the factors of production will be attractive to farmers using the medium levels, the high levels and a combination of fertilizer and manure.

Adoption of soil-fertility management technologies in Murewa

Household characteristics

There are more women heads of households in all the wards selected, with an average of 62% women-headed and 38% male-headed. The majority of farmers in Zimbabwe are women, as most men are in paid employment in urban centres or are in search of work. There is little variability in age and household size, with the average age of a household head being 54 and the average household size being seven. The average number of cattle in Ward 11 is

five, compared with the other wards, which range between seven and ten. Households in Ward 11 on average own one ox, compared with two in all the other wards, implying that most households in Ward 11 face a draught-power constraint. This difference could contribute to delayed ploughing.

Four logit regression models were run to analyse factors affecting the use of cattle manure; the combination of fertilizer and manure; and a basal- and top-dressing application of 250 kg ha^{-1} Compound D and 150 kg AN ha^{-1} in the Zimbabwe study and for compost use in the Malawi case-study (Table 17.6).

The results from the use of the manure model show that five independent variables were significant. The negative sign for the Compound D parameter estimate (COMPDHA) and the expected positive sign for the number of cattle owned (NOCATT) suggest that farmers with access to cattle manure are substituting chemical fertilizer with manure and those with more cattle apply more cattle manure or other forms of organic manure. However, there remains a problem of access to cattle manure, due to low cattle numbers in the study areas. Farmers with better extension contact (EXTCONT) and with a higher educational level of the household head (EDUC) have a higher likelihood of using manure for crop production. Contrary to our expectations, prices of Compound D (PCD) and ammonium nitrate (PAN) have the wrong signs but are significant.

The fertilizer and manure combination model indicates that only three of the variables are significant. Compound D is a basal fertilizer, which is normally applied at planting or just after crop emergence. Ammonium nitrate is applied as top-dressing. The two types of fertilizers are thus complementary. The positive and significant sign of ammonium nitrate use (ANHA) shows that it is used in the combination, while COMPDHA has a negative sign (significant), revealing a substitution effect of manure. Cattle number owned by households, as expected, had a positive and highly significant relationship with fertilizer and manure combination use (1% level).

Table 17.6. Results of logit regression for factors affecting farmers' use of soil-management technologies in Murewa and Zomba Districts, Zimbabwe and Malawi.

Variable	Manure (logit)	Fert. + Man. (logit)	Compound D (250 kg ha⁻¹) + ammonium nitrate (150 kg ha⁻¹) (logit)	Compost (logit)
Constant	−4.27*	−4.52*	−1.0603	−2.67**
ANHA	0.32	0.43****	−	−
COMPDHA	−0.43****	−0.48****	−	−
EXTCONT	0.56****	0.77	−0.5912	0.339**
NOCATT	0.43*	0.35*	0.1388*	
TSALES	−4.5E-06	−3.8E-06	0.0003	
TMZA	0.3	0.44	0.1495	
EDUC	0.24****	0.23****	0.314	0.080
FARMEXP	0.02	0.43	0.0011	0.63
MANUSE	−	−	−1.8219**	
PCD	−0.0048*	0.0028	0.0067	
PAN	0.0070*	−0.0033	−0.0016	
SITE	−	−	−	1.228*
CREDIT				0.930**

*, **, ***, **** indicate statistical significance at the 1%, 5%, 10% and 20% levels, respectively.

The third model (use of at least 250 kg ha⁻¹ and 150 kg ha⁻¹ AN) is assumed to identify factors that explain why farmers use these modest levels and very low levels of chemical fertilizer. Most of the independent variables are not significant except the related variable of manure use (MANUSE) and number of cattle. The large majority of farmers apply fertilizers at less than half the recommended rates. Less than 10% of the farmers responded as having any access to credit and availability to fertilizers at the right time.

Compost use in Zomba, Malawi[5]

One of the major constraints to improvement in maize productivity is the rising cost of factor inputs. In Malawi, the cost of fertilizer in the 1998/99 season rose by more than 100% and this is likely to substantially erode the profit margins of crops that depend on fertilizer, such as maize, unless it is matched by a corresponding increase in commodity prices. Although soil fertility and crop productivity can be increased by the application of appropriate levels of inorganic fertilizers, the high cost makes its availability and access difficult for

smallholder farmers. Smallholder farmers can improve soil fertility by incorporating organic manure in their farming practices. The collapse of the smallholder credit system in the 1993/94 season, led to the decline of fertilizer uptake to less than 70,000 tons, compared with 180,000 tons in 1992/93 (Kumwenda and Conroy, 1994). A study by the Malawi Maize Productivity Task Force (1996) concluded that inorganic-fertilizer use tends to be concentrated among the better-off farmers that have larger landholdings.

The majority of farmers (95%) applied fertilizer in the 1998/99 season. This study revealed that more than 90% of the fertilizer was applied to maize. A small proportion (37%) of the households in the sample has used compost manure as a soil-fertility management technology. Interestingly, more households in the NGO site (48%) have used compost than those in the Rural Development Project (RDP) site (24.7%). The history of both fertilizer and compost use indicates that some households started using compost, as well as fertilizer, as early as 1970. The number of people interested in using compost manure has been rising over the years since 1970, except for the period

1991–1994, when there was a sharp fall in the number of households adopting compost-manure technology.

There was a sharp rise in compost use after 1994, a situation that can be mainly attributed to a combined effect of removal of the input subsidy and limited access to credit. The study revealed that only 10% of the households had access to either formal or informal credit in the 1998/99 cropping season. Although differences in rates of adoption between the NGO and RDP appear to be centred on the issue of access to extension facilities, it was necessary to seek the views of farmers on why they do not apply compost to their crop. Results reveal that it is not just the intensity of extension, but also farmers' perception of a technology that influences adoption.

Lack of compost manure was reported as a major reason for non-adoption by 49% and 35% of the households from the NGO and RDP sites, respectively. It is not clear why farmers, particularly those at the NGO site, should lack compost when the majority have been trained in compost-making, while others have seen neighbours or friends using compost. Ninety-five per cent of the farmers responded that they used fertilizer in the 1998/99 season. The data indicate that a big proportion (45%) had applied very small amounts (less than 15 kg) of fertilizer to their fields. However, a significantly higher proportion from the RDP site (31%) than those from the NGO (8%) site indicate that their inability to use compost manure is attributed to the lack of knowledge on compost-making. About 25% and 17% from the NGO and RDP sites, respectively, mentioned that they could not use compost because they do not know its performance. Other reasons for non-adoption were reported as poor performance (3.5%) and high labour requirements (10.5%). Others (2.5%) were unable to use compost in the 1998/99 cropping season because the compost they had made did not decompose.

The logit regression model (Table 17.6) results, with the positive and significant signs for the parameter estimates of extension contact (EXTCONT), participation in the NGO program (SITE) and access to credit (CREDIT), imply that intensity of extension-service and NGO support activities at the site and better access to credit would increase the likelihood of compost adoption.

Conclusions and Implications

This chapter has documented the experiences of regional networking activities targeted to soil-fertility technology development, farm evaluation and farmers' assessment of the technologies and experiences with their promotion and adoption. The Soil Fertility Network is one of the institutional arrangements that is creating linkages and collaboration between researchers and extension to enhance the feedback to and from researchers and farmers. To this end the network has achieved its wider goal of bringing together technical scientists, extension staff, policy analysts, economists and farmers to plan, design and undertake research on soil-fertility-related problems that farmers consider are relevant. The network has contributed towards a better awareness of soil-fertility technologies and enhanced collaboration and the promotion of best-bet technologies generated by the partners. We are now beginning to see uptake by farmers of some of the results of this endeavour, but it is clear that many types of support appear necessary to help this process.

Very recently, several attempts have begun to understand the adoption constraints and challenges that farmers face. The case-studies reported (Chihota, Murewa and Zomba) identified lack of appropriate information about technologies, lack of input provision and inability to afford purchased inputs and lack of access to credit and output marketing as factors constraining the adoption of soil-fertility technologies. Farmers are willing to use the technologies provided they are made accessible to them. For this, the scaling up of technology transfer through on-farm trials and farmer participatory research requires more emphasis. In addition, socio-economics and policy-related studies are being initiated to complement the biophysical research efforts of the network.

While inorganic inputs are important, there is evidence that, if used alone, they are not enough to allow farmers to increase productivity sufficiently to meet food needs and to supply marketed surpluses. It is therefore imperative for research to try different combinations of organic and inorganic fertilizers, which also fit within the farming systems of smallholder farmers. Efficient methods of improving the quality of organic manure are also needed to improve the nutrient status of the manure. Other options, such as green manures, have to be screened and tested for their economic, technical and social feasibility. Extension efforts can be channelled to developing recommendations on combinations of organic and inorganic fertilizers.

From our experiences, special effort is required to address the policy environment, because the economic reforms, introduced since the early 1990s, have not favoured agriculture in general. Market imperfections persist within the economy with the rate of liberalization of input markets being faster than that of output markets, especially that of maize. The liberalization and deregulation of parastatals in Zimbabwe and Malawi have created a vacuum in the delivery of inputs and in output marketing because the private sector has not yet filled this gap. Smallholder farmers seem to be convinced about the importance of chemical fertilizer but the expense of the input due to the removal of subsidies has reduced the amounts being used by farmers. Access to inputs and credit is very limited.

New research needs to focus on the policy level to understand the private and social profitability of the various soil-fertility management options that are being suggested. Socio-economic research needs to understand the dynamics of soil fertility at different scales – the farm, village and community – and the agroecological, socio-economic, institutional, market and policy factors that affect the current trends in soil fertility.

Notes

[1] The views expressed in this chapter are those of the authors and do not necessarily reflect those of CIMMYT. The information summarized here has come from the work of many members of Soil Fert Net and MWIRNET. Any omissions or errors of interpretation are ours.

[2] The justification and background for the research work of Soil Fert Net is in Kumwenda *et al.* (1996, 1997) and Waddington and Heisey (1997). Much of the research on which the technology best bets are based is reported in Giller *et al.* (1998), Hikwa and Waddington (1998), Snapp *et al.* (1998) and Waddington *et al.* (1998).

[3] Into a relatively land- and rainfall-abundant area with N-deficient and responsive soils and a history of bush fallows.

[4] See Gatsi and Mekuria (2000).

[5] For detailed discussion on this see Mataya *et al.* (2000).

18 A Bio-economic Model of Integrated Crop–Livestock Farming Systems: the Case of the Ginchi Watershed in Ethiopia

B.N. Okumu,[1] M.A. Jabbar,[2] D. Colman[3] and Noel Russell[3]

[1]Department of Applied Economics and Management, Cornell University, 444 Warren Hall, Ithaca, NY 14850, USA; [2]International Livestock Research Institute (ILRI), PO Box 5689, Addis Ababa, Ethiopia; [3]University of Manchester, Manchester M13 9PL, UK

Unsustainable farming systems abound in much of the developing world, with severe consequences for the level of poverty and the scale of land degradation. It is now estimated that well over 1000 million people in the developing world just barely survive on US$1 day^{-1}. Many more with slightly better incomes lead lives of deep deprivation. For most of them, the biggest aspiration is to have an adequate diet and livelihood. But the resources, particularly the land area, needed to fulfil even these modest hopes are rapidly shrinking and their productivity is rapidly declining, resulting, inevitably, in social disintegration and a climate of conflict and unrest.

Ethiopia is a typical example of these problems. Over 46% of the country's gross domestic product (GDP) arises from the agricultural sector, which also accounts for about 80% of the total export revenue and contributes about 80% of the country's employment opportunities. Annual per capita income stands at US$100 (World Bank, 1995) and is among the lowest in sub-Saharan Africa. Agricultural production is concentrated in the highlands ecosystems, estimated at about 46% of the country's land

mass. Lying at an altitude of about 1500 m above sea level, the highlands are said to be some of the world's most highly degraded landscapes. Yet they are home to 88% of the country's 60 million population and have one of the highest population densities in the world (Srivastava *et al.*, 1993). Productivity is low and declining. Thus, although 80% of the population is engaged in agriculture, it generates less than 50% of the GDP (Shiferaw and Holden, 1997). A major reason cited for this low and declining productivity is land degradation.

This study was motivated by the observation that, although technological solutions to many degradation problems in Ethiopia exist, they are often not adopted by most farmers and hence degradation continues unabated. At the same time, many policy and institutional reforms have failed to produce rapid and sustained agricultural growth. One reason cited for this is the overemphasis on biophysical remedies in most of these solutions. Little consideration is given to the farmers' needs, perceptions and conditions. Moreover, whereas the individual impacts of various intervention technologies are known, there is little information on their combined impact or on the role of policy and institutional arrangements in conditioning their outcomes (Okumu, 2000).

Bio-economic models offer much promise as the kind of integrative tool that economists need to generate such information. First, they are capable of assessing the impact of policy and institutional reforms on incentives for the adoption of different types of technologies and natural resource management practices. Secondly, they can be used to analyse the impact of adoption on incomes, poverty and the condition of natural resources over time. These models rely mainly on quantitative procedures that simulate biophysical and socio-economic processes. They incorporate feedback mechanisms between individual farms and the landscape, between farms and regional markets and between resource-use decisions made in one period with the condition and productivity of the resource base in subsequent periods. In this way, bio-economic models enable deep insights to be drawn about the complex interactions that permeate an agricultural system (Hazell, 1998). Thus they form a powerful tool for *ex ante* analysis of various development options.

This chapter uses one particular bio-economic model, first, to explore some questions and hence generate information pertaining to the complex interactions existing in an integrated crop–livestock farming system in Ethiopia. Secondly, we examine various ways in which such a model may be used to assess, *ex ante*, the likely impact of multiple technology adoption on such systems under a set of policy scenarios. The Ginchi watershed in the Ethiopian highlands is used as a case-study.

The chapter is organized as follows. The next section describes the Ginchi-watershed integrated crop–livestock production system and highlights a watershed community as opposed to a purely household level of analysis. The third section explains some features of the Ginchi-watershed bio-economic model, while the fourth section discusses model results and policy implications. The final section outlines the main policy conclusions.

The Ginchi Watershed

The Ginchi watershed is located in the high-potential cereal–livestock ecological zone of the central highlands of Ethiopia. The area typifies an integrated crop–livestock production system common in many parts of Ethiopia and sub-Saharan Africa. Annual precipitation is high but variable, ranging between 1000 and 1500 mm and falling mainly in the *kiremt* or wet season (Andergie, 1994). Most of this rain is bimodal, with 70% recorded during the long-rains season (*meher*) and 30% during the short rains (*belg*).

Short rains are mostly inadequate for cultivation of major crops but are good for growth of pulses and regrowth of natural forages. The watershed is drained by the Lugo River, a sub-tributary of the Awash River. Table 18.1 summarizes the biophysical details and problems of the watershed. We note that based on slope gradient, land-forms, type of soil and other land features, the watershed may be delineated into four physiographic units (A, B, C and D).

Ginchi-watershed farming practices

Current, land-use practices in the watershed vary strongly with the physiography of the watershed. The highly fertile but poorly drained bottom lands (A and B) are used for teff and wheat cultivation, while other cereals, such as maize, barley, millet and sorghum, are grown mainly on the upper slopes (C and D). Land types B and C are also important for the cultivation of legumes (chickpea, rough pea and other pulses). The use of external, yield-increasing inputs is rudimentary and agricultural production relies heavily on technologies largely unchanged for centuries (Shiferaw and Holden, 1998). Landholdings average less than 2.5 ha per household in the watershed. This is a consequence of increasing family sizes and in-migration from the lowland areas of the country. Also, owing to the egalitarian allocation of land by the peasant associations[1] (PAs), land fragmentation is high, with most

Table 18.1. Watershed land categories, their characteristics and problems.

Physiographic unit	Characteristics	Problems
Land type A, 0–4% slope	Intensively cultivated flat lowland Soils are black vertisols (over 1 m depth) Major activities are cultivation of teff, wheat and pulses, and grazing	Waterlogging and other problems of poor drainage Rill erosion Nutrient mining Intense soil cracking Need for construction of a communal water channel
Land type B, 5–10% slope	Moderately cultivated moderate-slope land Main soils are black vertisols (0.5 m minimum depth) Main crops are teff, wheat, pulses, maize and pasture	Rill erosion resulting in gullies in some areas Poor drainage requiring construction of communal and feeder water channels Nutrient mining Soil cracking
Land type C, 11–15% slope	Highly degraded steep slopes with numerous gullies Alfisols and vertisols of less than 0.5 m depth Main source of potable water in the watershed Grazing and settlement area Main crops are maize, pulses and teff, and tree planting around homesteads is the main activity	Severe gully and rill erosion with some soil subsidence High livestock densities Nutrient mining
Land type D, over 16% slope	Very steep, poorly forested slopes Main soils are acrisols and alfisols of less than 15 cm depth Altitude about 2320 masl Firewood collection in nearby forest land is a major activity Restricted cultivation on the top slopes	Gully erosion Rill erosion Shallow soils Nutrient mining Deforestation

NB: about 5% (15 ha) of the watershed is under human settlements or homesteads.
masl, metres above sea level.

farmers having plots scattered all over the watershed. Incentives to conserve land in Ginchi and in most of the Ethiopian highlands are hence low, due to the prevailing land-tenure system, in which land may be reallocated with no compensation to the previous owner (Gryseels and Anderson, 1983; Yohannes and Holden, 2000a).

In terms of livestock keeping, both individual and communal pasture-lands are small and diminishing and they suffer from low productivity, due to high stocking rates and overgrazing. It is estimated that natural pasture (both private and communal) has the capacity to produce up to 6 tons of dry matter (DM) ha^{-1}. However, if continuously grazed, it yields only 2.5 tons DM ha^{-1} (Jutzi, 1987). Frequent grazing and trampling reduces this DM yield by a further 50%, i.e. 1.75 tons ha^{-1}. Thus, in many farm households in Ginchi and in the highlands generally, a 30–40% shortfall in animal-feed supplies is experienced every year (Jutzi, 1987).

Cattle are the most important livestock species. The common cattle breed kept is the traditional short-horned zebu. Integration of crop and livestock subsystems ensures that both systems complement and supplement each other. The principal livestock contributions to the crop system are draught power (for cultivation, transportation and threshing of crops) and manure. Animal dung is utilized in a number of ways in the watershed. For most of the year, dung is burned for fuel to complement wood, which is in short and declining supply. This signifies a rapid form of nutrient mining, since part of the biomass consumed by the animals does not re-enter the system as animal manure. The dry-season free roaming grazing practised in the watershed has resulted in household members going out to the fields to collect livestock droppings. Similarly, most crop residue is fed to livestock and its supply is erratic and seasonal, depending largely on the cropping pattern, harvesting practices, weeding intensity and frequency.

Family labour accounts for almost 80% of the farm labour requirements, most of which is provided by members of 15–65 years of age. In general, there is division of labour by gender and age. Labour substitutability is a very significant characteristic of labour supply in this region. Exchange labour (locally known as *debo*) plays a major role in mitigating labour shortages and allowing some critical tasks to be done on time. Despite all the above arrangements, shortages of labour during certain periods of the crop year are a common problem. The underlying reason is the adherence of most households (over 75%) to the Coptic Orthodox church practice of strictly observing church holidays. During such days (estimated at 10–15 days every month), no farm work is allowed. These cultural and religious practices provide a constrained boundary within which possibilities of introducing new technologies have to be examined at least for the foreseeable future.

Off-farm activities (e.g. trading) are also significant for some household members during non-holiday periods. Most of the trade is carried out in two local markets, Olenkomi and Ginchi. The markets are located about 3 km away – one to the east and the other to the west of the Ginchi watershed along the Addis Ababa–Ambo road. The area thus enjoys good market access, given the excellent tarmac road joining it to the two markets.

Land management problems

Scarcity of wood fuel and animal fodder contributes both directly and indirectly to the stripping of the landscape cover and exposing it to forces of erosion. It also means that the capacity to use crop residues as mulch and manure is diminished significantly, resulting in limited soil-nutrient recycling. Other cultivation techniques using the *maresha* (traditional plough) in the vertisol areas are geared towards minimization of waterlogging problems and involve the construction of ridges and furrows to facilitate drainage. On land of higher gradient, the flow of water through such furrows results in accelerated erosion and gully formation, as well as flooding of outlying fields in the bottom areas. However, in the absence of some larger system for channelling the water from the bottom land with a much lower gradient, much of the water remains between the ridges. This results in conflict among farmers in different parts of the landscape. What is needed is an integrated drainage system into a main drain.

An attempt to solve problems such as those arising from the poor resource management practices in the Ginchi area was the basis for the formation of the Joint Vertisol Project (JVP) in 1985. JVP is a multidisciplinary, multi-institutional project formed to spearhead farmer participatory vertisol management research (Tekalign *et al.*, 1993). The overriding objective of the JVP is to develop and verify technologies appropriate for improving land use in the predominantly vertisol areas of the watershed on a sustainable basis. It generated the following technology package.

1. The broad-bed and furrow maker (BBM), which is an improved animal-drawn implement for drainage.
2. New wheat varieties for early planting on drained beds.
3. Intercropping of cereals with forage legumes.
4. Grazing management techniques.
5. Agroforestry, mainly for fodder purposes.

The adoption and implementation of such a package, particularly the BBM and the wheat varieties, were to entail the drainage of the wetlands in the lower parts of the watershed, using both the traditional and modified ploughs, such as the BBM (among other equipment), drawn mainly by oxen. The construction and maintenance of drainage channels to drain off excess water from the individual farm plots to a communal drain was to be done by collaborative action at the community level by the watershed population. Such activities would, however, put pressure on the available amount of human resources, especially human labour, and on oxen draught power and cash endowment. It was observed further that the most probable way to offset such costs would be to use higher-yielding crop varieties and larger animal breeds. Such a package hence had to include improved animal breeds and new seed varieties as a means of boosting productivity in the watershed. But improved varieties would require the use of at least some minimum threshold amounts of chemical inputs (fertilizers, pesticides, herbicides, etc.), while new animal breeds called for more animal feeds with higher nutritive value than is locally available. To ensure an adequate supply of draught power for more intensive tillage, these animals had to be better fed than before. Similarly, new high-yielding crop varieties would require longer crop-growing periods, with perhaps different cropping patterns and most probably a general intensification of most farm activities.

Even with the package fully adopted, the moderate to steep slopes of the watershed could still suffer from extensive soil loss. To curb this, some agroforestry around the farmers' homesteads and afforestation across the slopes would be desirable. This would again have impacts on the various aspects of the farmers' activities. Where the trees planted could be used for forage, animal fodder supply to the households would increase, reducing the need for overgrazing or use of crop residues as animal feed. This in turn would increase the availability of residues for mulching, thus contributing to higher yields. Perhaps of greater implication would be the effect of agroforestry on fuel-wood supply. In an area like Ginchi, where animal dung cake (made from crop residue and animal waste) is the main source

of fuel, any increase in the supply of fuel wood would have a number of implications. It would enable increased use of animal dung as organic manure, which would improve the soil structure and fertility of the dominantly clayey vertisols, and it would reduce the rate of deforestation of indigenous forests, which have a better capacity to conserve the soils than exotic trees. Moreover, indigenous trees take longer to regenerate, while afforestation, in general, requires a more secure land-tenure policy than that currently observed in Ginchi.

Understanding how the above integrated crop–livestock system worked was central to the generation of appropriate solutions that might easily be adopted. Use of the BBM plough, for example, was experimentally known to improve drainage and hence could result in the cultivation of new wheat varieties, but it also resulted in flooding of the lowlands.

The Ginchi Watershed Bio-economic Model

To address these general problems of development in the Ginchi watershed, a bio-economic model was developed as an integrative analytical tool (see Fig. 18.1), with the aim of gaining some insight into what would happen if:

1. There was limited or no adoption of the JVP technologies under current policy conditions.
2. New technologies were adopted given certain policy and/or institutional changes.

Technologies are thus evaluated not only in terms of farm households' immediate and future needs but also in terms of their spatial and intertemporal impact on the landscape. A dynamic non-linear mathematical programming framework was used, mainly because of the amount of dynamic processes to be captured (see Okumu *et al.*, 2000 for model details).

The non-linear dynamic programming algorithm adequately captures the multifunctional and multidimensional interactions among the human, economic and

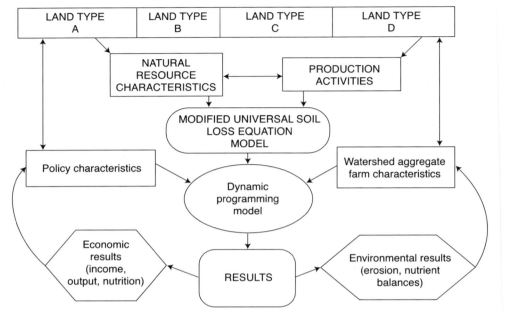

Fig. 18.1. Structure of the Ginchi watershed dynamic bio-economic model.

biophysical components of the watershed ecosystem. The model optimizes an aggregate watershed utility function (comprising production, consumption, profit and leisure, under the assumption of non-separability of these activities in the given situation, i.e. they are simultaneously optimized). The function is indirectly linked to the biophysical aspects of the watershed through an exponential soil-loss yield-decline equation with single-year time lags. Per hectare soil loss is estimated using the universal soil-loss equation and soil loss is estimated based on slope and the type of land-use activity chosen by farmers. Cumulative soil losses in previous periods determine yields in the next period, given the ameliorative effects of chemical and dung fertilizer. The framework takes into account seasonality in input and output supplies (conditioned mainly by the rainfall pattern, as well as by the cultural practices described above); labour substitutability; gender roles; crop–livestock constraints; minimum household food requirements; forestry activities; and the biophysical aspects of soil erosion and nutrient balances arising from these activities. Validation and calibration of the model

is reported in Okumu (2000). Both experimental fertilizer yield response and farm household cross-sectional economic data sets from the four land categories found in the watershed were used to test and run the model.

The model allows careful matching of ecological characteristics with socioeconomic characteristics and emphasizes the various levels of economic–ecological interaction. The result is a better characterization of geophysical settings of the study area, such as variation of soils across the landscape, steepness and geographical orientation of the terrain, predominant wind direction and the amount of precipitation. These factors are known to influence technology uptake. Similarly, the division of resources among individuals, including the formal and informal jurisdiction system, has some impact on the extent of technology adoption and must be taken into consideration. Decision-making is, however, assumed to be at watershed level. This is justified in part by a survey that revealed most farmers in the watershed to be members of the PA. The PA is a government-appointed council of local elders that oversees all communal affairs, including egalitarian allocation of resources, such as land, as well as the interests of the government. Thus, the existence of a PA implied decision-making at a higher level than the household, although not necessarily based on the watershed demarcation. These features are not unique to Ginchi but are common in most of the Ethiopian highlands (Gryseels and Anderson, 1983)

A maximum 12-year time horizon was chosen for the model. This was the length of time most farmers felt it would take before significant reallocation of plots by the Yebdo Keshina PA, which is in charge of the Ginchi watershed, could take place. As in many parts of Ethiopia, farmers considered investments in land (for periods beyond 12 years in Ginchi) to be too risky, as any of the plots could be given out to other PA members with no compensation (Yohannes and Holden, 2000a). The model captures this uncertainty by ignoring the 'end-game effects', i.e. assuming that everything comes to an end in the 12th year.

Similarly, despite the fact that most farmers belonged to groups known locally as *Mahabar*, which encouraged contributions to assist those members suffering calamities (including death and loss of animals (especially oxen) and crops), the capacity of such groups to handle risk arising from poor weather was nevertheless limited. Explicit modelling of risk was, however, difficult, due to the limited amount of data and also the already large size of the model. Again, the assumption of free flow of resources among households, such as high interhousehold interactions in terms of communal labour and draught-animal sharing, was justified by the need to reap some economies of scale in carrying out arduous activities, such as teff harvesting, through non-wage-based labour exchanges.

The model is used to investigate two scenarios. The first is a limited intervention scenario that assumes no change in current institutional structure. Following Holden *et al.* (1998) and Lapar and Pandey (1999), who emphasize a link between security of tenure and planning horizon, land-tenure issues are explored by investigating two

alternative planning horizons. This scenario also assumes that consumption is maintained at 1500 calories per adult equivalent (AE) per day.[2] The second scenario assumes multiple adoption of JVP technologies, a long planning horizon and associated institutional changes. Consumption is assumed to increase to 2000 calories AE^{-1} day^{-1}. Sensitivity analysis of the model was done through parametric analysis, in which market interest rates were varied and the effect on the model projected cash incomes and soil-erosion levels was recorded (Okumu, 2000). In both scenarios, prices were generally assumed to be invariant to watershed production levels due to the relatively small size of the watershed, i.e. 298 ha, and the very good access to the local markets, one of which is located in an urban (district town) centre.

The optimization procedure ensures that only those technologies whose per unit marginal returns are at least equal to their associated per unit marginal costs are considered in each period. For each location in the watershed, the model calculates the optimal fertilizer and dung application rates for every crop activity and selects the most profitable ones for cultivation. Relative prices, costs and yields are adjusted for the effects of erosion in each period. For each optimal farm plan, the model generates shadow prices for resources in the watershed, indicating marginal social costs. This is a unique feature of this modelling approach. Such shadow prices (not displayed here) may be used in designing natural resource management strategies and policies to correct both spatial and intertemporal externalities (Ehui, 1987). The approach thus represents a significant departure from past studies, which have been mostly diagnostic (Ehui, 1987; Omiti, 1995; Smaling *et al.*, 1996; Kassie, 1997; Jager *et al.*, 1998; Lapar and Pandey, 1999).

The study reports the likely impact of technology adoption on nutrient flows, cash income and human-nutrition indicators in the study area, given both the current and simulated policy environment and institutional settings. Trade-offs in the achievement of these indicators of human and biophysical welfare are similarly quantified.

Results and Discussion

Scenario 1: Land management under current policy conditions, with fertilizer as the only form of intervention

The limited-intervention scenario is based on the observation that farmers: (i) have some access to fertilizer, through their current cash-income levels; (ii) have the option of raising productivity by adopting explicit soil-conservation measures, such as planting grass-strip bunds along the slopes; and (iii) could leave some of their land fallow in some years and/or could practise crop rotation to ameliorate the effects of soil erosion and nutrient mining. Livestock numbers are held constant in order to simulate the static traditional livestock-keeping system, in which excessive numbers of cattle are kept for reasons other than their economic returns or animal-feed availability. In addition, tree planting is limited to homestead areas.

The results, presented in Table 18.2, display the observed land-use pattern in the base year, together with the 4-year (short time horizon) and 12-year (long time horizon) model projections.

Generally, model results suggest that explicit adoption of soil-conservation measures may not be a profitable venture both under the short and long time frames, given the existing policy conditions. Instead, crop rotation is likely to be the preferred option. Highly erosive but relatively more profitable and more culturally preferred crops, such as teff, are likely to be rotated with less erosive but less profitable crops, such as local wheat varieties. A unique land-use pattern is therefore likely to emerge. With a short time horizon, teff is rotated with wheat and maize on land type A and B, respectively. Teff may, however, increasingly take up more land in both land types as the end period approaches, resulting in a clear monoculture in the poorly drained bottom lands (A and B). The upper slopes (C and D)

are likely to be used for wheat and maize cultivation and haymaking.

If a longer 12-year time horizon is used (i.e. farmers are offered more secure land-user rights), the land-use pattern changes. Land type A could specialize in teff cultivation while wheat and teff could be continuously rotated on land type B, with the wheat area consistently increasing at the expense of teff land. Maize and wheat cultivation might be concentrated on the fragile slopes in land types C and D. An initial increase in local wheat planting on land type B may, however, change in later years. The model projects that, in the seventh year, the area under teff starts rising, while that under wheat declines. Thus by the tenth year, land under wheat cultivation is likely to be increasingly used for teff cultivation. The progressive increases in teff requirements for consumption purposes (due to in-migration), coupled with constantly declining yields of wheat and maize due to effects of cumulative soil loss, could be the main driving forces for this shift in land allocation. Wheat cultivation thus generates less income in each period, making market purchases of teff for consumption purposes unsustainable. Note that teff prices are 20% higher than wheat prices. An income-maximization strategy based on the production, consumption and sale of home-produced teff is, therefore, likely to be adopted in the last 5 years of the plan period. This is because the increased soil loss arising from teff cultivation at this later stage has fewer implications on productivity at this late stage than earlier in the plan period. It is also consistent with the observation that farmers tend to degrade the land more as the end of the time horizon approaches (Holden et al., 1998).

Fertilizer application closely mimics land allocations. Its use, however, is severely limited by the high fertilizer costs. The implied change in land-use strategy, as explained above, may be conditioned by the increasing failure of fertilizer application to mask the cumulative effects of soil loss on yields in the long run. For these reasons, wheat and maize are consistently replaced with teff in land types B and D, respectively, as the end of the plan period draws closer. By the 12th year, about 85% (or 90 ha) of land type B may be under teff cultivation.

Comparison of the short-term model land-use projections in the bottom land (type A) in year 1 with the base-year

Table 18.2. Actual and estimated figures of land use (ha), income (birr) and erosion (tons ha^{-1} year^{-1}) of the dynamic model, assuming limited use of JVP package.

Type of activity	1995 actual values	Four-year time horizon (1995–1998)			Twelve-year time horizon (1995–2006)	
		1995 (Year 1)	1996 (Year 2)	1998 (Year 4)	2001 (Year 7)	2006 (Year 12)
Production (by land type)						
Eucalyptus* A (no.)	–	–	–	–	2,966	2,910
Teff A	26.65	20.00	26.86	35.00	34.9	34.93
Wheat A	10.38	15.00	8.12	–	–	0.061
Others A	12.80	5.00	5.00	5.00	5.00	5.00
Hay A	–	13.00	13.00	13.00	13.00	13.00
Grazing A1	3.12	–	–	–	–	–
Grazing A2	54.00	54.00	54.00	54.00	54.00	54.00
Teff B	67.71	40.00	40.00	98.60	45.62	89.70
Wheat B	9.86	9.68	–	–	52.37	8.29
Maize B	1.47	48.32	58.00	–	–	–
Others B	20.96	2.00	2.00	2.00	2.00	2.00
Hay B	6.50	15.00	15.00	15.00	15.00	15.00
Grazing B1	8.5	–	–	–	–	–
Grazing B2	105	105.00	105.00	105.00	105.00	105.00
Teff C	15.31	–	–	1	–	–
Wheat C	7.67	1	1	–	–	0.15
Maize C	1.00	27.00	27.00	26.99	27.9	27.00
Others C	6.02	2.00	2.00	2.00	2.00	2.00
Hay C	2.50	7.50	7.50	7.5	7.50	7.50
Grazing C1	7.50	–	–	–	–	–
Grazing C2	25.50	25.50	25.50	25.50	25.5	–
Teff D	16.15	–	–	–	–	9.79
Wheat D	2.30	–	–	40.00	–	–
Maize D	5.77	40.00	40.00	–	40.00	30.205
Others D	15.78	–	–	–	–	–
Hay D	7.30	12.50	12.50	12.50	12.5	12.50
Grazing D1	5.00	–	–	–	–	–
Grazing D2	54.00	52.50	52.50	52.50	52.5	52.5
Cows (no.)	120	120	120	120	120	120
Oxen (no.)	240	240	240	240	240	240
Net teff buying (kg)	12,701	40,380	36,446	–12,111	28,979	6,770
Net wheat buying (kg)	7,106	0.00	–7,674	18,373.78	0	10,236
Cash income (in birr)	149,397	169,329	253,504	232,499	220,956	226,261
Cash income (in US$)	21,342	24,189	33,355	29,430	27,969	29,771
Erosion (tons year^{-1})	9,143	10,023	11,357	11,534	8,407	9,134
Erosion (tons ha^{-1} year^{-1})	(31.00)	(33.63)	(38.11)	(38.44)	(28.00)	(30.65)

*Until recently, eucalyptus trees were planted around homesteads and rarely on crop land.

observed values reveals them to be close. This suggests that current cultivation of these crops is close to optimal levels. This could be due to the significant amount of on-farm research on wheat that has been going on over the years in this land category. More probably, however, it could be an indication that farmers owning plots in this part of the watershed have a high time preference or high subjective discount rate of future returns and hence are more interested in short-term gains than otherwise. Such high time preference may be caused by the existing land-tenure policy, which gives limited land-user rights, with the possibility of frequent land redistribution and/or high levels of poverty (Holden et al., 1998).

By comparing cash income and erosion results of the 4-year and 12-year time-horizon runs of the dynamic model, one gets some indication of the impact of land-tenure policy on the natural resource base. Examination of estimated soil losses at the end of a 4-year period (short time horizon ending in the fourth year) and 12-year period (long time horizon ending in the 12th year), indicates soil losses in the fourth year to be 20% higher than in the 12th year. Surprisingly, model projected fourth-year cash incomes are only 2.6% higher than those in the 12th year. Thus, with a longer time horizon, farmers are likely to adopt a specific crop-rotation strategy to reduce soil erosion and sustain net cash incomes at an acceptable level, unlike the short-term situation, when monoculture of the staple food crop, teff, is likely to prevail. Model results thus suggest that, given farmers' limited resource endowments and especially the existing liquidity and land constraints, an insecure land policy is likely to create an income illusion that promotes land degradation. Giving farmers relatively more secure land-user rights may hence result in improved conservation of soils.

In conclusion, we note that the possibility of adoption of explicit land-conservation measures is slim under the current policy conditions. Instead, careful allocation of land resources among activities of different

soil-loss and income implications is likely to be observed. Fertilizer use is likely to play a central role in this crop-rotation strategy.

Scenario 2: Feasibility and possible impact of multiple adoption of JVP technologies given certain policy and/or institutional changes

This scenario is designed to investigate the feasibility of adopting the new high-yielding wheat variety supported by the BBM technology package. The simulation results, presented in Table 18.3, are discussed in terms of the resulting land-use pattern, economic returns, human nutrition, soil-erosion levels and soil-nutrient balances. It is assumed that households ensure adequate food supply (Yohannes and Holden, 2000b) and adopt a more commercially oriented livestock-keeping strategy (i.e. livestock numbers are allowed to vary over time). Results of this scenario are compared first with the 1995 actual watershed situation and then with the situation simulating limited intervention, discussed above. This 'multiple' intervention scenario is hence based on the assumption that farmers are able and willing to adopt either all or some of the JVP-generated technology package, given appropriate policy and institutional environments and based on the available resource endowments. The existence of high land-user rights is therefore assumed throughout the analysis and the simulations are based on a 12-year planning horizon. In addition, it is assumed that the watershed community raises its calorie intake levels by 33% from those consumed in scenario 1, as a response to a food-security policy emphasizing food self-sufficiency. Two tree varieties are available for planting, namely Sesbania (leguminous forage tree) and Eucalyptus (fuel wood with commercial value).

Crop land-use patterns

With the technologies in place, the watershed's capacity to produce most of the food crops necessary for meeting a higher daily calorie intake of 2000 calories AE⁻¹ is tested through model simulation. The main question to be answered is whether the new wheat variety, requiring better drainage in land type A, can be adopted along with the other technology components, as postulated in the discussion above. This scenario also

Table 18.3. Actual and estimated values of land use (ha), income (birr) and erosion (tons ha⁻¹) of the dynamic version of the model with multiple intervention and consumption at 2000 calories AE⁻¹ day⁻¹.

Type of activity	1995 actual values	Twelve-year time horizon 1995–2006				
		1995	1996	1998	2001	2006
Production (by land type)						
Eucalyptus A	–	–	–	–	–	–
Teff A	26.65	20.00	20.00	20.00	20.01	35.00
Wheat A	10.38	–	15.00	15.00	14.99	–
Others A	12.80	20.00	5.00	5.00	5.00	5.00
Hay A	–	13.00	13.00	13.00	13.00	13.00
Grazing A1	3.12	–	–	–	–	–
Grazing A2	54.00	54.00	54.00	54.00	54.00	54.00
Eucalyptus B	–	–	–	–	–	–
Teff B	67.71	40.00	40.00	40.00	40.00	59.58
Wheat B	9.86	–	11.63	13.32	16.06	40.42
Maize B	1.47	–	3.67	3.23	2.80	–
Others B	20.96	60.00	44.71	43.45	16.46	–
Hay B	6.50	15.00	15.00	15.00	15.00	15.00
Grazing B1	8.5	–	–	–	–	–
Grazing B2	105	105.00	105.00	105.00	105.00	105.00
Eucalyptus C	–	–	–	–	–	–
Teff C	15.31	–	–	–	–	–
Wheat C	7.67	–	–	–	–	–
Maize C	1.00	20.00	27.99	27.99	20.00	20.00
Others C	6.02	10.00	2.00	2.00	9.99	9.99
Hay C	2.50	7.50	7.50	7.5	7.50	7.50
Grazing C1	7.50	–	–	–	–	–
Grazing C2	25.50	25.50	25.50	25.50	25.50	25.50
Eucalyptus D	–	30.00	30.00	30	30	30
Teff D	16.15	–	–	–	–	–
Wheat D	2.30	–	–	–	–	10.00
Maize D	5.77	–	10.00	10.00	10.00	–
Others D	15.78	10.00	–	–	–	–
Hay D	7.30	12.50	12.50	12.50	12.50	12.50
Grazing D1	5.00	–	–	–	–	–
Grazing D2	54.00	22.50	22.50	22.50	22.50	22.50
Cows (no.)	120	26.00	34.00	34.00	33.00	48.00
Oxen (no.)	240	51.00	68.00	68.00	67.00	96.00
Net teff buying (kg)	12,701	40,380.00	−53,897.77	−31,343.79	−3,100.20	−4,885.35
Net wheat buying (kg)	7,106	72,000.00	−2,401.72	−2,544.29	−2,790.53	−3,646.63
Cash income (birr)	149,397	2,112,730.38	2,419,426.00	3,533,938.61	2,115,136.65	2,815,972.50
Cash income (US$)	21,342	301,818.63	301,418.25	441,742.33	264,392.08	351,996.56
Erosion (tons year⁻¹)	9,143	7,177.74	6,982.12	6,984.71	6,717.86	7,719.42
Erosion (tons ha⁻¹ year⁻¹)	(31)	(24.16)	(23.42)	(23.44)	(22.54)	(25.90)

allows for behaviour consistent with the notion that farmers wish to avoid market risk by reducing their reliance on market grains and pulses (Yohannes and Holden, 2000b).

Results show that adoption of the new wheat variety, ET 13, is possible but may not occur on land type A as expected. The most probable place for adoption of this new wheat variety is likely to be land type B, where explicit drainage efforts are not acutely required, due to the relatively higher slope of between 6–10% and hence a better drainage (see Table 18.1). Model results show further that cultivation of teff and local wheat varieties (known to be more tolerant to poor drainage conditions than ET 13) is more suitable on land type A. These land allocations suggest that, in order to achieve the higher intake of 2000 calories $AE^{-1}\,day^{-1}$, a specific combination of technologies should be adopted and such adoption should vary both across the landscape and across the years. Land-use activities on land type A, for instance, are likely to be unaffected by a higher consumption of grains and pulses in the watershed, while, on land type B, raising intake from 1500 to 2000 calories AE^{-1} is most likely to result in an increasing number of farmers with plots in land type B adopting the new wheat variety at the expense of areas under pulses/ spices and eucalyptus trees. Starting from no cultivation in the first year, the model projects that around 12 ha (11%) of total land type B in the second year may be put under the new cultivar of wheat. This area may rise further over the years to 23 ha and 40 ha by the eighth and ninth years (i.e. 23 and 38%), respectively, signalling a possible increase in the number of farmers adopting the new wheat cultivar over time. Such adoption could result in no cultivation of eucalyptus on land type B and could make fallowing of about 20% of land type B area in the seventh and eighth years profitable.

Thus inclusion of fallow in the crop-rotation pattern rather than tree planting or conscious land-conservation measures, such as planting grass strips on some of the plots with rapidly declining yields, signifies their high profitability. To see this, the difficulty of converting land under eucalyptus to food production has to be borne in mind. Thus, in spite of the high profitability of eucalyptus trees, the need for own-farm food self-sufficiency makes fallowing a preferred option. Explicit land-conservation measures, such as planting of grass strips or stone walls, are similarly not adopted, mainly because of the high labour requirements, the high opportunity cost of food production forgone and the limited immediate returns. Moreover, farmers know that any of their plots could be reallocated at the end of the plan period and hence their incentives to carry out explicit conservation measures are greatly reduced. Leaving some plots fallow for a maximum of 2 years before gradually bringing them back into production of cereals (wheat) is hence a worthwhile endeavour. Thus the model projects that, of the 20% land area left fallow in years 7 and 8, about 4% could be brought back into production in the ninth year, leaving 16% under fallow. In the tenth year, model projections show that a further 6% of the area under fallow could be cultivated, leaving the rest to be brought under cultivation in the 11th and 12th years.

From the above model projections, it may be concluded that fallowing is the most feasible alternative to tree planting as a means of soil and nutrient conservation in the vertisols and, more generally, in much of the highlands with similar conditions. This is likely to be the case especially when the food-consumption requirement from on-farm production is raised by 33% and when farmers are given more secure land-usufruct rights. We note that fallowing is commonly done in the watershed, although the area fallowed is being significantly reduced over the years. Admittedly, fallowing is only possible for those households with more abundant landholdings.

A further examination of model results (Table 18.3) shows that no changes in the allocation of land type C (highly eroded gully part of the watershed, with an average slope of 11–15%) may be expected, even when self-sufficiency constraint levels are raised. This is most probably due to the high suitability of land type C for maize cultivation both for consumption and as a cash crop as opposed to teff and wheat growing. The chances of this land area remaining under maize production are hence high, even when more own-farm-produced teff and wheat are required to meet consumption at higher levels of 2000 calories AE^{-1} day^{-1}. Such cultivation in the final year could also be a result of the 'end-game effects'.

On land type D (the hilly part of the watershed with slopes of over 16%), model projections suggest a 10 ha allocation of the land area to cereal production throughout the 12-year plan period. The allocation of this land to various crops across the years begins with sorghum cultivation in the first year, followed by maize in the second to the tenth year, teff in the 11th year and a local wheat variety in the final year. Thus, the least erosive crops are cultivated on the steep slopes initially but, as yields decline in the high-potential areas of the watershed, staples, such as teff and local wheat varieties, are cultivated in these parts as well.

These types of land use on the steep slopes appear to agree with the land use observed in the base year for most of the steep slopes that are dominantly used for the cultivation of cereals. We also note that the new wheat variety is not likely to be suitable for the steep slopes in Ginchi and elsewhere in the highlands, given the relatively shallow and less fertile soils. Crops such as sorghum and the traditional wheat variety are likely to be more suitable. With increasing demand for home-produced food, we note the way the new wheat variety forces traditional teff and wheat varieties out of the prime lands to the poorer and steeper slopes, such as land type D.

Tree land-use pattern

An examination of model results reveals no adoption of sesbania (forage trees) on any of the land types in the watershed. Eucalyptus trees are likely to be grown mainly on the high slopes of land type D (none is envisaged on land type A, B and C). Land type B is used, mainly, for cultivation of the new variety of wheat instead (as explained above). The positive effects of tree planting in the fertile bottom slopes are unlikely to be achieved and soil conservation is primarily to be expected through short fallows. On the steep slopes (land type D), however, 30 ha are put under eucalyptus trees throughout the plan period. This has two effects. First, the steep slopes are protected from extensive soil loss over the years and, secondly, watershed cash flow is greatly improved, especially after the 4-year tree gestation period. Both the first harvests and the tree regrowths that are harvested subsequently are sold at per ha profits that are almost ten times those of ordinary crops.

The optimal land-use practices generated by the model have some correspondence to observed watershed land-use patterns. Crop rotation is a common and central practice. On-farm trials reveal the BBM technology to be effective but too labour-intensive. Adoption of the ET 13 wheat variety in many parts of Ethiopia has been ad hoc and has tended to be location-specific.

Comparing the soil-loss levels of the two situations, with and without multiple interventions at recommended consumption levels, a widening gap in the amounts of soil lost over time is evident. In the scenario with multiple interventions, reduction in soil loss, in spite of higher consumption levels, is due to planting of eucalyptus trees on the steep slopes in the base year. This suggests that both natural resource conservation and improved human-nutrition goals could be achieved simultaneously through improved land-user rights that facilitate the multiple adoption of commercial trees, new high-yielding crop varieties and increased fertilizer application. This is especially the case when tree planting is done early enough, i.e. at the beginning of the plan period. Explicit soil-conservation measures, such as the construction of grass strips or even bench terraces or stone walls, are apparently not adopted even when a more secure food policy is pursued. The adoption of eucalyptus trees is mainly because of its dual (direct and indirect) impact on incomes and hence social welfare.

Similarly livestock are likely to play a key role in driving the whole system by providing the draught power required by some of the technologies and crop activities and also through nutrient cycling. Unlike in scenario 1, where it is assumed that current policy and institutional arrangements result in a static number of animals being kept, this assumption is relaxed in scenario 2 and a more commercially oriented live-stock-keeping strategy is simulated. Live-stock numbers are allowed to vary according to need and overall profitability, rather than for purely cultural reasons. Results tabu-lated in Table 18.2 show that livestock num-bers are likely to be reduced to less than a third their current numbers in the base year, but could rise gradually over the years as more and more of the new wheat variety is adopted. Adoption of new wheat varieties thus calls for a higher per ha oxen power for a finer tillage than is the case when a local wheat variety is cultivated. By the end of the 12-year planning horizon, livestock num-bers are projected to have risen by 86% from values estimated at the beginning of the plan period (small ruminants inclusive). Increases in stocking rates over the years could also be attributed to the rising demand for dung manure to replace some of the nutri-ents lost cumulatively through soil erosion, crop harvests and emissions. Again, it may be driven by the need to utilize the resulting crop residue arising from increased crop output.

Policy Conclusions

Under current policy conditions, the use of environmentally friendly animal and crop husbandry that fits the farmers' resource endowments and priorities (such as crop rotation) are likely to be preferred in the watershed to purely conservation-oriented technologies. Such practices may include application of scarce inputs, such as dung manure and chemical fertilizer. With improved land-user rights, multiple adoption of productivity-enhancing tech-nologies may prove to be more profitable than a piecemeal single technology-adoption approach. The former appears to benefit from some positive interaction effects among technologies. The importance of more secure land rights is emphasized by the model results. Only with security and a form of collective action will the optimal conservation measures occur, namely plan-tation of eucalyptus on a reasonable scale on land type D.

Overall, technology adoptions, input allocations, cumulative soil-loss levels, vari-ation in livestock numbers and increased net cash flows require a more secure land-user-rights policy. A change in any conditioning variable(s) that have an impact on any one of these issues is likely to have some effect on watershed sustainability indicators. For instance, with no policy change and hence low technology adoption, average per ha nutrient balances are −58 kg for nitrogen, −32 kg for phosphorus and −114 kg for potassium. With adoption, as demonstrated above, nutrient balances are likely to be as low as −25 kg, −14 kg and −68 kg of nitrogen, phosphorus and potassium, respectively. This is in spite of the higher human consumption levels.

It is also demonstrated that technology adoption is likely to vary across time and space. Thus, in defiance of initial expecta-tions by scientists that the BBM technology would result in cultivation of new wheat varieties on land type A, model results show this to be highly unlikely. Traditional wheat varieties are likely to be preferred in the immediate future and only when the cumulative effects of erosion set in will the new wheat variety be adopted. Again, this is likely to be on land type B, where the explicit need to drain the land using the BBM technology is low, due to a higher gradient. Cultivation of the new wheat variety on land type B rather than A is thus, among other things, a labour-conservation strategy. Dual-purpose technologies, such as eucalyptus trees, are also likely to be more appealing than initially envisaged and may require explicit policy control to ensure food self-sufficiency in the wake of market failure. The role of livestock in facilitating all the above changes is underscored.

Notes

[1] During the 1975 land-reform programme by the then socialist government, land was declared the property of the state and administrative units called peasant associations were formed to redistribute land equitably within its boundary. The new government that came into power in 1991 maintained land as a public property but has given slightly more user rights.

[2] This was the level observed in the watershed survey undertaken as part of the study reported here. A comprehensive nutritional study, completed in 2000–2001 in the community, indicates household consumption per adult equivalent to be higher than this value for the whole year but close to it during lean months (M. Jabbar, personal communication).

19 Nutrient Cycling in Integrated Plant–Animal Systems: Implications for Animal Management Strategies in Smallholder Farming Systems

Lindela R. Ndlovu[1,2,]* and Priscah H. Mugabe[2]

[1]*Department of Animal Production, University of the North, Private Bag X1106, Sovenga, South Africa;* [2]*Department of Animal Science, PO Box MP 167, University of Zimbabwe, Mount Pleasant, Harare, Zimbabwe*

Poverty and food insecurity characterize most of the agriculture in sub-Saharan Africa (SSA). This necessitates effective management of the natural resources on which African agriculture depends. The soil is a major resource base in this system and its interaction with plants and animals constitutes a potential entry point for the adoption of sustainable management strategies that contribute to the overall portfolio of adoption of natural resource management strategies. In most of SSA, the greatest opportunity for sustainable agriculture exists through close integration of plants and animals with each other and with the soil, each providing an input into the productivity of the other and making use of some output from the other system. The integrated agricultural systems could be animal–range, animal–crop or animal–range–crop. The interactions between plants, animals and soil that we emphasize in this chapter involve nutrient transfers.

The principal crops in less humid environments are maize, sorghum, millet, pulses (cowpeas, groundnuts, bambara nuts, beans, etc.) and the *Cucurbitae* family of crops. The major livestock are cattle, sheep, goats, poultry and, to a limited extent, pigs. In these integrated systems, there is an exchange of nutrients that involves the soil. The animal component consumes plant material, which undergoes digestion and absorption, processes that extract some (most) nutrients for metabolic uses in the animal. Those nutrients not digested and/or absorbed are passed out of the animal as excreta, which go to the soil. Excreta are acted upon by soil microorganisms to release nutrients, which are then available for use by plants, which may in turn be consumed by animals, and the cycle continues. This system is probably sustainable over the long term in low-output agriculture, but degradation of soil and loss of animal and plant productivity quickly ensue if intensification

---

* Current mailing address: National University of Science & Technology, Corner Gwanda Road/Cecil Avenue, PO Box AC 939, Ascot, Bulawayo, Zimbabwe.

©CAB *International* 2002. *Natural Resources Management in African Agriculture*
(eds C.B. Barrett, F. Place and A.A. Aboud)

is implemented without adoption of appropriate resource management practices.

This chapter will briefly review ruminant nutrition concepts related to soil-nutrient management in SSA farming systems, as most of the available traditional strategies involve ruminant animals. The discussion will then focus on the potential impact on nutrient cycling of some ruminant nutrition-related technologies that have been propagated by both national and international research organizations for integrated (mixed) plant–animal farming systems and finally the adoption patterns (or lack thereof) of these and related soil management techniques.

Ruminant-nutrition Concepts

Farmers of SSA keep ruminant livestock for multiple reasons. They produce essential food products throughout the year, contribute draught power and manure to crop production and provide financial reserves for periods of economic stress, as well as offering avenues for the investment of cash surplus for immediate needs in rural areas. They also act as a major source of food and cash income in rural areas, particularly in drought years when crops fail. However, livestock may compete with crops for resource inputs, including the supply of nutrients, especially nitrogen (N) and phosphorus (P), which are needed by both production components. The management of such crop–livestock competition is important for sustainable natural resource use in such systems.

The cycling of biomass from natural vegetation and crop residues through ruminant livestock into faeces and urine, which are used to ameliorate soils, is a widely used strategy in SSA (Powell et al., 1996; Snapp et al., 1998). Efficient cycling of nutrients within a farming system is a prerequisite for long-term sustainability (Romney et al., 1994). The effectiveness of livestock in recycling nutrients in the farming system depends on the types and numbers of animals kept by farmers and their feeding and watering regimes, as well as the temporal and spatial distribution of the livestock in the landscape. These factors are influenced by the management strategies that the farmers adopt. Four general livestock production systems are found in SSA: pastoral, agropastoral, mixed crop–livestock and commercial (often large-scale). Mohamed-Saleem (1998) detailed the major characteristics of each of the systems.

Faeces and urine are waste products of the ingestive and digestive processes of animals. The interaction between the form in which nutrients are ingested and the digestive processes in the animal will influence the concentration and balance of nutrients in excreta (Romney et al., 1994). This results in high variability in the amount and nutrient content of faeces and urine produced by animals. Faeces consist mostly of microbial and other endogenous components, plus that portion of the feed that is resistant to microbial and mammalian digestive enzyme action (Van Soest, 1994) and that passes out of the gastrointestinal tract with minimal alterations. Endogenous additions of enzymes, microbial cells, minerals and other macromolecules occur at variable rates (Orskov, 1982), depending on the feed, animal species and condition of animal. Urine, on the other hand, consists mostly of products of metabolism – especially protein metabolism – and consists mostly of urea. When applied to soils, most urinary N can be lost by volatilization and leaching, while faecal N is less susceptible (Powell et al., 1998b). In acidic soils, urine might be beneficial by increasing pH, which in turn increases the availability of P. P and N are often the main limiting nutrients in SSA soils (Powell et al., 1998a).

The level of food intake, the extractability of nutrients from the feed and animal requirements affect the production of waste from animals (Van Soest, 1994). In general, 50–60% of the nutrients ingested are retained by the animals (Van Soest, 1994). Increases in intake result in increased faecal output if digestibility and animal physiological conditions do not change. The feed intake of animals is affected by body condition, species, breed of animal, physiological state, environmental factors and feed availability and quality (Van Soest, 1994). Recent evidence suggests a major role for the previous experience of the animal with the feed in question, in addition to the above factors, in determining the feed intake of animals (Provenza, 1995). This implies that caution needs to be exercised when carrying out short-term stall-feeding experiments to allow for learning (and adaptation) if the results are to be extrapolated to extensive systems.

The quantity and quality of animal feed have a major impact on the amount and composition of animal excreta (Powell *et al.*, 1994; Romney *et al.*, 1994; Van Soest, 1994; Somda *et al.*, 1995; Gerdeman *et al.*, 1999), through effects on digestion and metabolism. In ruminant livestock, the major nutrient that affects the digestibility of feeds is protein because of its function in increasing microbial protein synthesis and thus increasing the population of microbes in the rumen (Van Soest, 1994). Consequently, increasing the crude protein content of the diet tends to reduce the amount of dry matter excreted in faeces unless the diet contains a high proportion of polyphenolic compounds.

The use of cereal grains and oil-seed cakes/meals in animal diets is prevalent in the commercial sector and is often recommended for intensive production by smallholders. Non-structural carbohydrates that are not digested are excreted in the faeces and are rapidly fermented by saccharolytic microbes, causing low pH in the faeces, especially soon after grain harvest, when livestock have access to a large amount of unprocessed grain (Murwira, 1995). The effect of this on manure quality has not been studied to any extent, although Murwira (1995) found that it had a minimal effect on the volatilization of N as ammonia in fresh faeces and when faeces are applied to the soil.

The partition of N excretion between urine and faeces is affected by diet composition (Romney *et al.*, 1994), especially the presence of N, fibre, lignin and tannins. Tannins are widespread in plants used for feed and food in tropical and subtropical areas and are abundant in feed resources readily available to ruminant livestock in smallholder farming systems. Tannins are complex plant secondary metabolites that are polyphenolic in nature and are soluble in polar solutions. Their main distinctive characteristic is their ability to precipitate macromolecules, such as proteins, carbohydrates and mineral complexes. The presence of tannins and related phenolic compounds has a major effect on N retention and the pathway of N excretion (Table 19.1). In general, a high content of condensed tannins (also called proanthocyanidins) results in a shift of N away from urine to faeces (Reed *et al.*, 1990; Powell *et al.*, 1994; Dube and Ndlovu, 1995; Reed, 1995; Woodward and Reed, 1997), although exceptions have been reported (Nherera *et al.*, 1998).

The brief review of animal nutrition concepts above has highlighted the importance of the composition of the animal's diet on the quality of faeces and urine that the animal produces. The section below will explore some issues related to the quantity of excreta produced by animals.

Production of Manure

A critical issue to sustainable crop production in SSA is the availability of manure for crop production. We define manure as the combination of faeces, urine, bedding, feed refusals, etc. that farmers apply to crop land. Of particular importance to manure availability are the number and types of livestock a smallholder farmer keeps. It has been estimated that a 500 kg ruminant in southern Africa produces 2–3 tons of recoverable manure per year if kraaled overnight (Schleik, 1986). Fernandez-Rivera *et al.* (1995) estimated corresponding production to be 2.6–3.5 tons for cattle, 0.01–0.33 tons for sheep and 0.01–0.20 tons for goats in West Africa. The estimated amounts of N and P in sheep manure range from 10 to 22 and 1.3 to 2.75 g kg^{-1}, respectively, and in cattle they range from 12 to 17 and 1.5 to 2.1 g kg^{-1}, respectively (Powell *et al.*, 1998b). Based on the content of these elements in manure and plant requirements, it is estimated that between 5 and 10 tons of

Table 19.1. Partitioning of feed nitrogen between faeces and urine in ruminants fed Acacia (A.) spp., Securinega virosa (S. virosa), Ziziphus mucronata (Z. mucronata), Leucaena (L.) spp., Calliandra calothyrsus (C. calothyrsus), Sesbania sesban (S. sesban) and vetch (Vicia (V.) dasycarpa).

Forage	Animal species	Condensed tannins	N intake (NI) (g day^{-1})	Faecal N (% of NI)	Urinary N (% of NI)	Reference
A. karroo	Goat	243 g kg^{-1}*	10.1	73.3	15.8	1
A. nilotica	Goat	67 g kg^{-1}	9.5	42.1	35.8	1
S. virosa	Goat	15 g kg^{-1}	18.0	26.7	30.0	1
Z. mucronata	Goat	46 g kg^{-1}	14.3	38.0	23.8	1
A. tortilis	Sheep	31.4 AU[†]	6.7	74.6	22.3	2
A. albida	Sheep	26.8 AU	5.4	75.9	22.2	2
A. nilotica	Sheep	89.2 AU	5.3	77.4	18.9	2
A. sieberiana	Sheep	37.4 AU	5.6	80.4	17.9	2
L. leucocephala	Goat	19.6 AU	7.5	53.6	2.5	3
L. diversifolia	Goat	19.7 AU	6.7	44.1	5.8	3
L. pallida	Goat	18.7 AU	6.8	44.0	5.4	3
C. calothyrsus	Goat	29.9 AU	6.7	43.5	4.7	3
A. brevispica	Goat	0.37 AU	10.9	41.0	51.0	4
S. sesban	Goat	0.01 AU	12.8	29.0	48.0	4
V. dasycarpa	Goat	0.01 AU	9.1	44.0	53.0	4
A. brevispica	Sheep	0.37 AU	11.2	46.0	41.0	4
S. sesban	Sheep	0.01 AU	13.2	34.0	48.0	4
V. dasycarpa	Sheep	0.01 AU	9.8	42.0	53.0	4

*g kg^{-1} as vanillin equivalent.
[†]Absorbance units at 550 nm g^{-1} of neutral detergent fibre (NDF).
1, Dube and Ndlovu, 1995; 2, Tanner et al., 1990; 3, Nherera et al., 1998; 4, Woodward and Reed, 1997.

manure are needed per hectare, annually, depending on the crop type. These estimates suggest that the amounts produced or the numbers of livestock required (and hence land) to produce adequate amounts are untenable if manure is used as the only soil amendment strategy. However, poultry manure consists of much higher proportions of N (48 g kg^{-1}) and P (18 g kg^{-1}) than ruminant manure, and thus only 1–2 tons of it are required to achieve a productivity equivalent to 7 tons of ruminant manure (Nandwa and Bekunda, 1998). In most smallholder crop-production systems in SSA, a combination of inorganic and organic fertilizers is recommended (and, even where such recommendations are non-existent, smallholder farmers use various combinations of organic and inorganic fertilizers). Despite the popularity of this practice in smallholder systems, optimum combinations for the different soil types and crops have not been authoritatively defined.

Estimates for urine production are few. Somda et al. (1995) estimated that, on average, cattle excrete 17–45 g of urine kg^{-1} of live-weight daily (approximately 8.5–22.5 kg for a 500 kg animal) and sheep and goats excrete 10–40 g kg^{-1} live-weight daily. However, the amount of urine voided is highly variable: intake of water, salt, N and polyphenolics influence it. Powell et al. (1998a) found that an average sheep voided about 64 g of urine day^{-1}, while Hove et al. (2001) reported results that ranged from 31 g per goat day^{-1} when fed 320 g leaves of Calliandra calothyrsus (a browse high in condensed tannins) to 236 g per goat day^{-1} when fed 320 g of cottonseed meal (an oil-seed meal with more than three times more N than in C. calothyrsus). Nsahlai et al. (1998) found that the use of sorghum stover from a bird-resistant variety in combination with noug oil-seed cake resulted in excretion of copious amounts of urine relative to animals fed the non-bird-resistant sorghum variety in combination with noug oil-seed cake. This was attributed to the content of tannins in the bird-resistant sorghum, which was higher than in the non-bird-resistant sorghum.

The amounts of both faeces and urine produced are also affected by the spatial and temporal fluctuations in feed availability and quality that are intrinsic to SSA farming systems (O'Reagan and Schwartz, 1995). Production of faecal matter is lowest in the dry season, when feed supplies are limited. N and P levels in the diet are much lower in the mid-dry season than in the early dry season or the early wet season (Powell *et al.*, 1996; Mpofu *et al.*, 1999). The age and production level of the animal also affect nutrient concentration in excreta. Young animals that are growing rapidly or dairy cows excrete manure with a lower nutrient content than fattening animals (Romney *et al.*, 1994).

Role of Animal Excreta in Nutrient Cycling

The main nutrients supplied by animal excreta in rangelands and crop lands of SSA are N and P. The supply of plant-available P due to manure application is a combination of release of soil endogenous P as a result of increased pH following application of urine, and P contained in faeces. Urine of ruminant animals is almost devoid of P (Ikpe *et al.*, 1999). An equally important contribution of animal excreta in the amendment of sandy soils is the increase in soil organic matter (SOM). Low SOM is probably the major cause of poor fertility in sandy soils (Snapp *et al.*, 1998). Manure increases SOM in soils, but sometimes this is not accompanied by increased productivity in the cropping systems due to temporary soil N immobilization (Nandwa and Bekunda, 1998). Manures from animals fed diets high in polyphenolics have shown both rapid mineralization of up to 50% of its N and P after application to soil (Somda *et al.*, 1995) and high resistance to any mineralization (Mafongoya *et al.*, 1997). The lack of mineralization implies decreased decomposition of manure and

plant material in the soil and hence increased SOM. These differences could be due to the different chemical structures of the phenolic compounds in the browses fed. This points to a need for more studies on the relationship between the chemical structure of polyphenolics and the physical and chemical attributes of manure if we are to more accurately predict the contribution of manure to crop nutrient demands.

Adopted Strategies and Impact on Nutrient Cycling

Use of crop residues as feed

This strategy has been widely adopted in its various forms (see African Research Network on Agro-Byproducts (ARNAB)[1] publications 1985–1990 and a subsequent series by the Animal Feed Resources Network (AFRENET, 1990–1995). It entails either harvesting the crop residues from the field, storing them in the kraal or elsewhere and feeding them to the whole herd or selected animals during the dry season or allowing the animals to graze the residues *in situ*.

The use of crop residues for feed involves removing nutrients from the cropping area and requires replenishing them. Manure produced in the kraal is collected and spread in the fields to achieve this. However, losses of up to 50% of N are incurred during storage of manure (Romney *et al.*, 1994; Reynolds and de Leeuw, 1995). Storage as slurry (Whitehead, 1990) or application when fresh accompanied by turning into soil (Romney *et al.*, 1994) reduces losses tremendously. In most smallholder systems these practices are not carried out because of cost and high labour demands.

In most of southern and eastern Africa, the use of crop residues as feed results in reduced weight losses, increased survival and improved draught-animal condition at the start of the ploughing season (Ndlovu *et al.*, 1996). Its adoption could be linked to the multiple benefits that accrue to individual households that practise it. However, it has a cost in terms of nutrient cycling, as most nutrients are not returned to the cropping land and, for it to be sustainable, there is a need for additional nutrients to be sourced external to the system. Indeed, in Zimbabwe, most smallholders use a combination of manure and inorganic fertilizer in their fields (Murwira et al., 1995), but the optimum rates of combination are not firmly established. In view of the variability in the quality of manure, there is need for definitive research on the 'best-bet' rates of combination for manures from various sources (animal species and diet types).

Crop residues have a low feeding value and can only meet maintenance requirements of animals with minimal excess nutrients for production. In order to increase their feed value, there is a need to process them or feed them with other feedstuffs as supplements. A variant of the strategy of direct feeding of crop residues has been to treat them with urea, ammonia or alkali to improve palatability and digestibility (Preston and Leng, 1987). The adoption of this variant has been minimal, due to costs of the chemicals and limited gains in terms of weight or milk produced. The use of urea or ammonia is likely to result in increases in urinary N excretion, which can be easily lost via ammonia volatilization.

Use of forage legumes as supplements

It has been widely recognized that the most limiting nutrient in ruminant diets in smallholder systems is N, especially during the dry season (Elliott and Topps, 1963). Various strategies to supply this nutrient have been suggested (Preston and Leng, 1987), but most have not been adopted by smallholders. The use of forage legumes is relatively recent in southern African smallholder systems (Dzowela et al., 1997). Most legumes are high in protein, minerals and readily soluble cell contents and low in fibre content (Van Soest, 1994). This makes forage legumes ideal supplements for ruminants fed on crop residues and dry grass, which are low in N and minerals.

Many tropical legume forages, however, tend to be high in polyphenolics and, as discussed above, this could affect their palatability, digestibility and partitioning of N excretion between urine and faeces. This could have large impacts on nutrient cycling if the partitioning favours N excretion through urine in a system where manure is collected from pens/kraals, stored and then applied to the fields, as manure collected from kraals is often devoid of urine since most urine will have evaporated during storage. On the other hand, allowing animals to graze crop lands overnight ensures that both urine and faeces are returned to the soil. While some urinary N will be lost through volatilization, this practice has been found to increase crop yield over use of manure alone (Ikpe et al., 1999). In some areas, this management option is untenable because of predators and livestock rustlers.

Various recent and ongoing studies have evaluated the use of tree leaves as browse and as a soil-fertility amendment, inter alia. While the two uses could be complementary, there is a real possibility of competition. Very few studies have examined the comparative advantages of cycling these forages through livestock versus direct application. Somda et al. (1995) concluded, from a series of experiments in acid soils in Niger, that feeding plant material to animals and applying manure to soil can accelerate the humification processes and nutrient turnover rates. However, the results of Mafongoya et al. (2000) suggest that the opposite might be the case for southern Africa.

Increased use of tanniniferous feeds in livestock diets will necessitate the management of tannins in the leaves of these feeds. This can be through drying (Ahn et al., 1997), mixing different forages (Hove et al., 2001) and use of polymers to bind tannins (Silanikove et al., 1997; Waghorn et al., 1997). The addition of polymers shifts N towards urine and increases the amount of intact and reactive polyphenols in the faeces. The impact of this on mineralization of the manure and on the ecology of the soil has not been determined but could be quite major.

Rangeland strategies

In an attempt to overcome the problem of deteriorating rangeland potential in SSA, much empirical work has been done on grazing management systems. Various systems have been recommended in which the frequency, intensity and seasonal timing of grazing are altered in different ways. A common feature of these systems is that management activities (movement of animals, changing of animal numbers, use of fire, bush clearing, etc.) are performed on a regular or constant basis. Each system has met with varying success at different times and in different places (Walker *et al.*, 1986).

Most research on grazing management technologies has not analysed effects on soil fertility or nutrient cycling. We shall attempt to deduce these effects on African rangelands from some of the published research results.

Lavado *et al.* (1996) found no significant differences in organic carbon (C) and total N between grazed and non-grazed range. The extractable P was slightly, but significantly, lower in the grazed area. The lack of differences between grazed and ungrazed areas in the spatial structure of organic C and total N mainly related to changes in the grassland structure. While the C return in the ungrazed area is spatially homogeneous because of the occurrence of litter, the patches of bare soil and the uneven distribution of the scarce litter in the grazed area could induce a higher heterogeneity of the C returns. It is possible that the spatial distribution of C input in the grazed area induced the observed spatial dependence of extractable P. The technological implication for African rangelands is that grazing management systems that result in patchy spatial animal distribution would result in patchy soil-nutrient cycling systems. For instance, Lavado *et al.* (1996) suggested that the apparent lack of excreta effect on the spatial nutrient variability may be attributed to the low livestock density in the studied grasslands, the large size of the plot (800 ha) and the distance to the water source (1200 m) which would 'dilute' the effect of animal excreta in the grazed area. Moreover, the spatial distribution of the animals could

lead to concentration of the excreted nutrients in non-productive areas of the range, such as along animal routes of movement, in resting areas and around watering points (Powell *et al.*, 1998b).

The pattern of soil and forage properties within a pasture influence cattle behaviour and urine deposition and so result in variations in the proportion of total deposition vulnerable to loss (Schimel *et al.*, 1986). The magnitude of losses from urine patches relative to potential losses from senescing vegetation suggests that the latter pathway is worthy of significant attention in grasslands, both as a loss vector for N and as a source of atmospheric NH_3 (Schimel *et al.*, 1986).

Frank *et al.* (1995) carried out a study of adjacent moderately and heavily grazed mixed-prairie pastures. Soil organic C by depth was not statistically different between exclosure and heavily grazed areas, but exclosure had significantly higher soil organic C than moderately grazed areas. These results suggested that moderate grazing slightly reduced soil organic C. These findings were attributable in part to C removal by grazing animals. Bauer *et al.* (1987) had also reported reduced soil organic C but not N content of grazed native grasslands compared with ungrazed grasslands.

The above results notwithstanding, high levels of grazing can accelerate the cycling of nutrients from plants back to the soil. The substrate chosen by herbivores generally has higher nutrient concentrations than the litter used by decomposer organisms, because herbivores are selective and because the leaves are eaten before nutrient reabsorption from senescing leaves can occur. High-intensity grazing sites, such as the Serengeti Plains of Africa, are characterized by high nutrient availability and rapid nutrient cycling through herbivores. In contrast, nutrient-poor ecosystems, such as the P-deficient grasslands of South Africa, do not support high levels of herbivory (O'Reagan and Schwartz, 1995). When grazing occurs in such ecosystems, it causes a reduction in nutrient uptake and plant growth (Chapin *et al.*, 1985) because of the slow release of nutrients from organic matter. Organic material produced in nutrient-poor sites decomposes and releases nutrients more slowly than that produced in nutrient-rich sites for several reasons. Firstly, the quality (inherent decomposability) of the organic C substrate is low; decomposers may be energy-limited. Secondly, phenolics and lignin can further alter decomposition and/or nutrient release through direct effects on microbes (Mafongoya et al., 1998). The high ratios of C to nutrients in the litter produced in nutrient-poor sites lead to a protracted immobilization of nutrients in microbial biomass and its by-products, and therefore to slow nutrient release (Swift et al., 1979). The land carrying capacities that are recommended in most SSA farming systems are based on beef production and are lower than the ecological capacity of the rangelands. This results in patchy spatial distribution of animals in the range and could have a negative impact on nutrient cycling. However, the potential for irreversible damage due to higher stocking rates cannot be completely discounted, as indicated below.

In Kenya, Mworia et al. (1997) showed that the ability of the range site to recuperate from shifts towards degradation depended mainly on two factors: the level of grazing intensity and the variation in rainfall. In their study, stocking densities greater than four heifers ha^{-1} produced soil bulk density changes that were irreversible during a 2-year rest period. These results are of relevance to the SSA rangeland situations in that most rangelands are managed under very high stocking intensities.

Among the management practices, the increase in nutrient availability of the grassland system is viewed as a key factor (Lavado et al., 1996). Undoubtedly, fertilization and increasing stocking rate will have a large effect on the SOM and nutrient dynamics. Lavado et al. (1996) showed that, after 13 years of protection from grazing, significantly lower extractable P was found in the soil of the grazed area than in the ungrazed area. This indicates a non-steady-state situation for this nutrient under

grazing. To achieve a sustained level of forage production from the natural vegetation, P fertilization was suggested. Such a technology would have limited success in African peasant farming systems because of prohibitive costs and competition for fertilizer requirements with cropping systems.

In general, while there are few empirical data on the effect of grazing management technologies on rangeland nutrient cycling in SSA, it can be postulated that nutrient cycling would be negatively affected by the patchiness of grazing, harsh abiotic conditions and high stock numbers per unit area common in most situations. There is a need for more quantitative research in this area in order to provide meaningful inputs in bio-economic models being developed as tools for sustainable natural resource management.

Whither Technological and Policy Initiatives?

The previous sections outlined some natural resource management technologies with a potential for adoption in smallholder farming systems, but the reality on the ground is that most are not adopted at any meaningful scale. In this section we highlight some of the constraints to adoption of technically feasible natural resource management technologies based on crop–livestock integration.

Labour requirements

The majority of the strategies (e.g. cut-and-carry browse systems, moving dry manure from kraals, etc.) have high labour demands. In SSA, males often seek employment away from their homes and this reduces labour availability. Even in situations where the male members of the households are not in full-time employment they often seek seasonal employment on commercial farms or the lands of wealthy rural dwellers, further depleting labour available in their own agricultural enterprises. This is further exacerbated by the decline of the social networks upon which depended the cooperative work parties that alleviated household-level labour shortages (Scoones *et al.*, 1996). A potential exists for small-scale mechanization for some of these processes in order to alleviate the drudgery associated with them. However, such strategies should not be costly and this calls for the imaginative use of design and material by researchers, in combination with communities that will use the technologies.

In crop–livestock systems, herding of livestock during the cropping season is labour-intensive, especially as school attendance becomes prevalent. In past years, teenagers did this chore but, with increased access to education, teenagers stay in school longer and sometimes move to boarding-schools quite distant from their homes, thus becoming unavailable for herding work. This affects where cattle are grazed, often around homesteads while waiting for children to come back from school, leading to denudation of the area. Fodder-bank technologies would seem to offer an opportunity for retaining cattle longer in kraals without causing degradation to the environs of the homestead. However, labour demands for cutting and carrying the fodder to the animals has made the system of limited success. Paddocking and/or purchase of good-quality feed are generally not suitable options because of limited access to investment capital and credit opportunities in smallholder agriculture in SSA.

Access to credit

The uncertainty of smallholder agriculture, especially under dryland conditions, limits the possibilities of commercial credit facilities. Non-profit organizations and governments regularly investigate alternative credit schemes that would encourage the adoption of technologies in such areas. While non-farm income has been suggested as a substitute for credit (Scoones *et al.*,

1996), state and donor investment in such endeavours needs to be examined against the trade-offs and benefits of large-scale natural resource management efforts that exclude community participation.

Grazing management technologies

There has been very limited adoption of grazing management technologies that aim at the prevention of overgrazing. Most of the technologies proposed for grazing land management are premised on the pending calamity that will ensue if stocking rates are not limited to fixed numbers based on beef-production objectives (Scoones, 1993). These range management strategies assume a steady state (equilibrium state) of vegetation – the climax – which, if perturbed, will lead to a reduction in both plant and animal production. The theory assumes that it is livestock that alter the balance: if their numbers exceed the ability of the plant population to replenish itself, then degradation will occur. The numbers that need to be retained are calculated to ensure maximal growth rates of individual animals. In a range that is in disequilibrium, the vegetation cover is determined by rainfall more than by animal consumption, and rates of stocking can be maintained at levels higher than is predicted from beef models during rainy seasons. The use of key resources, such as wetlands and riverine areas, during the dry season ensures that populations do not collapse, as predicted by the classical range management strategies (Scoones, 1995). Given the non-terminal benefits, such as draught power, milk and manure, which are valued by rural dwellers, it is not surprising that management techniques based on reducing livestock numbers have not been widely adopted. This calls for efforts that are more sensitive to indigenous knowledge and technologies, plus better involvement of communities in determining which technologies are appropriate for their circumstances.

Conclusion

Soils are an important resource in small-holder agriculture, and interactions with plants and animals determine its sustainability in the absence of external interventions. Ruminants play an important role in nutrient cycling in SSA agroecosystems and offer opportunities for the adoption of sustainable soil management strategies, which feed into broader natural resource management efforts. However, very few such technologies are adopted in SSA, largely due to labour and credit constraints.

Note

[1]   ARNAB was coordinated and funded by the International Livestock Centre for Africa (ILCA); it later became AFRENET by amalgamating with the Pasture Network of East and Southern Africa (PANESA) and West African pasture networks. ILCA is now the International Livestock Research Institute (ILRI).

20 Natural Resource Technologies for Semi-arid Regions of Sub-Saharan Africa

Barry I. Shapiro[1] and John H. Sanders[2]

[1]International Crops Research Institute for the Semi-Arid Tropics (ICRISAT),
BP 320, Bamako, Mali; [2]Purdue University, 1145 Krannert Building,
West Lafayette, IN 47907-1145, USA

As population pressure and soil degradation have proceeded in the semi-arid zones of sub-Saharan Africa, farmers have responded with a series of labour-intensive measures intended to conserve soil resources. However, on closer inspection, yield effects are also clearly important objectives for the low-income farmers exhibiting high time discount rates, because natural resource management (NRM) investments improve the utilization of available water and are generally combined with soil-fertility improvements. This chapter reviews information available on the diffusion of a number of these technologies and evaluates the constraints to their more rapid diffusion.

The semi-arid zones are largely neglected sectors in the agriculture of these countries and have become concentrations of rural poverty. But there is an important efficiency objective to be gained from greater concern with the semi-arid regions. When water availability and soil fertility increase, semi-arid regions can have a comparative advantage over many higher-rainfall zones due to their lower plant-disease incidence and longer sunlight hours. This has been well recognized in California, Israel, Australia, the Iberian Peninsula and South Africa. The highest crop yields in the world generally come from formerly arid or semi-arid regions with irrigation. Part of these large effects from irrigation can be obtained as a result of improved water use from a series of techniques that better utilize available water. Fortunately, there are a large number of these NRM and other water-retention technologies besides irrigation.[1]

The chapter begins with a brief differentiation of two principal cereal-based production systems in semi-arid regions. Then comes the chapter's main section on natural resource technologies, first considering the heavier and then the lighter soils. We draw lessons learned from the available information on the diffusion of the major types of water-retention technology and then address the two related issues of soil fertility and new cultivar development. Finally, the conclusions section reviews the characteristics of natural resource technologies for the semi-arid zones and makes some public policy recommendations to accelerate their diffusion.

Farming Systems in the Semi-arid Regions

After Australia, Africa is the continent where agriculture is most subject to

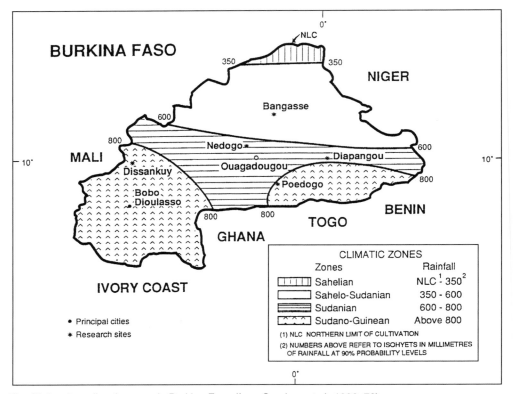

Fig. 20.1. Agroclimatic zones in Burkina Faso (from Sanders *et al.*, 1996: 73).

drought. Thirty sub-Saharan countries depend heavily upon drylands for their crop area and the support of their rural populations (see Appendix to this chapter). This chapter focuses on the semi-arid zone, though the dry subhumid zone is also subject to periodic droughts.

The semi-arid region is divided into Sudanian and Sahelo-Sudanian, corresponding approximately to sorghum- and millet-based cropping systems, respectively (Fig. 20.1). This division is based upon rainfall at the 90% probability level and is similar to divisions resulting from the aridity index and crop-growing season (see Appendix to this chapter), but allowed a further distinction between semi-arid zones (Gorse and Steeds, 1988).

Outside the Sahel, other cereals (and grain legumes) also become important in semi-arid regions. In East and southern Africa, maize often pushes into the semi-arid zone, as does teff in Ethiopia. Nevertheless,

these farming systems generally continue with sorghum as another insurance crop. Besides rainfall, there are important soil distinctions, since millet is not only more drought-resistant but also more tolerant of low soil fertility and is found on sandier soils. Compared with millet, sorghum prefers heavier soils, with some clay or silt.

Natural Resource Management Technologies for the Semi-arid Zone

Farmers in the semi-arid regions engage in many activities to control erosion. Bunds or dykes of dirt, stone or living vegetation are placed on the contour. Terracing has the same effect of reducing erosion on more steeply sloped areas. Holes are dug in the field, *zaï* in Burkina, trenches in Ethiopia. All these techniques slow up and thereby hold water behind or in them. Catchments of various types also retain and then utilize the available water.[2]

With a little more water available from the NRM technique, the returns to soil-fertility improvement are increased and the risk reduced (Viets, 1962). The yield and risk-reducing effects of a typical water management technique, tied ridges,[3] are illustrated with data from researcher-controlled on-farm trials in Burkina Faso and Ethiopia (Tables 20.1 and 20.2).

NRM techniques differ between the soil types, corresponding to the sorghum- and millet-based systems. On the crusting soils,

Table 20.1. Farmers' performance with fertilizer and tied ridges in sorghum production in Burkina Faso villages, 1983 and 1984 (from Sanders *et al.*, 1996).

			Yields				% of farmers with cash losses	
Year/village	No. of farmers	Traction source	Control	Tied ridges	Fertilization	Tied ridges and fertilization	Fert.	Tied ridges and fert.
1983								
Nedogo	3	Manual	430	484	547	851	56	0
Nedogo	11	Donkey	444	644	604	962	58	42
Bangasse	12	Manual	406	493	705	690	21	17
Diapangou	24	Manual	363	441	719	753	8	8
Diapangou	25	Donkey	481	552	837	871	12	16
Diapangou	25	Ox	526	578	857	991	20	12
1984								
Nedogo	11	Manual	157	416	431	652	27	9
Nedogo	18	Donkey	173	425	355	773	50	0
Bangasse	12	Manual	293	456	616	944	8	17
Dissankuy	25	Ox	447	588	681	855	28	0
Diapangou	19	Manual	335	571	729	1006	26	0
Diapangou	19	Donkey	498	688	849	1133	21	0
Diapangou	19	Ox	466	704	839	1177	5	0

Cash expenditures were only for inorganic fertilizer. Tied ridges alone never increased cash expenditures. The only additional input for tied ridges was a substantial increase in the use of family labour.

Table 20.2. Effects of tied ridges on yield of sorghum, mung bean and maize in research stations in the semi-arid areas of Ethiopia (from Kidane and Rezene, 1989).

	Average grain yield (t ha^{-1})		
Soil conservation method	Kobba	Melkassa	Mean
Sorghum			
Flat planting (farmers' traditional practice)	1.6	0.8	1.2
Tied ridges (planting in furrow)	2.9	3.0	2.95
Mung bean			
Flat planting (farmers' traditional practice)	0.4	–	0.4
Tied ridges (planting in furrow)	0.7	–	0.7
Maize			
Flat planting (farmers' traditional practice)	1.2	–	1.2
Tied ridges (planting in furrow)	2.7	–	2.7

Ridge height: 35 cm.
Ridge spacing: 80 cm for mung bean, 75 cm for sorghum and maize.
Ridges tied at 6 m intervals.

the first priority is to reduce runoff. On the sandy dune soils of the millet system, the water-retention technique needs to respond to the infiltration or percolation problem, in which the water enters the soil but passes rapidly beyond the roots of the plant. As with runoff, the movement of the water needs to be slowed.

Water-retention technologies where crusting is a problem

Water-retention techniques on the crusting soils can be separated into types I and II. Type I techniques include the rock, dirt and vegetative bunds on the contour, *zaï*, terracing, mulching and catchment basins. Type I technologies are constructed all over the semi-arid regions of Africa outside the crop season (Reij, 1983; Roose, 1990; Université AM de Niamey et Ecole Polytechnique Fédérale de Zurich, 1995). For farmers to utilize these methods, opportunity costs have to be low, since these techniques are very labour-intensive and most produce only small absolute yield gains. The yield gains are relatively large to farmers with depleted soils offering only very low yields.

In the early 1980s, David Wright of Oxfam noticed in his village in the Ouaha-gouya region of Burkina Faso that crusting was an extremely serious problem and absolute yields very low. Rather than continue his tree-replanting programme, he adapted a plastic hose to identify the contour lines and showed farmers how to build dirt and rock bunds along the contour to reduce the runoff. On these very degraded soils, yields were increased from 155 to 290 kg ha⁻¹, in 1984 an 87% increase (Wright, 1985: 57). This technique was picked up by a World Bank programme and extended all over the central plateau of Burkina Faso in the 1980s. By 1986, there were 60,000 ha covered with these dirt bunds on the Mossi Plateau of Burkina Faso (Sanders *et al.*, 1996).

Farmers noticed a downside to the dirt bunds. Water could collect behind them in sufficient quantities to break through and then cause erosion with this rapid release of larger quantities of water. So farmers began

shifting to the more porous rock bunds. With less water retained, the yield effect across both degraded and undegraded soils was only 11%, while the average on degraded soils over 3 years was 47% (Wright, cited in Hulagalle *et al.*, 1990: 149; Kabore *et al.*, 1994). Often the necessary stones were not on the farmers' fields. The World Bank-supported programme was critical, as trucks were utilized with workforces from the communities to bring in the stones. Vegetative barriers with and without the bunds were also utilized.

Farmers discovered the water retention immediately behind the bunds and also put manure and/or mulch there to get the combined effects of water and soil fertility. Both manure and mulch increase water-retention capacity and provide nutrients. The yield effect was only observed in the second and third years after bund construction (Kabore *et al.*, 1994: 79).

Burkinabe farmers also increased their use of *zaï*, an old Mossi technique, in the 1980s and 1990s. The *zaï* are holes in which water is captured and into which manure and/or mulch are commonly added. *Zaï* are thus a manual type of tied ridges. Both the *zaï* and the addition of organic matter are performed before the crop season (Kabore *et al.*, 1994: 67). In two villages, average yields on control plots were 49 and 111 kg ha⁻¹, increasing to 191 and 513 kg ha⁻¹ over three *zaï* treatments (Kambou *et al.*, 1994: 51). Despite high percentage-yield increases of 290 and 362%, these still represent very low absolute yield increases and final yields, as these soils were highly degraded.[4] The implicit returns for the additional yields were less than the agricultural wage of 50 FCFA (Kabore *et al.*, 1994: 80, 81).

The *zaï* combined with compost is an impressive type I technology, where there is substantial labour and compost available. Farmers have been observed reaching very respectable 1.2 t ha⁻¹ mean yields in non-degraded regions with 20,000–25,000 holes ha⁻¹ and 4 to 18 t ha⁻¹ of compost. For the digging of the holes alone, 780 h ha⁻¹ of labour are required (Maatman *et al.*, 1998: 127). The natural evolution of this system is to also put inorganic fertilizer into the holes to reduce the compost requirement. This is currently being done in Burkina (Kabore *et al.*, 1994). The next step is to shift to the analogous mechanized operation, tied ridges, as the opportunity costs of labour increase. So the *zaï* is an exception among type I technologies; it offers substantial potential absolute yield increases at a cost of considerable labour inputs for the digging and composting.

By the mid-1990s, Burkina claimed that 700,000 farmers were using some of these type I techniques (Sanchez *et al.*, 1997b: 35). This is undoubtedly an overstatement but these techniques were being widely disseminated by non-governmental organizations (NGOs) and community-based organizations (CBOs) as the government scaled down its involvement. Extension agents in Senegal, Niger and Mali, influenced by the Burkina successes, put the diffusion of type I technologies on their agenda in the 1990s.

The complementary fertilization of type I techniques with organic fertilizers was given an incentive by the structural adjustment programmes (SAPs). As inorganic-fertilizer subsidies were gradually eliminated in the SAPs, a new technology was introduced to extend the quantity and improve the quality of manure. Covering and watering compost heaps was promoted all over the Sahel (Sanders *et al.*, 1996). Crop residues were put into the corrals and this material, along with the manure, was then added to compost heaps. Some even added rock phosphate to the compost heaps. This case demonstrated the principle of induced innovation: higher inorganic-fertilizer prices encouraged the development and diffusion of new techniques to collect and improve the quality of organic fertilizers.

In the second half of the 1990s, Ethiopia became a model of demonstration and diffusion of new agricultural technologies (Quinones *et al.*, 1997). Once the Ethiopian programme was successful in the higher-rainfall regions, demonstration programmes were expanded into the semi-arid zones (IGAD/INTSORMIL/USAID/REDSO, 2001). The initial emphasis for the semi-arid region was on the combination of new sorghum varieties, a tied-ridger and inorganic

fertilizers. The tied-ridger implement turned out to be heavy and awkward and was returned to the experiment station for further adaptation.

Farmers shifted to the type I technologies after they were introduced in 1999. In 2001, in a sample of 90 farmers in Tigray, 72% of the farmers used stone bunds, 33% used dirt bunds, 16% dug trenches (holes in the field similar to *zaï*), 30% dug ditches for runoff collection and later utilization, and 63% used manure (Wubenah and Sanders, 2001). So this was an extremely rapid diffusion process for two crop seasons.

The big absolute gains in yields come from the shift to type II technologies. These techniques include ridging, tied ridging and better soil preparation (Tables 20.1 and 20.2). Type II technologies are done within the crop season. These techniques are generally used before the soil has been completely degraded. The yield gains are relatively smaller than type I since the yields have not yet plummeted. Better soil preparation resulted in yield gains between 22 and 103% (Nicou and Charreau, 1985). The yield gains from type II technologies are expected to be larger in absolute terms due to the use of better water-retention techniques[5] plus the inorganic fertilizers (Rodriguez, 1987). Since these type II water-retention techniques need to be implemented during periods of high labour demand (planting or first weeding), the operation needs to be mechanized with animal traction. So successful introduction depends upon the field performance of the implement.

In the mid-1980s, Burkina Faso introduced approximately 250 tied-ridgers. Heavier soils crust more, so the importance of breaking these crusts is greater. However, the heavier soils also put more stress on the implement. There was no attempt to develop supporting services, such as the local blacksmiths who supported the diffusion and repair of tube-wells in the Punjab of India. These tied-ridgers thus ended up in heaps in the villages.

In Ethiopia, a new model tied-ridger has now been released: a simple metal plate added to the traditional plough (IGAD/INTSORMIL/USAID/REDSO, 2001). It was

released to a private company by the
Ethiopian Agricultural Research Organization (EARO) in 2000.

In contrast to the failure to date of
tied-ridging implement adoption by Burkina
farmers, the biggest type II technology success story is the ridging[6] done in Mali-Sud.
An estimated 85% of farmers there use oxen
traction (Coulibaly, 1995) and ridging on the
contour is pervasive. During the last decade,
the more successful farmers in the semi-arid
region, an estimated 15%, have begun combining the ridging with inorganic fertilizers
(Vitale, 2001; also on Mali, see Kelly et al.,
Chapter 15, this volume).

Lessons learned about natural resource technologies on the heavier soils

The type I technologies are often referred
to as soil-conservation techniques. These
techniques are a widespread response to
soil degradation. These techniques capture
a little water and then manure is added. The
ridging or bunds are very labour-intensive
and would only hold small quantities of
water immediately behind the bunds. The
zaï holes in the field capture more water,
as does terracing or combining vegetative
barriers with dirt or rock bunds. More comprehensive water retention is provided by
ridges, tied ridges or better land preparation
– the type II technologies – and they are
often combined with inorganic fertilizers.
These measures respond to the same constraints of water and soil fertility as do type
I techniques. But they do it more efficiently
and on a larger scale and generally require
animal traction because of the labour scarcity in agricultural operations at the time
these measures are required.

At extremely low yields from soil degradation, type I technologies provide a relatively large yield boost, but absolute yield
increases are small and yields remain low
after implementation. Type II technologies
provide more water and therefore the use of
low or moderate levels of inorganic fertilizers becomes more viable (more profitable
and less risky). Absolute yield gains can be
much higher but the relative gains are less,

since the soils have not yet been depleted.[7]
The natural evolution of the system, with
higher opportunity costs for labour and the
continuing difficulty of obtaining adequate
organic matter, is then expected to be
towards the use of these animal-traction
implements and inorganic fertilizers with
the type II technologies.

The principal hypothesis for the slow
diffusion of both type I and type II techniques in the semi-arid zones of sub-Saharan
Africa is that farmers have not seen them
widely enough in the field. The Tigray example cited above indicates that, when farmers
with crusting soils see the type I innovations, adoption can be very rapid.[8] Type I
technologies now have been widely adopted
on the heavier soils of Burkina Faso and,
more recently, in Ethiopia. In both cases
there was substantial public-sector involvement to encourage community labour and to
offset some of the costs, such as hauling
stones for the dykes in Burkina Faso. The
public sector can definitely facilitate this
process.[9] In Burkina Faso, the diffusion continued even after the public sector scaled
down and phased out its involvement.
Diffusion of type I technologies has
occurred in some villages without public-
sector involvement.

Why have the type I technologies been
so successful while there has been so little
adoption to date of type II technologies?
Our principal hypothesis is that the type
I technologies have been simpler and easier
to perform out of season, when labour was
more abundant. But farmers would be interested in these technologies only when they
had degraded their soils (except for zaï) and
their opportunity costs are low because the
returns to their labour are very low. In order
to respond to the substantial cropping-
season labour requirements, type II technologies require animal traction and specific
implements. Since the semi-arid regions
tend to suffer disproportionate concentrations of rural poverty, many farmers do not
have access to animal traction.

Where crusting is a serious problem,
soils are heavy and implements often break
down. The lack of local capacity to repair
and the lack of national capacity to adapt

these implements seem important factors in the failed diffusion of tied-ridgers in Burkina Faso. Since little machinery other than hand-tools is being repaired in the villages, blacksmiths are rare in rural areas. Typically one must go to regional towns for such work. A sustained extension-service effort could probably help make these services more widely available, perhaps through travelling mechanical repair operations. Support in adapting implements is apparently a requirement at the start of the process, as has occurred in Ethiopia.

Improved technology for millet production systems on sandy soils

On sandy dunes, rather than reducing runoff on the surface, water must be held in the soil longer. The principal cereal in these soils is millet, with some sorghum planted in areas with better access to water (water-recession areas) or higher soil fertility (under the acacia trees) (Sanders *et al.*, 1996). The principal natural resource technologies evaluated for these sandy soils are the use of manure and of crop residues (Bationo and Mokwunye, 1991b; Bationo *et al.*, 1993, 1998; Williams *et al.*, 1993). Both manure and crop residues hold water and nutrients and thereby increase the nutrients available to crops. At high levels, they can provide principal nutrients. However, manure suffers problems of availability and crop residues are frequently used for grazing. Moreover, with the high temperatures in these soils, organic matter burns up rapidly so the carry-over between crop

seasons is sharply reduced. Inorganic fertilizer by itself has also shown a significant effect on yields in these sandy dune soils (Mokwunye and Hammond, 1992). With fertilizer, there is more plant biomass below and above ground (Sivakumar and Salaam, 1999), which also increases water-use efficiency (Shapiro and Sanders, 1998). Inorganic fertilizer alone can increase yields here, but eventually yields will decline as micronutrient deficiencies occur. Hence, high sustainable millet yields require the combined effects of moderate levels of organic and inorganic fertilizers (Bationo *et al.*, 1993: 252–253).

We simulated farm-level decision-making by Niger farmers using a programming model incorporating risk (Shapiro, 1990; Shapiro and Sanders, 1998; Table 20.3). At present, farmers in the model adopt short-season cultivars and P alone.[10] Farmers would use the N and P combination only if they have access to improved longer-season cultivars. However, much of the breeding activities in the Sahel after the prolonged 1968–1973 drought have concentrated on short-season cultivars, which have insufficient time in the field to respond well to fertilizer in average- or good-rainfall years. As a consequence of cultivar availability, joint N and P use has a low pay-off and thus uptake remains low.

Short-season cultivars offer a good risk-reduction technique for adverse years, but they are generally adopted only on small areas or when the early rains fail, because in most years they are less profitable than intermediate or long-season cultivars.[11] In Ethiopia, the short-season, *Striga*-resistant

Table 20.3. Effects of various technologies and a fertilizer-subsidy fertilizer on a representative farm in the Sahelo-Sudanian zone in Niger (from Shapiro and Sanders, 1998: 478).

Policy or programme	Fertilizer use (ha)	Rain-fed crop income (US$)	% Change in crop income
Current practices	n/a	486	–
Improved short-cycle cultivars	0	631	30
Phosphorus only	2.1	685	41
Long-cycle cultivars*	1.5	651	34
Input subsidy (10%)	1.2	657	35

*Combined with both N and P fertilizers.
Exchange rate: 273 FCFA = US$1.

sorghums accompanying the other new type I technologies in the Tigray region were introduced on only 8% of the crop area. Farmers specifically identified the problems of lack of response and growing time for the short-season cultivars in normal- and good-rainfall years. When the rains are late – an estimated 38% of the time according to farmers – they plant the short-season cultivars (Wubenah and Sanders, 2001). There is an important difference between a portfolio strategy to reduce risk and avoid the loss of the cereal crop and a strategy to increase incomes when risk is reduced by providing more water. Farmers in semi-arid Tigray clearly understand this distinction.

Increasing soil fertility in sorghum- and millet-based cropping systems

With the disappearance of fallows and encroachment on to grazing lands, very large quantities of organic fertilizer are necessary to replace basic soil nutrients. However, sources of organic fertilizer, such as manure, are severely limited in most of semi-arid sub-Saharan Africa. For most farmers, sufficient manure is available only for small areas surrounding family compounds (Williams et al., 1993). While manure can be managed more effectively by improved corralling and composting, thereby extending the quality and quantity of nutrients, the overall increase and absolute levels of essential soil nutrients (nitrogen, phosphorus and potassium) remain small (Williams et al., 1993).

Increases in the relative prices of inorganic to organic fertilizer in the late 1980s and 1990s, due both to the elimination of input subsidies and to currency devaluation, have encouraged the expansion of manure use and the diffusion of techniques (corralling and composting) for improving manure quality. With the increased use of composting in Mali and Burkina Faso, manure use has increased and inorganic fertilizer use decreased (Sanders et al., 1996). However, limited ability to expand manure supply, even with increased composting, limits the potential for this substitution strategy. Manure's chief effect is thus as a complement to inorganic fertilizers (McIntire et al., 1992).

Because foreign-exchange constraints have discouraged fertilizer imports, some Sahelian governments have been developing their own sources of natural rock phosphate. Research in Niger, however, indicates that high-quality rock phosphate from Tahoua, even when partially acidulated, would be competitive with imported super-simple phosphate only if supplied to farmers at approximately 25% of the price of imported super-simple phosphate (Jomini, 1990). At present solubility levels, rock phosphate is unlikely to compete with imported inorganic sources even after the 1994 50% devaluation of the CFA franc (Mokwunye and Hammond, 1992). Moreover, acidification is an expensive process in sub-Saharan Africa; the industrial raw materials for the acidification process are available in only a few countries.

In summary, soils are deficient in N and P over most of the semi-arid zone in sub-Saharan Africa. The cheapest sources of these two essential nutrients (by cost per nutrient unit of N and P) appear to be inorganic fertilizers.[12] There are other important effects from organic fertilizers, such as increased water and nutrient retention and increased biological activity from manure. In situations where soil organic matter is too low to retain the necessary water and nutrients, organic fertilizer is an essential complement to inorganic fertilizers (Bationo and Mokwunye, 1991a,b; Bationo et al., 1993; Sedogo, 1993).

One important question for the semi-arid region is: what can be produced that will yield a sufficiently high return to justify inorganic fertilizer use? Traditional food crops, such as sorghum, millet and cowpeas, face sufficiently price-inelastic demand for prices to collapse with either good weather or a new technology introduction. When good weather and rapid technological change are combined, as for maize in Ethiopia in 1996 and 2001, price collapses cause major crises for producers.

So a central concern of any intensification strategy has to be the market potential.

In a recent programme to extend new pigeon-pea cultivars, the agency introducing the cultivars made it a priority to connect farmers' groups to exporters of processed 'dhal' (Jones *et al.*, 2000). Among the drought-tolerant grain legumes, there is undoubtedly similar potential for obtaining higher prices from exploiting other niche markets.

Moreover, in the next 5–10 years, the shift to higher-quality foods with increased incomes will substantially increase the demand for feed grains for broilers and egg production, as occurred in the 1970s in Brazil and in the 1980s in Honduras. This transition has begun in Botswana, Kenya, Senegal and Zimbabwe. No developing country has been able to respond to these diet shifts without substantially increasing net cereal imports. With their drought-resistant characteristics, sorghum and millet are in an excellent position to supply a large part of the cereal component of rations in many countries with substantial semi-arid zones (see Appendix to this chapter).[13] The drought-resistant cereals compete with maize, but maize is limited by its much greater susceptibility to drought and soil-fertility stress.

Developing new cultivars

Traditional cultivars for the semi-arid zone have been selected over time by farmers for a stable response to a wide range of conditions. The introduction of new cultivars alone in semi-arid regions of sub-Saharan Africa has not been successful in increasing aggregate yields for sorghum and millet (Ahmed *et al.*, 2000). Moreover, a cultivar-alone strategy is not sustainable, since it would draw down essential plant nutrients, which are already often deficient in semi-arid regions.

Local cultivars tend not to respond well to improved conditions, as they have evolved to give stable yields under adverse conditions. They often lodge with higher soil fertility. They have not been selected for the various characteristics necessary to respond well to a higher-potential-yielding environment. Once agronomic conditions change by providing more water and higher soil fertility, there is a new environment for improved cultivars.

In the USA, sorghum yields tripled over approximately 20 years, from the mid-1950s to the mid-1970s (Miller and Kebede, 1984). Only 33–39% of these gains came from genetic improvements. The introduction of hybrids in 1956 was combined over time with the increased use of water, fertilizer and herbicides to improve the growing environment, making these improved cultivars more attractive to farmers.

Local cultivars respond well to small changes in water availability and soil fertility from type I technologies. However, the introduction of new cultivars will be very important to accompany and stimulate adoption of type II technologies. Breeders need to anticipate these agronomic improvements and develop new cultivars to be responsive to these changes. In particular, cultivars with biotic resistances that can respond to moderately higher inputs without lodging appear highly desirable.

Since new cultivars take at least 5–10 years to develop and diffuse, it is critical that national and international organizations dedicate sufficient resources now to breeding activities. There is a continuing emergence of new biotic problems in agriculture. Moreover, breeding has been very successful in the last three decades for developing new higher-yielding cultivars, using higher input levels, and resistances to diseases and insects in the cereals (rice, wheat, maize and sorghum) in developed countries, Asia, and Latin America. National and international systems of agricultural research need to take advantage of these gains (new cultivars and breeding techniques) and more aggressively produce and exchange improved materials.

Conclusions

Farmers in the semi-arid zones of sub-Saharan Africa are justifiably concerned with short-run yield increases, because these households are under pressure to survive in a harsh environment. As their soil resources are depleted, they adopt measures to conserve soil, but the short-run effect is also to harbour a little more water. With this water behind the bunds or in the *zaï*, the returns to manure increase for type I technologies. Diffusion has been rapid in Burkina Faso and Ethiopia for these type I techniques. Type II technologies exhibit considerable potential for greater water conservation and higher soil fertility. Mali is in the middle of this process, with large-scale diffusion of ridging and the initiation of inorganic fertilizer use in the semi-arid zone.

In both Burkina Faso and Ethiopia, the shift to type II technologies has been constrained by problems with the implement. There has been a public response to this failure in Ethiopia, but not in Burkina Faso. As animal-traction implements improve and become more accessible to smallholders, more local repair facilities will also need to develop, as well as better public- or private-sector support from agricultural engineers.

With the adoption of type II technologies, new cultivars will offer a higher yield response in the new agronomic environment than will local cultivars. Research stations need to be anticipating this shift in their breeding work now. Incorporation of biotic resistances in intermediate and long-season material continues to be a necessary priority, as well as higher yielding characteristics.

Once the agronomic environment improves moderately in the semi-arid zone, a range of commodities can take advantage of the semi-arid zone's naturally lower disease incidence and increased sunlight. In order to take advantage of this potential, it will be essential to study the feasible market outlets. As incomes increase and consumers demand better diets, the traditional cereals have substantial potential as feed grains for supporting rapidly increasing poultry production. Moreover, the higher quality characteristics of these traditional cereals may enable some regaining of their loss of domestic urban market share to rice or maize.

Appendix: Importance of Drylands in Sub-Saharan Africa

Both crop area and human populations are concentrated in the semi-arid zone of the Sahelian countries (Tables 20.A-1 and 20.A-2). In Burkina Faso, Mali, Niger and Senegal, 69–100% of their available crop area is in the semi-arid zone. Forty-five per cent of Niger's population lives in the semi-arid zone, with the other three countries ranging from 69 to 75%.

In east and southern Africa, the population is less concentrated in the semi-arid zones, except in Botswana, Eritrea, Namibia and South Africa. In some of the driest countries, such as Sudan, Djibouti and Somalia, there has been significant irrigation development in the extremely dry regions, so the population concentration is less than might be predicted based on the proportion of crop area in the semi-arid zone. In Zimbabwe and Angola, a higher percentage of the population lives in the drylands than the proportion of the crop area there. In Zimbabwe, this occurs because the prime lands were taken by the colonial settlers, pushing much of the rural African population into the drylands. South Africa also has a high population proportion in semi-arid regions (60%), in part because South Africa has been relatively successful at developing technologies for these semi-arid zones.[14] Perhaps more importantly, under apartheid, blacks without urban jobs had to return to the Homelands, creating densely populated semi-arid regions that have now been reincorporated into South Africa. In Angola, Ethiopia, Kenya, Mozambique, Sudan and Tanzania, the concentration of the population is not as high in the semi-arid zone, but there is a greater concentration of poverty and welfare problems there.

In most regions of the world, drought-tolerant crops would not be found in the dry, subhumid regions. This is approximately the Sudano-Guinean zone in Fig. 20.1. But low soil fertility, low use of fertilizer and little use of water-retention devices make drought-tolerant crops very important in most dry subhumid zones of Africa. For example, in the cotton/maize zones of

Burkina Faso and Mali, sorghum and millet are regularly found on half of the crop area (Sanders *et al.*, 1996; Vitale, 2001). Drought-resistant crops are more tolerant of low soil fertility than the more demanding principal cash crops of the dry subhumid zone, maize and cotton. So they offer useful responses to microvariability in soil fertility on farmers' fields and are often rotated with the principal cash crops, thus taking advantage of the residual effects of the fertilization.

Notes

[1]   Irrigation of various types is the preferred alternative when viable. But it is often either too expensive or not technically possible. Hence, for most of the drylands, we need to concentrate on alternative technologies to capture available rainfall before it can run off or to hold it better in the soil within access to plant roots.

[2]   For World Bank reviews of the different techniques utilized in sub-Saharan Africa, see Reij *et al.* (1988) or Critchley *et al.* (1994).

Table 20.A-1.   Dryland crop area (1000 ha) and population (1000) in West and Central Africa (calculated from Murray *et al.*, 1999: 12, 13).

Countries	Dry subhumid (% of crop area*)	Semi-arid (% of crop area*)	Dry subhumid population (% of total)	Semi-arid population (% of total)
Benin	3,009.0 (26%)	1,595.6 (14%)	355.0 (7%)	146.0 (3%)
Burkina Faso	5,850.8 (22%)	18,899.3 (72%)	2,128 (20%)	7,957 (76%)
Cameroon	2,137.3 (5%)	3,246.9 (7%)	894 (7%)	2,060 (16%)
Central African Republic	2,662.5 (4%)	1,325.0 (2%)	37 (1%)	13 (0.4%)
Chad	12,219.6 (29%)	2,539.5 (60%)	1,873 (30)	2,188 (35%)
Gambia	746.1 (69%)	337.4 (31%)	940 (85%)	170 (15%)
Mali	6,165.9 (17%)	26,947.6 (75%)	1,205 (11%)	7,501 (70%)
Mauritania	0 (0)	2,577.6 (100%)	0 (0)	216 (10%)
Niger	0 (0)	7,792.2 (100%)	0 (0)	4,070 (45%)
Nigeria	13,533.5 (15%)	29,864.7 (34%)	9,610 (9%)	32,171 (29%)
Senegal	2,992.1 (18%)	11,623.0 (69%)	432 (5%)	6,424 (77%)

*Crop area did not include arid or hyperarid. This would understate the crop area for that part of these zones being irrigated or in oases.

The semi-arid zone had an aridity index of 0.2–0.5. In the dry subhumid zone, the aridity index was 0.5–0.65. The aridity index is the mean annual precipitation (total moisture) divided by the mean annual potential evapotranspiration (moisture loss). The weather data were collected between 1920 and 1990. The aridity indices for the two zones correspond with the length of growing period of 60–119 days (semi-arid) and 120–179 (dry subhumid) (Murray *et al.*, 1999: 2, 3). To evaluate water available to the plant, it would be necessary also to adjust for soil types, cultural practices and rainfall distribution. Even this more complicated aridity index does not give us this insight.

Table 20.A-2. Dryland crop area (1000 ha) and population (1000) in dry subhumid and semi-arid regions in East and southern Africa (calculated from Murray *et al.*, 1999: 12, 13).

Countries	Dry subhumid (% of crop area*)	Semi-arid (% of crop area*)	Dry subhumid population (%)	Semi-arid population (%)
Angola	17,583.8 (14%)	26,256.3 (21%)	851 (8%)	2,572 (24%)
Botswana	0 (0)	49,462.4 (100%)	0 (0)	1,396 (97%)
Djibouti	0 (0)	26.6 (100%)	0 (0)	1 (0.2%)
Eritrea	0 (0)	41,769 (100%)	0 (0)	2,306 (73%)
Ethiopia	15,927.6 (19%)	34,237.8 (42%)	10,137 (18%)	7,706 (14%)
Kenya	4,821.9 (13%)	23,947.2 (66%)	4,672 (17%)	4,655 (17%)
Lesotho	1,423.1 (47%)	224.3 (7%)	1,298 (64%)	181 (9%)
Madagascar	4,500.9 (8%)	8,714.9 (15%)	413 (3%)	1,297 (9%)
Malawi	6,035.5 (51%)	302.0 (3%)	5,448 (56%)	335 (4%)
Mozambique	21,383.1 (27%)	19,567.0 (25%)	6,511 (38%)	2,947 (17%)
Namibia	0 (0)	41,069.6 (100%)	0 (0)	1,245 (81%)
Somalia	0 (0)	12,282.0 (100%)	0 (0)	3,325 (35%)
South Africa	7,847.7 (19%)	15,067.2 (71%)	6,404 (15%)	25,022 (60%)
Sudan	14,269.2 (15%)	16,119.7 (72%)	1,520 (6%)	9,881 (37%)
Swaziland	375.9 (50%)	859.4 (29%)	429 (50%)	147 (17%)
Tanzania	34,902.6 (37%)	35,099.5 (24%)	10,117 (34%)	5,493 (18%)
Uganda	17,106.5 (17%)	4,187.6 (8%)	1,867 (9%)	261 (1%)
Zambia	38,415.6 (32%)	24,138.3 (16%)	3,214 (40%)	1,000 (12%)
Zimbabwe	798.5 (14%)	5,584.3 (8%)	4,018 (36%)	6,713 (60%)

*Crop area did not include arid or hyperarid. This would understate the crop area for that part of these zones being irrigated or in oases.
See footnote to Table 20.A-1 for definitions of these two agroclimatic zones.

³ Tied ridges, also called furrow dykes in the USA, are ridges crossed perpendicularly so that there is a space between them (often about 1 m in each direction). Here the runoff water collects and the crop is planted on top of the ridges.

⁴ Twenty-four per cent of the Burkina land area is highly degraded. Extemely dense bare surfaces, with a 1–2 cm thick crust, are known as white places (*zi-pele*), where part of the fields is like a paved basketball court (Kambou *et al.*, 1994: 43). This is the common form of extreme degradation.

⁵ The *zaï* and tied ridges are essentially the same technique of creating holes in the field to hold water. The *zaï* are done by hand and the tied ridges are done by animal or mechanical traction. The difference is in the labour requirements. The *zaï* are also done out of season and combined with organic inputs – manure and/or mulch – rather than in season combined with inorganic fertilizer.

⁶ According to unpublished data from the Cinzana experiment station in Mali, if the ridging is done on the contour, much of the effect of tied ridging was achieved.

⁷ Table 20.1 offers an example of the differential effects on degraded and undegraded soils, even though this is only for type II technologies. Note the extremely low yields of the control in degraded Nedogo, as opposed to the higher initial yields in Diapongou. With tied ridges, sorghum yields in Nedogo increased 146–164% but remained only 416–425 kg ha⁻¹. In contrast, in Diapongou yields increased 38–70% but reached 654 kg ha⁻¹ average. The main point about type II technologies is the big absolute yield gains with the combination of tied ridges and inorganic fertilizer. Yield increases from around 400 kg ha⁻¹ to 1 t ha⁻¹ are impressive gains. With a new cultivar, 1.2 to 1.4 t ha⁻¹ would be expected. These are still a long way from the 3–4 t ha⁻¹ of US yields (Miller and Kebede, 1984: 7, 8) but, nevertheless, substantial potential-yield gains.

⁸ The concentration on Tigray by the present Ethiopian government and the traditional top-down extension activity in a region that has been at war for most of the last two decades both need to be noted (editors' contribution; Hagos *et al.*, 1999). In a war zone, extension advice is more than a suggestion to farmers. Besides, the extension-support input use was partially subsidized in early years of the programme. So there was undoubtedly a much more rapid diffusion process in Tigray than would be expected in most regions.

⁹ The Sasakawa 2000 experience in various African countries of introducing new technologies by promoting demonstration trials and improvements in input markets has shown that diffusion can proceed rapidly after widespread successful demonstrations (Quiñones *et al.*, 1997; IGAD/INTSORMIL/USAID/REDSO, 2001). In the discussion below, the focus is on the Global 2000 programmes in Ethiopia and Mozambique. In 1995 the government of Ethiopia took over most of the Global 2000 programme and has been implementing and expanding this programme to the present.

There has been a continuing interdisciplinary conflict between the agronomists managing Global 2000 and economists analysing the field data of these programmes. Global 2000 programmes assume that technologies are available for the food crops used in their demonstrations, generally make blanket high-input recommendations across regions and do not provide data on the farm-level performance of the technologies. Global 2000 now needs to adjust their recommendations regionally based on trials on input levels, analyse better the economic performance of their technologies and support accompanying policy for critical related problems of food-crop intensification, including the postharvest price collapse and the between-year price collapse with good weather and/or technological change. So the interaction between disciplines has been useful, as Global 2000 and related programmes are starting to do all of the above in Ethiopia and Mozambique (J. Low, personal communication).

The advantage in the debate is still with the pro-activist position of Global 2000 in getting things done by extension services and governments for the food crops, while economists spend their time pointing out the high inelasticity of demand for traditional food crops and the difficulty of getting credit for input purchases for them. Many economists are also concerned with the sustainability of the Global 2000 programmes, as initially their programmes subsidize the availability of inputs, including seeds, fertilizer, extension services and credit, before proceeding to phase out the subsidies over time. Clearly the Global 2000 programmes will become more sustainable if they continue to pay attention to the economists' critiques (Howard *et al.*, 1998, 1999; Reardon *et al.*, 1999; Kelly *et al.*, Chapter 15, this volume).

¹⁰ Farmers in Niger have been adopting the early cultivars on small parts of their millet areas as a risk-reduction strategy. Farmers who have been able to observe demonstration trials in these sandy dune soils are often now using micro-doses (approximately 20 kg ha⁻¹) of inorganic fertilizers applied in the seed pocket. The effect here is to improve germination and early vigour. There

are better results for diammonium phosphate (DAP) than the compound inorganic fertilizer (unpublished data, International Crops Research Institute for the Semi-Arid Tropics (ICRISAT), Sadore Station). Obviously, increasing yields imply more than good germination, so higher rates of inorganic fertilizer will be required. But this is a good start in the use of inorganic fertilizers on the traditional food crops.

[11] This draws on a synthesis of studies of the impact of the introduction of new sorghum and millet cultivars in seven sub-Saharan countries (Ahmed et al., 2000).

[12] For an illustration in Burkina Faso, see Sanders (1989). For policy discussion of the future importance of inorganic fertilizers in sub-Saharan Africa, see Bumb and Baanante (1996) and Larson and Frisvold (1996). For the alternative approaches of Burkina and Ethiopia to increasing soil fertility, see Sanders and Ahmed (2001).

[13] At present, new higher-quality sorghum and millet cultivars are available. Moreover, there have been advances in processing and preparation techniques, which reduce women's time requirements for preparing traditional cereals. Hence, millet and sorghum can make a comeback from the loss of urban market shares to rice in the Sahel. Millet couscous and porridge are now available in ready-to-boil plastic sacks in the markets of Dakar and Bamako and are even exported (B. Ouendeba, personal communication, 2001). Similar products are being developed for sorghum. One local manufacturer is also beginning to use higher-quality white sorghums in local biscuit production in Bamako. Note that, besides having a better taste and higher- quality food characteristics, these new white sorghum cultivars are also more susceptible to bird damage, and birds have been an intractable research problem. The high-tannin cultivars developed by Doggett in the 1960s in Uganda are still the predominant technology response in regions with high bird pressure, such as most of the Rift Valley.

[14] In much of South Africa, the soils have an impermeable layer 1–2 m down. To accumulate water, the soils are kept cultivated after the harvest. Then more rainfall can be absorbed and stored in the soil above the impermeable layer before the start of the next production season (Ahmed et al., 2000). With this technique, sandy soils are preferred, as they are more permeable, so they will absorb more rainfall.

21 Lessons for Natural Resource Management Technology Adoption and Research

Frank Place,[1] Brent M. Swallow,[1] Justine Wangila[1] and
Christopher B. Barrett[2]

[1]International Centre for Research in Agroforestry (ICRAF), PO Box 30677, Nairobi,
Kenya; [2]Cornell University, 315 Warren Hall, Ithaca, NY 14853, USA

This final part of the book highlights key implications from the preceding chapters. This chapter focuses on lessons for the study of natural resource management (NRM) technology adoption in smallholder African agriculture. The next chapter draws out implications for agricultural and economic policy.

An improved understanding of smallholder needs, constraints and practices is a prerequisite for international and national agricultural research systems being effective in improving the livelihoods and natural environments of African farmers. The development of appropriate technologies and policies rests upon a foundation of sound biological and social science knowledge derived from improved working relationships with smallholders and understanding of their decision-making. We therefore begin this summary section by presenting a NRM technology research and development framework that depicts the ways that researchers engage in participatory processes of problem identification, technology development and dissemination. Subsequent sections present major findings regarding adoption of and research on NRM.

Rethinking the Questions and Stakeholders in Studies of NRM Practices

Stakeholders in NRM

Social scientists concerned with agricultural and natural resource technologies have often felt that they were asked to address the wrong questions. Consider an extreme example: an agronomist or engineer develops a new production or conservation technique, publishes the results in the best peer-reviewed journals and then a year or two later asks an anthropologist 'Why haven't farmers adopted this new technique?' Such situations are thankfully less common now than they once were. With the advent of farming-systems research in the early 1980s, social scientists were instead asked to consider questions like 'What kinds of technologies do farmers need?' A decade later, the rise of participatory research led to widespread consideration of the question 'What do farmers want?'

Certainly, it is useful to learn the answers to all three of these questions. But the exclusive focus on farmers – indeed, on

just the agricultural practices of rural households commonly pursuing diversified livelihoods – has perhaps obscured other important questions. The presence of externalities
provides a standard justification for public
investment in research on and extension of
new NRM technologies; many of the benefits
(or costs) of improved NRM accrue to
resource users who do not bear the costs (or
receive the benefits). Since farmers are only
one group of stakeholders with interests in
agriculture and NRM, it is appropriate
also to ask, 'What do other stakeholders in
the food-production system (e.g. consumers,
wholesalers, service providers, input distributors, regulatory agencies) want?' 'What
do other stakeholders in NRM want, including those who use resources downslope and
public agencies concerned with resource
conservation?' 'Finally, who stands to gain
or lose from a change in NRM or agricultural
production?'

Such an expanded rethinking of the
questions and stakeholders relevant to the
study of NRM practices has several important implications. First, few specific NRM
practices will benefit or be desired to the
same extent by all stakeholders. Indeed,
benefits to some may imply costs to others.
As such, socially optimal levels of NRM
technology adoption may well differ from
private levels. So, although plot- and farm-
level research must be done, it needs to be
combined with larger-scale analysis. Such
analysis is still quite uncommon. Larger-
scale analysis requires more than simple
aggregation across smaller units whenever
externalities do not occur in a smooth, linear
fashion. Examples of non-linear externalities include the effects of organic-nutrient
systems on pests and conservation systems
on erosion. With the introduction of new
plants, pests may only accumulate after a
relatively large number are growing in the
landscape. Similarly, the introduction of a
few fragmented conservation structures may
actually exacerbate soil erosion by channelling water flows.

Secondly, progress can be most easily
achieved by identifying NRM problems that
most stakeholders can agree upon. For example, most stakeholders, ranging from farmers

to traders, to processors, to policy-makers,
will benefit from reducing soil-fertility problems in Africa. Soil fertility has therefore
become the primary focus of NRM technology development efforts in Africa, as plainly
reflected in this book. On the other hand,
efforts to improve forest management in
Africa have proved less successful, due
to divergent interests and, consequently,
differences of opinion between distinct
stakeholder groups.

Adoption research in a
research–development continuum

The development, adoption and diffusion
of NRM technologies are interrelated processes that are underpinned by sound
problem analysis. Adoption studies need
to take account of these interrelationships.
The concept of a research–development
continuum is a useful way to capture those
interrelationships.

Figure 21.1 depicts this continuum,
along with the types of adoption and impact-
assessment studies that are relevant at
the different stages. Starting at the top and
moving in a clockwise direction, the outside
sequence reflects the common stages in
NRM technology development and dissemination. Of course, there are feedbacks
between stages, but these are omitted from
the figure for ease of viewing. On the inside
of the circle, we identify useful types of
research related to adoption at the corresponding stage of technology development
and dissemination. The main point to be
made is that adoption research should
not be seen as a single type of study that
occurs at a particular stage of the technology
development and dissemination process.

Studies of technology adoption must
inform other stages of technology development and dissemination to be of maximum
benefit. These studies must also integrate
biophysical and socio-economic variables/
analyses. For example, studies at the problem identification stage should contain a
strong emphasis on ex ante adoption and
help focus research on identifying NRM
practices that address key problems of

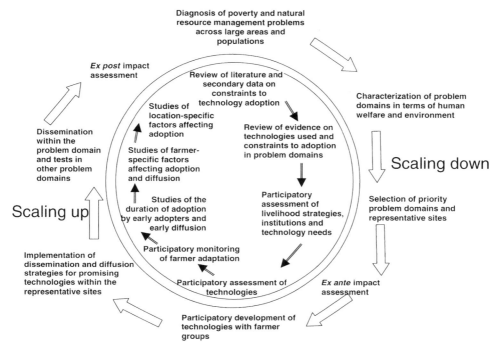

Fig. 21.1. Adoption and impact studies in a research–development continuum.

concern to farmers. *Ex ante* analyses should be relatively rapid, so as to identify major constraints (e.g. policy) and opportunities and to generate hypotheses that will advance technology development. On the other hand, *ex post* studies of farmer or location specific factors affecting adoption need to be more thorough to determine whether impact has been achieved and to identify second-generation issues that arise from wide-scale dissemination (e.g. issues regarding the production and distribution of germ-plasm to meet increased demand).

The preceding discussion underscores that there exist multiple reasons for conducting adoption studies (adapted from Place and Swallow, 2000):

1. Providing input into discussions of policies related to adoption.
2. Defining recommendation domains for existing technologies or those under development.
3. Identifying traits that will make new technology attractive to farmers.

4. Assessing the impacts of technology on objectives such as production and poverty alleviation.
5. Identifying groups (e.g. women) that may not be able to adopt a technology for various reasons.

The objective(s) of the research needs to be clarified at the outset because different objectives sometimes require different research methodologies. For example, objective 2 favours the building of predictive models based on strong associations between adoption and characteristics of the landscape and population that are easy to measure and analyse. On the other hand, objective 1 requires more careful development of structural models to identify causal relationships. The study of NRM technology adoption is complicated by the nesting of individual adoption decisions in a multi-scale, dynamic context involving interactions among many actors. This creates challenges for developing conceptual models of farmer decisions about NRM technology and for conducting empirical studies of such decisions. It is often difficult to identify underlying causal processes that condition adoption behaviour and to separate these from more spurious correlations.

The complexity of adoption behaviours necessarily limits the capacity of any single method or discipline to address the full range of constituent issues satisfactorily. The studies presented in this book therefore draw upon a range of methods and disciplines. None the less, most of the adoption studies in the literature and in this book are quantitative, economic analyses at the household or plot levels. Further, most of the studies in the literature examine cross-sectional variation in adoption behaviour rather than the dynamic processes of technology sequencing, adaptation and disadoption. The strengths of the cross-sectional models are the inclusion of a large number of households, the simultaneous analysis of several important explanatory variables and the co-adoption of various NRM practices. Several of the studies presented in this book utilize new methods to overcome the limitations of static, cross-sectional studies – introducing dynamics (Wyatt, Chapter 10, and Okumu *et al.*, Chapter 18), looking at farmer adaptation (Adesina and Chianu, Chapter 4), examining variation in adoption across communities or regions (Kristjanson *et al.*, Chapter 13, Adesina and Chianu, Chapter 4, Gebremedhin and Swinton, Chapter 6, and Freudenberger and Freudenberger, Chapter 14), and paying more attention to details of decision-making through more qualitative analyses (Gladwin *et al.*, Chapter 9, Peters, Chapter 3, and Pretty and Buck, Chapter 2). Several core findings none the less reappear across different sites, technologies and analytical approaches.

Major Findings Concerning NRM Practices and their Adoption

This section briefly summarizes major findings regarding the adoption of NRM practices in sub-Saharan Africa. The first subsection presents a number of generalizations that emanate from the studies presented in this book and, to a lesser extent, in the more general literature (what we know). The following subsection presents a number of propositions about the most important knowledge gaps that still remain (what we need to know).

What we know about the adoption of NRM practices

Farmers have different needs/constraints according to the external conditions they face and their internal characteristics. Therefore, the identification of desirable attributes or functions of NRM technologies or baskets of NRM technological options is critical for reaching a large number of farmers and communities.

Rural African landscapes are home to a wide diversity of farming households, most of whom engage in non-farm activities as well as agricultural production. In recent years, a broader livelihood perspective has begun to pervade the literature on agricultural development, as reflected in many of this book's chapters. Recognition that rural households have different reasons and incentives to engage in agricultural production as a part of their livelihood strategy naturally gives rise to greater attention to the heterogeneous needs of distinct rural subpopulations. As Wyatt (Chapter 10) points out, even households with similar endowments may demand different technologies because of differences in preferences, objectives, constraints or incentives.

Knowing the technology attributes that farmers want is essential to the process of technology development. In the case of dual-purpose cowpeas, for example, Kristjanson *et al.* (Chapter 13) found that desirable attributes for communities in northern Nigeria were most importantly high grain yield, followed by pest resistance and then fodder yield. Ndlovu and Mugabe (Chapter 19) found that methods that reduce labour intensity could enhance the adoption of livestock intensification techniques and thus improve livestock productivity.

There is an inherent dilemma between deliberate targeting of technologies to areas and social groups most likely to adopt and benefit from those technologies and the desire to make technology dissemination more 'demand-driven'.

Some of the studies presented in this book indicate apparent gains from targeting of regions and households with technologies that are most likely to be adopted and beneficial. The extension service will quickly lose credibility if it recommends that farmers use livestock breeds that are likely to succumb quickly to disease. But this logic cannot be taken too far. Farmers invariably consider factors that outsiders could not anticipate. The way out of this dilemma is to present farmers with a range of technology adoptions that pass an initial screening test. Studies presented in this book show that a greater percentage of a heterogeneous population are likely to adopt NRM practices when all households in the population are presented with a range of options (Kelly *et al.*, Chapter 15, Tarawali *et al.*, Chapter 5, and Shapiro and Sanders, Chapter 20).

The adoption of innovation processes by individual farmers and farmer groups is often more important than the adoption of individual technologies.

The transfer-of-technology approach, in which technology is developed by scientists and then passed to farmers through extension agents, has been gradually replaced over the past generation by more participatory approaches of innovation, diffusion and adaptation. This transition has proved central to most successful cases of improved NRM technology adoption and impact, as shown by several of the preceding chapters. As Peters (Chapter 3) points out, farmers and researchers have complementary knowledge – and knowledge deficiencies – so that integration of this knowledge through participatory processes typically has the most profound effects.

Indeed, Adesina and Chianu (Chapter 4) find that farmers by themselves made several adaptations of the alley-farming system in Nigeria. Tarawali *et al.* (Chapter 5) highlight the advances that can be achieved when farmers and researchers combine efforts in livestock feeding systems in West Africa. Pretty and Buck (Chapter 2) urge the facilitation of such processes more generally to advance improved NRM. The increased emphasis on smallholder innovation processes, whether on their own or in partnership with research scientists, calls for greater emphasis on building up human and social capital assets of communities, rather than focusing solely on technology development.

NRM practices that improve soil fertility, raise agricultural production and prove profitable to small farmers do exist. Soil-fertility and soil-conservation practices are being widely adopted in some areas, with integration of inorganic and organic fertilizers and fertility and conservation techniques often proving most attractive to farmers.

Investment in soil fertility has been shown to be attractive and profitable for rural Africans in at least some areas of each of the major agroecological zones in Africa. This holds for inorganic fertilizers (e.g. Mekuria and Waddington, Chapter 17, and Shapiro and Sanders, Chapter 20), as well as organic techniques, such as improved fallows (Place *et al.*, Chapter 12). Significant uptake of organic and inorganic practices for improving soil fertility has occurred in the highlands (Gebremedhin and Swinton, Chapter 6, Place *et al.*, Chapter 12, and Clay *et al.*, Chapter 8), the humid zone (Tarawali *et al.*, Chapter 5), the subhumid zone (Mekuria and Waddington, Chapter 17, and Kristjanson *et al.*, Chapter 13) and the semi-arid areas (Freeman and Coe, Chapter 11, Shapiro and Sanders, Chapter 20, and Kelly *et al.*, Chapter 15). Studies conducted in study sites in Central, East and West Africa (Gebremedhin and Swinton, Chapter 6, Hatibu *et al.*, Chapter 16, Clay *et al.*, Chapter 8, and Shapiro and Sanders, Chapter 20) have found that significant proportions of households were using soil- or water-conservation measures on their farms. These studies show significant pay-offs from the integration of conservation and fertility techniques (Gebremedhin and Swinton, Chapter 6, and Shapiro and Sanders, Chapter 20), as well as the integration of inorganic and organic sources of nutrients (Place et al., Chapter 12, Kelly et al., Chapter 15, Shapiro and Sanders, Chapter 20, Freeman and Coe, Chapter 11, Peters, Chapter 3, and Mekuria and Waddington, Chapter 17). It is important to note, however, that most studies of technology adoption are located in places where specific soil-fertility and soil-conservation technologies have been developed and disseminated to farmers through fairly intensive participatory research and extension inputs.

Farmers who recognize natural resource problems are not always induced to invest in improved NRM practices.

African farmers typically have a good understanding of the nature and extent of natural resource problems on their farms (Gebremedhin and Swinton, Chapter 6, and Gladwin et al., Chapter 9). However, this does not always translate into investment to address those problems. High rates of discount for future costs and benefits, compounded by high rates of poverty, are identified as key constraints to investment by Wyatt (Chapter 10), Holden and Shiferaw (Chapter 7) and Okumu et al. (Chapter 18). Consistent with this, several studies identified the need for investments that generate greater returns in the short term (Tarawali et al., Chapter 5, and Gebremedhin and Swinton, Chapter 6). Lack of secure land tenure in some sites, such as much of Ethiopia (Okumu et al., Chapter 18, and Gebremedhin and Swinton, Chapter 6), tenant farmers in Rwanda (Clay et al., Chapter 8) and female-headed farms in Nigeria (Adesina and Chianu, Chapter 4) further compound the problems of poverty and high discount rates.

Working-capital constraints or high opportunity costs of capital commonly limit investment in improved NRM practices. The linking of

high-value cash crops to cash investment therefore helps make such investments more attractive.

Households in many rural areas face chronic financial-liquidity problems. Scarce financial resources must meet multiple household demands for both consumption and investment in farm or non-farm activities (Wyatt, Chapter 10). Holden and Shiferaw (Chapter 7) found that poverty leads to very low willingness to pay for land-conservation investments in Ethiopia. Similarly, other studies found that greater wealth is associated with adoption of more costly investments, such as fertilizer or terracing (Place et al., Chapter 12, Kelly et al., Chapter 15, and Wyatt, Chapter 10). For roughly the same biomass response per unit investment, high-value cash crops generate greater economic returns than do low-value cereals, so systems that include higher-value crops exhibit markedly higher rates of adoption of NRM methods per unit land area than do semi-subsistence systems based on extensive production of staple grains, roots and tubers (Kelly et al., Chapter 15, Shapiro and Sanders, Chapter 20, and Freeman and Coe, Chapter 11).

Farmers commonly find ways to accommodate new NRM technologies into their farming systems when incentives are sufficiently high.

While a few of the studies found that the size of the family labour endowment was positively linked to the probability or level of adoption of certain types of NRM practices, there is contrary evidence (even from the same studies) that farmers find ways to accommodate practices that generate very high returns, no matter what the size of their family. For example, small households have adopted stone terraces at lower levels in Ethiopia (Gebremedhin and Swinton, Chapter 6) and conservation techniques in Burkina Faso by investing during the dry season (Shapiro and Sanders, Chapter 20). Households in western Kenya have modified initial fallow systems to greatly reduce the labour time required (Place et al., Chapter 12). These findings indicate that agricultural intensification in Africa does not always follow a simple Boserupian process, wherein increasing population density fosters adoption of more labour-intensive methods that conserve the natural resource base and generate higher crop yields. While labour availability matters, all else held constant, the empirical evidence across multiple sites demonstrates that processes of improved NRM can be initiated and undertaken gradually, despite perceived seasonal labour shortages.

Improved NRM technologies, like agricultural technologies more generally, fail to be adopted by women farmers and poor farmers at the same rate as male farmers who enjoy greater wealth, education and socio-economic power. Situations in which high percentages of women and disadvantaged groups have adopted NRM techniques almost always included concerted dissemination efforts to reach those groups.

A key motivation for participatory NRM practice development lies in the perceived inability of traditional top-down research methods to reach disadvantaged groups, such as the poor and women. These groups none the less remain more difficult to reach with even low-cost, improved NRM practices, even if participatory development and extension processes are used, unless special care is taken to target and reach identifiable subpopulations. For example, wealthier farmers proved more likely to use soil-fertility amendments, such as manure (Mekuria and Waddington, Chapter 17, and Place *et al.*, Chapter 12) and inorganic fertilizer (Kelly *et al.*, Chapter 15, and Freeman and Coe, Chapter 11). Adesina and Chianu (Chapter 4) found that alley farming in Nigeria was more likely to be adopted by males than by females, echoing findings with respect to fertilizer in Kenya (Place *et al.*, Chapter 12) and Rwanda (Clay *et al.*, Chapter 8). On the other hand, the two studies of improved fallows show that women were equally likely to adopt as males in Zambia (Gladwin *et al.*, Chapter 9) and that the poor used the technology at an early stage of dissemination in western Kenya (Place *et al.*, Chapter 12). However, in both cases, concerted efforts have been made by projects and dissemination partners to

involve these disadvantaged groups, at relatively high costs of dissemination and support. There does not seem to be an inexpensive short cut to reaching subpopulations that traditionally lag in the adoption of improved technologies.

What we need to know about the adoption of NRM practices

Despite its importance, there has been relatively little research on the adoption of water-management practices in Africa.

With the exception of the chapters by Hatibu *et al.* (Chapter 16) and Shapiro and Sanders (Chapter 20), this volume has not addressed the issue of water management. This is not because the issue is not important, but rather because national and international agricultural research centres have placed far greater emphasis on soil-management research in the past decade or so. Given that African agriculture remains largely rain-fed and that water-scarcity issues are receiving much more prominence, we anticipate much more work on technology development and adoption studies in this area. Given the complementarity between soil fertility and water management, as strikingly demonstrated by recent experiences with the system of rice intensification (SRI) developed initially in Madagascar (Uphoff, 2000; Moser, 2001), extending research on NRM in agriculture to water shows great promise.

Research and extension on improved NRM practices tend to be based on presumptions about the social desirability of those practices. At this point, however, there have been few satisfactory studies of the social costs and benefits of resource degradation or improvement.

Information on the social costs of resource degradation and comparisons between the private and social costs of degradation and restoration are crucial for justifying public investments and for identifying key constraints to the use of socially beneficial technologies. The few studies that have attempted to assess the external costs or total social costs of land degradation have tended to assume very simple processes of scaling between the plot, farm, landscape and national levels. The results are apt to overestimate the social costs of farm-level degradation and ignore the costs of the negative effects that emerge at the landscape or watershed scales. For example, soil that is a loss to one plot may be a gain to another plot further down a hillside. As a result, there are few hard data on which to justify subsidies or public investments in NRM.

One of the key challenges for African agricultural development is the widespread use of participatory methods that have proven to be effective in pilot research and development projects.

The pilot research and development projects in which participatory research and development techniques are developed and tested frequently involve intense interactions between well-qualified researchers and a modest number of farmers. Scaling up these experiences to millions of farm families presents major challenges. Even large national-level programmes still reach a small percentage of all farmers each year. For example, the National Agriculture and Livestock Extension Programme in Kenya has a goal of engaging 100,000 new farm families in participatory problem definition and technology adaptation each year. Even under the best of circumstances, however, this still amounts to less than 5% of the total farm population. Diffusion outside project areas appears to be minimal.

Main conclusions about research methods on NRM technology adoption

Most studies are conducted at the household level. There is considerable value added to complementary research at community, landscape and watershed levels.

As indicated previously, most adoption studies, including most chapters in this book, focus on the household as the unit of analysis. This has indeed provided much insight into the importance of household characteristics (e.g. labour, land and capital endowments), which vary considerably across households within even small geographical areas. Even smaller-scale analysis can be valuable, as Clay *et al.* (Chapter 8) demonstrate in studying conservation investments at the plot level in Rwanda, where landholdings are highly fragmented. Although plot and household characteristics are important and still needed, larger-scale factors are increasingly recognized to be of equal and, in some cases, perhaps even greater importance. Three studies purposefully sampled over larger areas (Kristjanson *et al.*, Chapter 13, Hatibu *et al.*, Chapter 16, and Adesina and Chianu, Chapter 4) and found village-level variables to be very important in predicting the adoption patterns of households. Moreover, Freudenberger and Freudenberger (Chapter 14) show how regional factors, such as transport infrastructure, can overwhelm more micro-level determinants of cropping and NRM patterns. The rise of spatial analysis exploiting geographical information systems should facilitate a useful expansion of adoption research to larger scales of analysis.

Perhaps the greatest remaining puzzle about NRM in African agriculture relates to the spatial clustering of adoption and adaptation. Why do some pockets exhibit high rates of innovation while others with nearly identical household and agroecological characteristics do not? So long as location-specific dummy variables commonly explain the largest proportion of predictable variation in adoption patterns in regression analyses, our understanding of farmer behaviour remains insufficient. Macro- and meso-level issues related to input and output market access, financial systems, social capital, the political economy of public-services delivery and project siting, etc. probably play key roles, but the causal mechanisms remain poorly understood, impeding the development of reliable policy recommendations. Various chapters find positive relationships between NRM technology adoption and more capital-intensive cash crops or between communities' capacity to organize cooperatives and rates of acquisition of new information and investment in improved practices. We do not yet have a solid grasp on precisely why such associations exist or how those associations can be replicated elsewhere.

Static quantitative analyses provide some insight into adoption processes, but the greatest gaps in knowledge revolve around adoption dynamics. While quantitative methods may be useful in filling this gap, such studies are costly and can miss important features more readily captured through historical and qualitative case-study methods.

Cross-sectional, quantitative analyses of dichotomous adoption behaviours (adopter/non-adopter) using limited dependent-variable regression methods continue to generate valuable information as to: (i) who is adopting different technologies; (ii) whether NRM technologies are reaching the poor, women or other groups of particular interest to donors and governments; (iii) what technology attributes farmers most want; and (iv) underlying constraints to adoption that may be amenable to remediation through policy interventions. Findings from such studies can therefore be useful in helping to expand the use of promising technologies or develop next-generation methods that will appeal more to a broader range of farmers. These methods are well developed and widely understood, can be done relatively quickly and cheaply and suffer few measurement problems.

Nevertheless, cross-sectional studies of inherently dynamic adoption processes are prone to biased inference due to bidirectional causality between dependent and independent variables that coevolve over time. They may also miss important differences in intensity of adoption. Although dynamic adoption studies are inherently slower and more costly to undertake, the main empirical gap in our understanding of NRM practice adoption patterns is with respect to adoption, expansion, adaptation and disadoption dynamics. As shown in Fig. 21.1, this gap jeopardizes a proper assessment of the impact of technology. Okumu *et al.* (Chapter 18) and Wyatt (Chapter 10) employ dynamic simulation models to predict such patterns, but there remains a dearth of useful statistical inference using duration models and panel-data methods. Given the cost and limited coverage of such methods, in the near term, the study of adoption dynamics may be best dealt with through qualitative research methods (e.g. oral history, ethnography) to understand the evolution of decision-making processes at plot, farm and community level. The chapters by Peters (Chapter 3), Freudenberger and Freudenberger (Chapter 14) and Gladwin *et al.* (Chapter 9) follow this approach. By probing more into social and cultural issues, these analyses shed much useful light on the unobserved heterogeneity that commonly plagues quantitative analyses, particularly static ones. An accurate understanding of the history of past interventions and the reasons why they succeeded or failed is especially valuable to development practitioners, since no one wishes to reinvent the wheel, much less the flat tyre. Ultimately, adoption dynamics may be best understood through studies that successfully integrate qualitative and quantitative methods, but this remains unexplored territory as yet.

Analysts increasingly acknowledge that technology and NRM practice choice are but a part of household decision-making under uncertainty, although this broader livelihoods perspective remains insufficiently pervasive.

Implicitly, most adoption studies assume that the more adoption, the better, as if adoption of a NRM practice or an agricultural technology were a farm household's objective in its own right. Since agricultural households have different needs and objectives and because there are many stakeholders in addition to farming households, more adoption is not always better. Decision-makers have optimal adoption levels; in some instances, non-adoption or disadoption may be optimal. For example, Tarawali *et al.* (Chapter 5) note that disadoption of improved mucuna fallows in West Africa after 2–3 years reflects not rejection of the cover crop by local farmers, but rather the success of the NRM practice in controlling noxious weeds. The chapters by Holden and Shiferaw (Chapter 7) and Wyatt (Chapter 10) similarly demonstrate that even profitable technologies may not be attractive to farmers if alternative, off-farm investments offer superior returns or if these technologies also increase households' exposure to risk. Indeed, the empirical relationship between risk aversion and preferences for and adoption of NRM practices has received insufficient attention in the literature, a particularly critical information gap in relation to households facing chronic food insecurity. What types of technologies are feasible and attractive to the very poorest of households with both very short-term planning horizons and high aversion to risk? Lack of understanding in this area is perhaps the most serious gap needing attention if technology-development research is to contribute directly and substantively to the reduction of poverty and food insecurity in Africa.

NRM technology-adoption studies have handled the farmer–user ('demand') side reasonably well. Supply-side issues, such as the role of social capital, extension services, private traders and community organizations in information flow and adaptation of on-the-shelf technologies to local conditions, are increasingly recognized as important, but remain understudied.

The adoption of NRM practices results from interaction between providers of information and farmer decision-makers who act upon the information, and this is conditioned upon other factors, such as incentives and access to inputs. Most of the studies in this volume provide careful, detailed descriptions of how and why farmers are motivated to adopt NRM practices (the demand side). Consequently, most empirical tests of adoption patterns have emphasized demand-side variables at the household level. Whether such demand-side factors are overwhelmed by more pervasive problems of information, incentives or market failure are generally not well studied.

Taking the case of information dissemination as an example, researchers and non-governmental and community-based organizations and projects have played an increased role in spreading information on NRM principles and practices. As a result, the flow of information to and awareness of farmers and other natural resource users is highly variable in coverage, timing and quality/reliability, so the 'supply side' of NRM technology is also important for explaining observed adoption patterns. However, this aspect has received far less systematic attention in the literature. Pretty and Buck (Chapter 2) and Kelly et al. (Chapter 15) do highlight these issues from the perspective of development efforts, while Peters (Chapter 3), Tarawali et al. (Chapter 5) and Mekuria and Waddington (Chapter 17) highlight these issues from a research point of view. Even these studies pay insufficient attention to questions of costs, fiscal or political sustainability or capacity to scale up information delivery. Increased attention to institutional design and performance, pathways of information flow and the processes by which farmers become aware of and interested in new NRM practices would seem warranted in future adoption studies, especially given the rapid change both in information technology and in the organization of public-services delivery across rural Africa.

There is a relative lack of quality data from farmers' fields on the investment requirements of NRM technologies and the social, economic and ecological benefits that they generate. This would be extremely useful, complementary information to the adoption analyses.

In order to be attractive to farmers, NRM technologies must provide short-term pay-offs, as well as improving the long-term sustainability of the natural resource base on which agriculture depends. Yet few studies, in this book or in the broader literature, provide conclusive data on even the profitability and payback periods of NRM practices (Mekuria and Waddington, Chapter 17, Okumu et al., Chapter 18, Kelly et al., Chapter 15, and Place et al., Chapter 12, are exceptions). It is not easy to obtain accurate information on profits or other impacts, such as food consumption, due to recall and measurement problems. Assessing such impacts at a community or national level requires considerable resources. None the less, such information is hotly sought after by African policy-makers and its absence prolongs the myth that NRM practices tend to be low-input, low-output and hardly, if at all, economically attractive. As Hatibu *et al.* (Chapter 16) point out, promotion of improved NRM practices must become more prominent on the development agenda, rather than remaining entrenched only on the environmental agenda. Until analysts convincingly demonstrate that improved NRM contributes to economic growth and poverty alleviation, however, the probability of such a policy transition will remain low

Conclusions

The chapters in this volume draw on diverse experiences across the African continent, affording an unprecedented opportunity to identify general tendencies that hold across agroecological zones and agricultural and economic policy regimes. This chapter has highlighted key findings useful to those working with farmers to develop NRM practices that are more productive in the short term, as well as the long term. There is evidence that recent movements to more participatory approaches have had a positive impact on the adoption of improved soil-fertility and conservation practices. These should be strengthened.

Technology development and dissemination systems must continue to emphasize practices that require little capital and methods of scaling up improved processes and techniques to wider communities. The next chapter, on lessons learned for agricultural and economic policy for rural Africa, focuses heavily on the prerequisites for effectively scaling up improved NRM adoption.

As NRM technology development processes become more complex, the study of adoption will also become more complicated. Understanding the types of processes, approaches, methods and tools that can lead to greater fulfilment of farmer and community demand for NRM practices will be more difficult, but more imperative, as donor funds for such research have become scarcer in recent years. More work is required to understand farmer behaviour in the context of a complex livelihood strategy and the way in which risk and time preferences bear on his or her decision-making. In addition, significant value added would emerge from more qualitative historical studies, especially those focused on larger community or landscape scales. Finally, it is important once again to consider that adoption and impact can be enhanced by undertaking a range of adoption-related research at different stages along the NRM technology development and dissemination continuum.

22 Towards Improved Natural Resource Management in African Agriculture

Christopher B. Barrett,[1] John Lynam,[2] Frank Place,[3] Thomas Reardon[4] and Abdillahi A. Aboud[5]

[1]Cornell University, 315 Warren Hall, Ithaca, NY 14853, USA; [2]The Rockefeller Foundation, PO Box 47543, Nairobi, Kenya; [3]International Centre for Research in Agroforestry (ICRAF), PO Box 30677, Nairobi, Kenya; [4]Michigan State University, 211F Agriculture Hall, East Lansing, MI 48824-1039, USA; [5]Egerton University, PO Box 536, Njoro, Kenya

This final chapter draws together lessons learned from the preceding chapters as to how governments and donors might stimulate necessary investment in improved natural resource management (NRM). The mass of evidence clearly demonstrates that improved NRM practices can contribute significantly to increased agricultural productivity, environmental sustainability and reduced poverty and vulnerability. Since these are strategic objectives in all sub-Saharan African countries, NRM in agriculture plainly merits attention. Moreover, widespread uptake of improved NRM, on a scale sufficient to have a significant impact on aggregate productivity and income measures, is not occurring spontaneously and remains unlikely in the near term without external stimulus. Continued neglect of NRM in agriculture therefore comes at a significant cost.

The key issues revolve around the 'what' and 'how' questions of support to improve NRM: what principles and priorities need to be followed and how can governments and other influential stakeholders take practical steps to follow those principles and priorities? Although the dearth of past policy interventions to support improved NRM limit existing knowledge, some clear principles and priorities exist for fostering the accelerated uptake of improved NRM in African agriculture, as do a few practical findings concerning their implementation.

This chapter highlights these findings, which cumulatively recommend a 'five ins' strategy built on the fundamental point that NRM is an investment choice.[1] Investment is the first and biggest 'in', the strategic objective. Investment depends on four supporting 'ins' – incentives, information, inputs and institutions – just as a table rests upon its four legs. Individuals invest only when adequate information supports the conclusion that the investment will probably prove profitable within the relevant planning horizon and when they have the resources to put into it and confidence in the rules and organizations (the institutions) that ensure they will reap their just returns. As this concluding chapter goes on to describe, each of these four supports is individually necessary, but not sufficient, to stimulate investment in improved NRM and, with it, the dynamic of sustainable agricultural intensification and rural development in Africa.

The Policy Imperative: the 'Five Ins' Strategy

Global economic history teaches us that investment in agriculture lays the foundation for economic growth, industrialization and improved health and nutrition. Agriculture continues to account for the largest sectoral share of employment in sub-Saharan Africa (SSA), where poverty remains primarily a rural phenomenon. Africa is the lone continent where per capita agricultural productivity and the incidence of undernourishment have stagnated over the past 40 years. Plainly, agricultural development is a prerequisite for poverty reduction in Africa and yet no significant, widespread and sustained progress has been made since independence. Over the past decade or two, agricultural researchers and rural communities have jointly concluded that the poor state of the natural resource base on which agriculture depends is a primary factor limiting agricultural development and, derivatively, rural economic growth, poverty reduction and food security, both in the near term and for future generations. Unfortunately, however, policy-makers and donors have been slow to invest in improved NRM for agriculture.

The preceding chapters document that improved NRM is feasible and can be economically profitable throughout the continent. Improved fallows, inorganic and organic fertilizers and soil- and water-conservation structures indeed increase yields, returns to labour and cash income in systems as diverse as the semi-arid areas of Niger, Tanzania and Zimbabwe, highland Ethiopia and Rwanda and the subhumid zones of western Kenya and Nigeria. The challenge stems from the limited scale of uptake of improved NRM practices thus far. In order to stimulate aggregate agricultural productivity and rural incomes, adoption rates need to increase by several orders of magnitude, to millions and tens of millions of African farmers. So there is both a pressing need and a demonstrable potential for improved NRM.

The challenge is the dearth of generalized private investment in NRM – the 'in' at the apex of the strategy we advocate – due to problems of incentives, information, inputs and institutions – the remaining four 'ins' that support investment. The constituent problems of this challenge are surely familiar to most readers, but the studies assembled in this volume point towards a new, holistic view that satisfies the crucial test of providing both a useful descriptive lens on recent history and a prescriptive model for moving ahead. Perhaps more importantly, this collection of studies offers key insights on how to promote improved incentives, information, inputs and institutions. Before moving to these issues, we first consider a very brief historical narrative to underscore the under-recognized interrelation between the five 'ins'.

One can usefully oversimplify the history of agricultural development efforts in post-independence Africa as divisible into three distinct periods. The first, post-independence era, roughly from the early 1960s to the early 1980s, emphasized state provision of inputs, such as subsidized credit and fertilizer, and the establishment of institutions intended to support agriculture through parastatal marketing bodies and national agricultural research and extension services. Unfortunately, heavy-handed government intervention generally proved fiscally unsustainable, involved top-down designs that restricted information flow and badly distorted farmer incentives to invest in agriculture, particularly by depressing output prices so as to make farming unprofitable for those with alternative opportunities. This sowed the seeds of institutional collapse in African agriculture.

The failures of the first-generation strategy fed the macroeconomic crisis of the early 1980s and the ensuing era of liberalization and structural adjustment. The policy rhetoric turned almost entirely to incentives – 'getting prices right', as the famous injunction termed it – and macroeconomic reform programmes across the continent emphasized scaling back the state and letting market allocation mechanisms take over. Unaccompanied by ancillary investments in the physical and institutional infrastructure necessary to support markets, these changes often merely exposed the underlying structural weaknesses that had previously spawned state intervention in rural Africa. The virtual institutional collapse that beset African agriculture reduced the availability of inputs, slowed the flow of information and ultimately undermined the profitability of all sorts of crops, thereby reducing incentives to invest in soil or water conservation or in integrated nutrient or pest management.

Most recently, the failure of market-oriented reforms to stimulate a robust supply response or to reduce rural poverty appreciably has prompted a new-found emphasis on democratization and civil society, moving the focus from incentives to institutions and information. The virtues of participatory approaches to development, of a free press and of social capital have become celebrated by governments, donors and scholars alike. Attention has rightly returned to the need to build capacity in community-based organizations for improved management of common property resources, to reduce information costs and increase information and financial flows through farmer field schools, farmer research committees and microfinancial institutions and to carry out authentically participatory research on poverty and technologies. None the less, the natural resource base continues to deteriorate, as African smallholders respond to weak NRM incentives and scarce essential inputs by divesting their natural capital through the harvesting of nutrients without adequate replenishment.

By its nature, improved NRM in agriculture requires widespread private investment. The absence of widespread, spontaneous adoption of improved methods indicates that conditions prevailing in rural Africa, at least outside intensively supported pilot projects, do not support farmers making essential investments in NRM. The only feasible path forward requires concerted public investment in providing the necessary incentives, information, inputs and institutions. In so far as these pieces of the puzzle have each been the subject of extensive reflection in the past, much appears familiar in the strategy we advocate. It is their necessary integration into a whole, as a foundation for broad-based investment in improved NRM, that is new, as well as urgent.

Core Principles, Priorities and Practical Next Steps

The 'five ins' strategy rests upon several core, interrelated principles that emerge clearly from the preceding chapters. Each principle implies certain priorities for policy-makers who are serious about stimulating improved NRM for accelerated agricultural development in Africa. Where both theory and empirical evidence provide support, we also offer practical suggestions as to the appropriate next steps in policy.

Knowledge-intensive integrated natural resource management

The first core principle around which policy must be designed is that improved NRM practices are knowledge-intensive. They are management practices, not discrete inputs like those that underpin agricultural or industrial production technologies commonly embodied in seed, chemicals or machinery. In part, the knowledge intensity of NRM arises because practices are inherently interrelated. Economists accustomed to thinking about inputs as substitutes would do well to heed the cautions of biological scientists that natural inputs are primarily complements to crop and livestock production. Plants need sufficient minerals and soil organic matter, water and sunlight to grow well. Substitution possibilities among these essential inputs are somewhat limited. So farmers must manage multiple resources well in order to attain and maintain high productivity. This basic observation is too commonly overlooked in the agricultural and development policy communities.

This principle carries several implications for policy priorities and practices. First, the agricultural community needs to move more vigorously towards integrated NRM that tackles the simultaneous problem of soil, water and biomass management, as distinct from promotion of individual practices or technologies (e.g. alley farming, irrigation, terracing, tied ridges). Appropriate packages of practices have been insufficiently identified and extended in most of Africa. Most individual elements of these packages exist already, but they are scattered. In business terms, the issue is less production than packaging and distribution. A high research priority needs to be placed on identifying and promoting best-bet packages of practices and technologies, much as the Soil Fertility Network is doing in southern Africa (Makuria and Waddington, Chapter 17). The economic pay-off appears to be high. The collaborative report *Can Africa Claim the 21st Century?* (World Bank, 2000b) finds a 37% internal rate of return on agricultural research, and recent International Food Policy Research Institute (IFPRI) research shows similar, spectacular expected returns to agricultural research (Alston *et al.*, 2000).

Secondly, and relatedly, it is time to end the artificial conflict between so-called 'modern' methods, based on chemical fertilizers, irrigation and improved cultivars, and 'traditional' or 'agroecological' methods, based on intercropping, rotations, cover crops and organic nutrient supplements. Most of the chapters emphasize the existence of crucial complementarities between inorganics and organics. Although farmers will try to substitute one sort for the other, as occurred in much of Africa over the past decade as fertilizer prices rose sharply in the wake of structural adjustment policies (see Barrett *et al.*, Chapter 1, Gladwin *et al.*, Chapter 9, and Shapiro and Sanders, Chapter 20), such substitution mainly limits the rate of productivity decline. Productivity improvements depend on recognizing and reinforcing complementarities. Governments and donors might do well to enact explicit interlinkage and cross-compliance policies, such as providing farmers who undertake and maintain soil- or water-conservation investments with coupons to subsidize fertilizer purchases through commercial distributors – that is, subsidies that foster NRM investment and simultaneously provide demand stimulus for the development of private fertilizer markets. This same basic design has proved remarkably successful in other settings, such as in respect of investments in early childhood nutrition through the USA's Women, Infants and Children (WIC) programme (Barrett, 2002).

Thirdly, information flow must improve between and within national agricultural research and extension systems (including universities), rural communities, non-governmental organizations (NGOs), private traders and individual farmers. The issue extends well beyond familiar prescriptions to encourage participation. More effort needs to go into conceptualizing and implementing institutional and organizational frameworks within which participation occurs, so as to harness the comparative advantages of inherently complementary groups in a resource-starved environment. There is, as yet, no accepted analytical model for how to effectively integrate universities and agricultural research institutes, which are better positioned to undertake *de novo* applied research and to conduct both *ex ante* impact assessment and *ex post* evaluation of interventions, with NGOs, extension services and community-based organizations, which are relatively more able to engage farmers in an ongoing dialogue about research and policy priorities, and with cooperatives and commercial traders, which are effective at distributing new materials (e.g. germ-plasm, fertilizer, lime) to those who can best use them. Entities' capabilities vary across functions and user groups, so there need to be multiple, complementary channels for the production and dissemination of information for farmers.

Information is central to the ongoing research problem surrounding NRM (Place *et al.*, Chapter 21). Although some technology gaps remain – especially in water management, less so with respect to soil conservation and fertility management – the immediate scaling-up problem reflects mainly insufficient farmer demand for investment in improved NRM under prevailing conditions. Although essential technological components exist amid the wealth of modern and traditional practices observable on the African continent, the agricultural research and extension community still lacks a clear understanding of how best to combine techniques to suit different agroecological and market conditions. As several chapters establish, where truly practised, farmer participation is realizing much of its potential by accelerating the identification of which among the many potentially problematic factors limits farm productivity and the uptake of existing techniques in a given location, thereby accelerating the development of more suitable practices. The experiences Tarawali *et al.* (Chapter 5) report from West Africa are especially encouraging.

Fourthly, knowledge intensity places a premium on education, not just for literacy or numeracy, but for analytical, observational and communication skills. Rural schools have suffered across Africa over the past decade, as central government budgets have been cut and local government revenues have proved insufficient to sustain public schools. The introduction of user fees has caused many poorer families to pull children out of school, at least in times of financial stress, although households commonly go to great lengths to try to get and keep at least some of their children in school. In high-potential areas, education often provides not only the capacity to respond better to changing technologies and environmental conditions (Schultz, 1975; Barrett *et al.*, 2001e), but also access to non-farm income-earning opportunities, which are essential to on-farm investment in improved NRM (Tiffen *et al.*, 1994; Barrett *et al.*, 2001d; Barrett *et al.*, Chapter 1, Clay *et al.*, Chapter 8, and Wyatt, Chapter 10, this volume). In lower-potential areas, the relationship between education and NRM appears more complex, as the educated commonly disinvest from agriculture (Wyatt, Chapter 10). Donors and NGOs can contribute materially here since investment in education remains low. For example, less than 8% of World Bank lending since 1994 has gone to education projects. Doubling this would help improve the quality of instruction for existing students, as well as making education accessible to children from poorer families and more remote regions. One appealing possibility is to couple this with the global school feeding programmes being advocated in high-level policy discussions currently, paying for education partly through food-aid deliveries, so as to pursue integrated human-capital formation linking education, health and nutrition.

Farmer-centred policy and research design

The second core principle evident in the preceding chapters – and closely related to the first – is that the development, extension and evaluation of NRM innovations and policies must be farmer-centred. The extraordinary biophysical, cultural and economic variability of rural Africa makes it difficult to identify effective local solutions without early, active involvement of local farmers and communities. The undistinguished history of top-down technology development, input distribution and extension services in Africa stands in stark contrast to encouraging cases of rapid adoption and resourceful adaptation of researcher-developed techniques in places where farmers have been fully involved as co-developers and co-evaluators of NRM practices from the early, problem-identification stage onwards (Pretty and Buck, Chapter 2, Adesina and Chianu, Chapter 4, Tarawali *et al.*, Chapter 5, Kelly *et al.*, Chapter 15, Shapiro and Sanders, Chapter 20, and Place *et al.*, Chapter 21). An insufficiently nuanced understanding of the local context by outsiders and the (often prudent) distrust of outsiders' 'science' account for much of the low adoption rates of many 'improved' practices and technologies among African smallholders. Quite apart from ethical and political-economy questions of empowerment, participatory methods are necessary in order to identify and promote locally appropriate practices. Relatedly, there is an inherent complementarity between 'indigenous' and 'scientific' knowledge (Peters, Chapter 3).

The disconnection has been not only between bench scientists developing supposedly 'improved' NRM techniques and farmers who reject these innovations, but also between policy-makers, who set national-level policy, and their constituents, who pursue strikingly different priorities. Hatibu *et al.* (Chapter 16) offer an especially vivid description of the policy failure of top-down approaches in semi-arid Tanzania.

Three policy priorities emerge from this second core principle. First, although both the agricultural and development communities have largely embraced participatory approaches, there remains little institutional structure for expanding participation to scales beyond specific research or intervention sites. The establishment of effective local governance complemented by competent, specialized central government agencies must be made a priority. In recent years, much emphasis has rightly been placed on decentralization of authority from the centre to localities in response to past malfeasance or misfeasance. Too often, however, decentralization and the broader roll-back of the public sector have emasculated the central government's capacity to perform essential functions. As Krueger (1990) emphasized, past government nonfeasance is as much the problem as malfeasance or misfeasance.

There are important, unresolved issues surrounding the appropriate division of responsibility between different levels of government authorities and between state and extrastate institutions (Pretty and Buck, Chapter 2, Gebremedhin and Swinton, Chapter 6, Freudenberger and Freudenberger, Chapter 14, Kelly *et al.*, Chapter 15, and Hatibu *et al.*, Chapter 16). These issues revolve largely around reconciling the need to: (i) internalize environmental externalities associated with NRM – a classic common-property management problem; (ii) achieve a minimum efficient scale in activities characterized by significant fixed costs – a coordination problem familiar to students of cooperatives; and (iii) respect different institutions' comparative advantage in

performing distinct, but complementary tasks. For example, communities commonly prove more effective at enforcing access and use rules (Ostrom, 1990; Baland and Platteau, 1996; Gebremedhin and Swinton, Chapter 6, this volume), but are ineffective at providing essential infrastructural services, in which case regional or national governments must be involved (Freudenberger and Freudenberger, Chapter 14). The appropriate level for control depends fundamentally on the scope of the ecological externalities, as well as on the capacity of institutions at different levels to gather information, to make and enforce sound judgements and to raise the necessary financing for essential operations (Barrett *et al.*, 2001c; Gjertsen and Barrett, 2001).

Secondly, and relatedly, civil strife impedes the effective functioning of government and extragovernmental institutions responsible for coordinating resource use and policy. Divisive politics and recurrent violence pose hazards beyond and more serious than poor NRM. Much as macroeconomic factors tend to have a greater effect on agriculture than do sectoral interventions (Krueger *et al.*, 1988), so too do broader societal conditions trump narrower NRM policies in conditioning the use of soils and water. There is, at best, limited potential for progress in stimulating improved NRM, agricultural intensification and rural economic growth and poverty alleviation if governments and donors fail to effectively address the widespread problems of uncivil society and violations of the rule of law. Indeed, these fundamental problems undermine the functioning of rural institutions to support agriculture and NRM. Although agricultural specialists have little expertise – and sometimes interest – in inherently political questions about social stability, its importance must be acknowledged and supporting efforts to establish a civil social and political discourse must be made a priority, even by those of us with expertise elsewhere.

Thirdly, and perhaps paradoxically, a farmer-centred approach to NRM in agriculture must transcend traditional sectoral specificity in order to take seriously the broader livelihood objectives of rural

Africans. Priority must be placed on developing the rural non-farm economy along with agriculture. Most rural Africans farm because, given the assets they hold, the opportunities and constraints they face and their location, farming is an attractive piece of a broader strategy to take care of themselves and their families. Most African farmers none the less undertake non-farm activities, and non-farm earnings are positively related to subsequent upward income mobility in rural Africa (Reardon, 1997a; Barrett *et al.*, 2001b,d).

Beyond introducing the non-farm rural economy into the debate about agricultural development policy, a broader livelihoods perspective also implies a caution against mistaking adoption of NRM methods as an end in itself. Indeed, as Tarawali *et al.* (Chapter 5) show, disadoption of NRM – in their case, of green-manure cover crops – may indicate attainment of the livelihood objective that motivated the initial adoption. Furthermore, as Wyatt (Chapter 10) emphasizes, for some rural Africans the most economically and agroecologically appropriate investment opportunities lie in non-farm sectors, not in agriculture.

In the medium to long term, economic growth will inevitably spawn a disproportionately rapid expansion of the rural non-agricultural sector, as has been the case in all agricultural transformations in history (Timmer, 1986). Growth in the rural non-farm economy can fuel a virtuous circle of improved NRM management by resolving liquidity constraints (Holden and Shiferaw, Chapter 7, Clay *et al.*, Chapter 8, and Hatibu *et al.*, Chapter 16) or it can induce disadoption of improved NRM practices, or even accelerated resource degradation, as the returns to labour and capital outside agriculture outcompete investments in improved NRM (Barrett *et al.*, 2001a; Wyatt, Chapter 10, this volume). At present, we have insufficient understanding of the factors behind the empirically ambiguous relationship between NRM and the rural non-farm economy in SSA. Much seems to depend on whether input and output market conditions make investment in agriculture and supporting natural capital attractive.

Improved natural resource management must pay

Farmers incur real costs to undertake NRM investments. They must dedicate time that could be devoted instead to other farm or non-farm activities, and they must often also use land or cash having considerable opportunity costs. Investment in immovable natural resources also exposes them to considerable risk – of poor harvests, low prices, asset appropriation – which weighs heavily on vulnerable people. No reasonable person would incur such costs unless the broader economic environment makes it pay, and within a reasonable time span. The third core principle of these studies is consequently that the widespread uptake of improved NRM practices depends on commercially viable agriculture or significant subsidies. While long fallow rotations worked in the distant past, in the face of low population densities, semi-subsistence agriculture cannot support widespread improved NRM in contemporary Africa.

Improved NRM consistently appears among high-value cash crops. Farmers apply chemical fertilizer and invest in conservation structures, organic-matter application, cover crops and improved fallows at much higher rates on areas planted in commercial crops than they do on areas devoted to subsistence production. Farmers see the return to the quite tangible costs of investing in improving productivity (and sustaining improved productivity) when they get an obvious pay-out from the market. If the returns to agriculture cannot compete with those from other activities, the empirical evidence clearly indicates that farmers do not invest scarce investment resources in natural capital.

The tough challenge then revolves around how to make NRM investment pay among the broad mass of smallholders producing cereals, tubers and roots under rain-fed conditions for home consumption or purely local markets. Simply put, this requires subsidies – information production and dissemination, institutions to organize input procurement and output marketing, or complementary inputs (not just fertilizer or only terraces). Improved NRM takes place in the production of semi-subsistence staples almost only where one finds localized, *de facto* subsidization through temporary projects offering institutional and informational support and inputs (Gebremedhin and Swinton, Chapter 6, Place *et al.*, Chapter 12, Kristjanson *et al.*, Chapter 13, Freudenberger and Freudenberger, Chapter 14, Kelly *et al.*, Chapter 15, Hatibu *et al.*, Chapter 16, Mekuria and Waddington, Chapter 17, and Shapiro and Sanders, Chapter 20). Across the continent, such mini-packages have replaced and replicated policies terminated 15–20 years ago in the context of structural adjustment programmes. Scant NRM investment takes place outside these implicitly subsidized intervention zones and commercially viable operations. Improved NRM practices can render staples production profitable without ongoing subsidies to farmers, but adoption does not seem to occur without an initial stimulus through subsidized, public investment in information and institutions. Poor smallholders dependent on low-value crops cannot afford to invest much in experimentation, so the upfront costs of establishing the efficacy of a technique must be borne more broadly, as are the later benefits of improved NRM. Thereafter, improved NRM adoption may boost agricultural profitability, rather than the other way around, by either or both of two pathways. Cereals, roots and tubers can become profitable, as demonstrated by the studies of improved fallows in western Kenya (Place *et al.*, Chapter 12) and soil- and water-conservation structures in the Sahel and Ethiopia (Shapiro and Sanders, Chapter 20). Or improved NRM can encourage the adoption of higher-value crops, as in the case of biomass transfer that was demonstrated on maize to western Kenyan farmers, who then began to use the technique on vegetables grown for market (Place *et al.*, Chapter 12).

Generalized subsidization of investment is fiscally infeasible in sub-Saharan Africa. Even for rich-country governments in North America and Europe, conservation programmes involving direct payments or subsidies to farmers pose significant budgetary burdens. The practical question of how to make NRM investment pay must therefore be approached more obliquely: what feasible public investments can increase the returns to agriculture enough to make investment in improved NRM pay? Put differently, what public goods and institutions 'crowd in' private investment in improved NRM practices and soil- and water-conservation structures? This is in marked contrast to pre-structural-adjustment policies, which crowded out private investment.

Physical infrastructure and associated technologies are an important ingredient. Freudenberger and Freudenberger (Chapter 14) emphasize the importance of transport infrastructure if smallholders are to play to their agroecological comparative advantage (e.g. cultivating a commercial and sustainable banana crop rather than an extensive rice crop in the forests of Madagascar's eastern escarpment). One might also emphasize communications infrastructure, since informational deficiencies appear to be an important part of rural market imperfections in Africa (Omamo, 2001). The African continent has only as many phone lines as does the Borough of Manhattan within New York City. Cellular and wireless communication technologies make it possible to establish and maintain service to areas too costly to serve with wire-based technologies. Such innovations are taking place in rural Bangladesh already, under the leadership of Mohammed Yunus, the founder of the celebrated Grameen Bank, a pioneer in the area of microfinance. Credible government or donor guarantees of infrastructural investments in the event that the local community undertakes the minimum necessary levels of private investment in improved NRM, whether through collective or individual efforts, can be used effectively as inducements, as has been the experience in India (Chopra, 1997).

The second priority area associated with establishing an enabling market environment relates to the institutional setting, or the rules of conduct necessary for individuals to contract with confidence and to feel secure in their claims to durable assets, such as land, livestock and water. Most of the preceding chapters emphasize the importance of clear, durable property rights. This includes security against seizure by invaders, by the state or by powerful individuals, which can include husbands and brothers who prey upon women's relative powerlessness in resource control in some cultures (Gladwin *et al.*, Chapter 9). Building up land markets can help to capitalize soil quality in land values, thereby making investment in NRM more attractive. But, in the presence of binding financial constraints, increased investment may be concentrated mainly among the wealthy, exacerbating rural wealth and productivity inequality (Carter and Olinto, forthcoming).

The institutional issues extend well beyond the security of productive assets – especially land and livestock – to ensuring that farmers indeed receive what a package's label or a trader or another farmer promises, or else they can reasonably expect compensation for breach of contract. Trust is an essential ingredient of market exchange, but is itself a scarce commodity. Individuals' repeated interaction provides a modest level of contract enforcement capacity, but economic history clearly shows that public and private order institutions are central to the minimization of transactions costs and the development of commerce (Platteau, 1997a,b). Establishing or restoring the rule of contract law in agricultural markets can substantially extend the reach of commercially viable agriculture in rural Africa.

The third priority area relates to rural financial systems, which are notoriously underdeveloped throughout Africa. Insufficient credit, insurance and savings impede investment in improved NRM, just as in other forms of productive capital, and thereby trap rural Africans in long-standing cycles of poverty and vulnerability. As the old adage has it, 'it takes money to make money', and African smallholders too often lack the funds necessary to invest in remunerative (and sometimes risk-reducing) livelihood strategies based on improved NRM (Barrett *et al.*, 2001d; Holden and Shiferaw, Chapter 7, Clay *et al.*, Chapter 8, Wyatt, Chapter 10, and Hatibu *et al.*, Chapter 16, this volume). The preceding chapters none the less underscore that low-cost

investments are increasingly available and adopted by small farmers. This includes the use of fertilizers now distributed in small sachets (Freeman and Coe, Chapter 11) and the adoption of inexpensive improved-fallow seedlings (Gladwin *et al.*, Chapter 9, and Place *et al.*, Chapter 12).

The fourth and final priority area concerns organization for collective action, echoing a point made in the previous subsection in respect of farmer-centred approaches to agricultural development. Significant fixed costs can make market participation unremunerative at the individual level. When farmers can cooperate in purchasing fertilizer, lime and other inputs, in building community nurseries to cultivate seedlings not readily purchased on the market and in selling marketable produce, they can thereby achieve the minimum efficient scale of production or distribution necessary to make investment in agriculture, and derivatively in improved NRM, pay. Cooperatives have unfortunately had a largely undistinguished history in independent Africa, but often that relates to top-down organizational designs and rules established outside the group itself, as in Tanzania's experience under *ujamaa*. Collective action can be highly complementary to market development, as amply demonstrated by the plethora of small self-help groups that have formed group marketing arrangements across the continent.

Conclusion

The agricultural development community has gradually come to recognize improved NRM as fundamental to sustainable agricultural intensification, which is itself a necessary, albeit not sufficient, condition of economic growth, poverty alleviation and environmental conservation (Lee and Barrett, 2000). However, the relatively slow rate of uptake of improved NRM practices by small farmers operating under harsh agroecological conditions and considerable socio-economic stress underscores the magnitude of the challenge of stimulating private investment on a capital-starved continent. Meeting this challenge requires the proper incentives, information, inputs and institutions to support widespread investment in improved NRM, or what we term the 'five ins' strategy. For too long, policy-makers and donors have focused on just a subset of these, to the effective exclusion of the others, ultimately undermining African farmers' ability or willingness to invest in natural resources to support agriculture. All four pieces of the puzzle are necessary to promote investment in improved NRM. Donors and governments have searched for short cuts for too long.

The varied experiences reported in this volume reveal three core principles that must underpin an effective strategy of improved NRM for African agriculture: (i) improved NRM practices are knowledge-intensive; (ii) the development, extension and evaluation of NRM innovations and policies must be farmer-centred; and (iii) the widespread uptake of improved NRM practices depends on a commercially viable agriculture or significant subsidies. Each of these principles implies a few policy priorities, which this chapter has briefly outlined and the preceding chapters have elaborated on in some detail. Cumulatively, the

resources required to advance this agenda will doubtless prove substantial. But these pale beside the cost of failing to pursue a strategy up to the task of stimulating NRM investment by African farmers. If the agricultural and development communities follow these principles and priorities, the encouraging improvements in NRM already evident on thousands of farms in communities ranging across Africa can be multiplied rapidly, thereby improving livelihoods for the current generation of rural Africans and sustaining the resource base on which their children and grandchildren will depend.

Note

[1]   Just before this volume went to press, Tom Tomich helpfully pointed out the similarity between our 'five ins' strategy for NRM and the 'four Is' strategy for agricultural development in the 1986 Food and Agricultural Organization (FAO) report, *African Agriculture: the Next Twenty-five Years*, its expansion to a 'five Is' approach by the Kenyan agricultural economist Bill Omamo, and the later 'six Is' strategy in Tomich *et al.* (1995: 166–177). This volume reinforces those earlier analyses.

References

Aboud, A., Sofranko, A.J. and Ndiaye, S. (1996) The effect of gender on adoption of conservation practices by heads of farm households in Kenya. *Society and Natural Resources* 9, 447–463.

Adaptive Research Group (1990) *Fertilisation of Maize Varieties: Fertiliser Rates*. Machinga ADD, Liwonde, Malawi.

Adesina, A.A. (1996) Factors affecting the adoption of fertilizers by rice farmers in Côte d'Ivoire. *Nutrient Cycling in Agroecosystems* 46, 29–39.

Adesina, A.A. and Baidu-Forson, J. (1995) Farmers' perceptions and adoption of new agricultural technology: evidence from analysis in Burkina Faso and Guinea, West Africa. *Agricultural Economics* 13(1), 1–9.

Adesina, A.A. and Zinnah, M.M. (1993) Technology characteristics, farmers' perceptions and adoption decisions: a Tobit model application in Sierra Leone. *Agricultural Economics* 9, 297–311.

Adesina, A.A., Chianu, J. and Mbila, D. (1997) Property rights and alley farming technology adoption in West and Central Africa. Paper presented at the International Workshop on Property Rights, Collective Action and Technology Adoption, International Centre for Agricultural Research on Dryland Areas, Aleppo, Syria, 22–25 November.

Adesina, A.A., Mbila, D., Nkamleu, G.B. and Endamana, D. (2000) Econometric analysis of the determinants of adoption of alley farming by farmers in the forest zone of southwest Cameroon. *Agriculture, Ecosystems and Environment* 80, 255–265.

Agarwal, A. (1995) Dismantling the divide between indigenous and scientific knowledge. *Development and Change* 26, 413–439.

Agarwal, B. (1997) Editorial: Re-sounding the alert – gender, resources, and community action. *World Development* 25(9), 1373–1380.

AGERAS (1999) *Synthèse de l'Analyse Diagnostic du Corridor Forestier Ranomafana–Andringitra*. Cellule Technique d'Appui, Fianarantsoa, Madagascar.

Ahmed, M.M., Sanders, J.H. and Nell, W.T. (2000) New sorghum and millet cultivar introduction in sub-Saharan Africa: impacts and research agenda. *Agricultural Systems* 64, 55–65.

Ahn, J.-H., Elliott, R. and Norton, B.W. (1997) Oven drying improves the nutritional value of *Calliandra calothyrsus* and *Gliricidia sepium* as supplements for sheep given low quality straw. *Journal of the Science of Food and Agriculture* 75, 503–510.

Aina, P.O., Lal, R. and Roose, E.J. (1991) Tillage methods and soil and water conservation in West Africa. *Soil and Tillage Research* 20(2–4), 165–186.

Akinola, A.A. and Young, T. (1985) An application of the Tobit model in the analysis of agricultural innovation adoption processes: a study of the use of cocoa spraying chemicals among Nigerian cocoa farmers. *Oxford Agrarian Studies* 14(1), 26–51.

Ali, M. and Byerlee, D. (1991) Economic efficiency of small farmers in a changing world: a survey of recent evidence. *Journal of International Development* 3(1), 1–27.

Alston, J.M., Chan-Kang, C., Marra, M.C., Pardey, P.G. and Wyatt, T.J. (2000) *A Meta-Analysis of Rates of Return to Agricultural R&D: Ex Pede Herculem?* International Food Policy Research Institute, Washington, DC.

Andergie, T. (1994) *Characterization of Lugo River Watershed*. Consultant Report, March.

Anon. (1984) Soil and water. *Crops and Soils Magazine* 37, 26–27.

Argwings-Kodhek, G. (1996) The evolution of fertilizer marketing in Kenya. Paper presented at Conference on Fine-tuning Market Reforms for Improved Agricultural Performance, Policy Analysis Project, Egerton University, Kenya.

Argyris, C. and Schön, D. (1978) *Organisational Learning*. Addison-Wesley, Reading, Massachusetts.

Arrow, K., Solo, R., Portend, R., Leader, E.E., Radar, R. and Schulman, H. (1993) Report of the NOAA panel on contingent valuation. *Federal Registry* 58(10), 4602–4614.

ASB (1997) *The Alternatives to Slash-and-Burn. Report of the 6th Annual Review Meeting, 17–27 August 1997*. Bogor, Indonesia.

Attah-Krah, A.N. and Francis, P.A. (1987) The role of on-farm trials in the evaluation of composite technologies: the case of alley farming in southern Nigeria. *Agricultural Systems* 23, 133–152.

Atwood, D. (1990) Land registration in Africa: the impact on agricultural production. *World Development* 18(5), 659–671.

Badiane, O. and Delgado, C. (1995) *A 2020 Vision for Food, Agriculture, and the Environment in Sub-Saharan Africa*. Food, Agriculture and the Environment Discussion Paper 4, International Food Policy Research Institute, Washington, DC.

Baland, J.-M. and Platteau, J.-P. (1996) *Halting Degradation of Natural Resources: Is There a Role for Rural Communities?* Clarendon Press, Oxford.

Baland, J.-M. and Platteau, J.-P. (1998) Division of the commons: a partial assessment of the new institutional economics of land rights. *American Journal of Agricultural Economics* 80(3), 644–650.

Baland, J.-M. and Platteau, J.-P. (1999) The ambiguous impact of poverty on local resource management. *World Development* 27(5), 773–788.

Balfour, E.B. (1943) *The Living Soil*. Faber and Faber, London.

Bannister, M.E. and Nair, P.K.R. (1990) Alley cropping as a sustainable agricultural technology for the hillsides of Haiti: experience of an agroforestry outreach project. *American Journal of Alternative Agriculture* 5(2), 51–59.

Barrett, C.B. (1997a) How credible are estimates of peasant allocative scale or scope efficiency? A commentary. *Journal of International Development* 9(2), 221–229.

Barrett, C.B. (1997b) Food marketing liberalization and trader entry: evidence from Madagascar. *World Development* 25(5), 763–777.

Barrett, C.B. (1999) Stochastic food prices and slash-and-burn agriculture. *Environment and Development Economics* 4(2), 161–176.

Barrett, C.B. (2002) Food security and food assistance programs. In: Gardner, B.L. and Rausser, G.S. (eds) *Handbook of Agricultural Economics*. Elsevier Science, Amsterdam.

Barrett, C.B. and Carter, M.R. (1999) Microeconomically coherent agricultural policy reform in Africa. In: Paulson, J. (ed.) *African Economies in Transition, Vol. 2: The Reform Experiences*. Macmillan, London, pp. 288–347.

Barrett, C.B. and Reardon, T. (2000) *Asset, Activity, and Income Diversification Among African Agriculturalists: Some Practical Issues*. Project report to USAID BASIS CRSP, March. Ithaca, New York.

Barrett, C.B., Barbier, E.B. and Reardon, T. (2001a) Agroindustrialization, globalization, and international development: the environmental implications. *Environment and Development Economics* 6(4), 419–433.

Barrett, C.B., Bezuneh, M. and Aboud, A. (2001b) Income diversification, poverty traps and policy shocks in Côte d'Ivoire and Kenya. *Food Policy* 26(4), 367–384.

Barrett, C.B., Brandon, K., Gibson, C. and Gjertsen, H. (2001c) Conserving tropical biodiversity amid weak institutions. *BioScience* 51(6), 497–502.

Barrett, C.B., Reardon, T. and Webb, P. (2001d) Nonfarm income diversification and household livelihood strategies in rural Africa: concepts, dynamics and policy implications. *Food Policy* 26(4), 315–331.

Barrett, C.B., Sherlund, S. and Adesina, A.A. (2001e) Macroeconomic shocks, human capital and productive efficiency: evidence from West African farmers. Unpublished working paper.

Basu, K. and Foster, J.E. (1998) On measuring literacy. *Economic Journal* 108(451), 1733–1749.

Bationo, A. and Mokwunye, A.U. (1991a) Alleviating soil fertility constraints to increased crop production in West Africa. The experience in the Sahel. *Fertilizer Research* 29, 95–115.

Bationo, A. and Mokwunye, A.U. (1991b) Role of manures and crop residues in alleviating soil fertility constraints to crop production with special reference to the Sahelian zones of West Africa. *Fertilizer Research* 29, 117–125.

Bationo, A., Christianson, C.B. and Klaij, M. (1993) The effects of crop residues and fertilizer use on pearl millet grain yields in Niger. *Fertilizer Research* 28, 271–279.

Bationo, A., Rhodes, E., Smaling, E. and Visker, C. (1996) Technologies for restoring soil fertility. In: Mokwunye, A.U., de Jager, A. and Smaling, E.H.A. (eds) *Restoring and Maintaining the Productivity of West African Soils*. Miscellaneous Fertilizer Studies 14, IFDC, Lomé, Togo, pp. 61–82.

Bationo, A., Lompo, F. and Koala, S. (1998) Research on nutrient flows and balances in West Africa: state-of-the-art. *Agriculture, Ecosystems and Environment* 71, 19–35.

Batz, F.J., Peters, K.J. and Janssen, W. (1999) The influence of technology characteristics on the rate and speed of adoption. *Agricultural Economics* 21(2), 121–130.

Bauer, A., Cole, C.V. and Black, A.L. (1987) Soil property comparisons in virgin grassland between grazed and ungrazed management systems. *Soil Science Society of America Journal* 51, 176–182.

Behnke, R.H., Scoones, I. and Kerven, C. (eds) (1993) *Range Ecology at Disequilibrium: New Models of Natural Variability and Pastoral Adaptation in African Savannas*. Overseas Development Institute, London.

Beinart, W. (1984) Soil erosion, conservationism and ideas about development: a southern African exploration, 1900–1960. *Journal of Southern African Studies* 11(1), 52–83.

Bekunda, M.A., Bationo, A. and Ssali, H. (1997) Soil fertility management in Africa: a review of selected research trials. In: Buresh, R., Sanchez, P. and Calhoun, F. (eds) *Replenishing Soil Fertility in Africa*. Soil Science Society of America (SSSA) Special Publication 51, Soil Science Society of America, Madison, Wisconsin, pp. 63–180.

Bellon, M.R., Gambara, P., Gatsi, T., Machemedze, T.E., Maminimini, O. and Waddington, S.R. (1999) *Farmers' Taxonomies as a Participatory Diagnostic Tool: Soil Fertility Management in Chihota, Zimbabwe*. Economics Working Paper 99–13, CIMMYT, Mexico, DF, 18 pp.

Benson, T. (1996) A geographic and economic analysis of past nutrient response trial results in Malawi. In: *Fifth Regional Maize Symposium for Eastern and Southern Africa*.

CIMMYT Maize Programme, Arusha, Tanzania.

Bergeret, A. (1993) Discours et politiques forestières coloniales en Afrique et Madagascar. In: *Colonialisations et environnement*. Etudes, Vol. 13, Société française d'histoire d'outre-mer, Paris.

Bernheim, B.D. (1994) A theory of conformity. *Journal of Political Economy* 102(5), 841–877.

Besley, T. (1995) Property rights and investment incentives. *Journal of Political Economy* 103, 913–937.

Bishop, J. (1995) *The Economics of Soil Degradation: An Illustration of the Change in Productivity Approach to Valuation in Mali and Malawi*. Discussion Paper DP 95-02, International Institute for Environment and Development, London.

Blackie, M.J. (1994) Maize productivity for the 21st century: the African challenge. *Outlook on Agriculture* 23, 189–195.

Blake, R.O. (1995) *The Challenges of Restoring African Soil Fertility. An Open Letter to the World Bank*. Committee on Agricultural Sustainability for Developing Countries, Washington, DC.

Blarel, B. (1994) Tenure security and agricultural production under land scarcity: the case of Rwanda. In: Bruce, J. and Migot-Adholla, S.E. (eds) *Searching for Land Tenure Security in Africa*. Kendall/Hunt, Dubuque, Iowa, pp. 71–95.

Blarel, B., Hazell, P., Place, F. and Quiggin, J. (1992) The economics of farm fragmentation: evidence from Ghana and Rwanda. *World Bank Economic Review* 6(2), 233–254.

Bojo, J. (1994) The cost of land degradation from a national perspective: an assessment of African evidence. Paper presented at the Eighth International Soil Conservation Conference, 4–8 December, New Delhi, India.

Bosch Van den, H., Gitari, J.N., Ogaro, V.N., Maobe, S. and Vlaming, J. (1998) Monitoring nutrient flows and economic performance in African farming systems (NUTMON). III. Monitoring nutrient flows and balances in three districts in Kenya. *Agricultural Ecosystems and Environment* 71, 63–80.

Boserup, E. (1965) *The Conditions of Agricultural Growth*. Aldine Publishing, New York.

Boserup, E. (1981) *Population and Technical Change*. Blackwell, Oxford.

Bourdieu, P. (1986) The forms of capital. In: Richardson, J. (ed.) *Handbook of Theory and Research for the Sociology of Education*. Greenwood Press, Westport, Connecticut.

Bourn, D., Wint, W., Blench, R.M. and Woolley, E. (1994) Nigerian livestock resources survey. *World Animal Review* 78, 49–58.

Boyd, C., Turton, C., Hatibu, N., Mahoo, H.F., Lazaro, E., Rwehumbiza, F.B., Okubal, P. and Makumbi, M. (2000) *The Contribution of Soil and Water Conservation to Sustainable Livelihoods in Semi-arid Areas of Sub-Saharan Africa*. Agricultural Research and Extension Network Paper no. 102, Overseas Development Institute, London.

Boyte, H. (1995) Beyond deliberation: citizenship as public work. Paper delivered at PEGS conference, 11–12 February 1995. Electronic publication on Civic Practices Network at: www.cpn.org

Braun, A. (2000) *The CIALs (Comité de Investigación Agrícultura Tropical) at a Glance*. CIAT, Colombia.

Breman, H. and Sissoko, K. (eds) (1998) *L'Intensification Agricole au Sahel*. Karthala, Paris.

Brokensha, D.W., Warren, D.M. and Werner, O. (eds) (1980) *Indigenous Knowledge Systems and Development*. University Press of America, Lanham, Maryland.

Bromley, D. (1993) Common property as metaphor: systems of knowledge, resources and the decline of individualism. *The Common Property Digest* 27, 1–8. IASCP, Winrock and ICRISAT, Hyderabad.

Brown, L. and Wolf, E. (1984) *Soil Erosion: Quiet Crisis in the World Economy*. WorldWatch Paper no. 60, WorldWatch Institute, Washington, DC.

Bruce, J.W. and Migot-Adholla, S.E. (1994) *Searching for Land Tenure Security in Africa*. University of Wisconsin Press, Madison, Wisconsin.

Brunner, J., Henninger, N. and Deichmann, U. (1995) *West Africa Long Term Perspective Study (WALTPS). Database and User's Guide*. World Resources Institute, Washington, DC.

Brush, S.B. and Stabinsky, D. (eds) (1996) *Valuing Local Knowledge: Indigenous People and Intellectual Property Rights*. Island Press, Washington, DC.

Buck, L.E. (1988a) *Agroforestry Extension Training Source Book*. CARE International, New York.

Buck, L.E. (1988b) Participatory monitoring and evaluation in agroforestry extension. In: Lilewe, A.M., Kealey, K.M. and Kebaara, K.K. (eds) *Agroforestry Development in Kenya. Proceedings of the 2nd Kenya National Seminar on Agroforestry held in Nairobi, Kenya,* 7–16 November. National Council for Science and Technology and the International Centre for Research in Agroforestry, Nairobi, pp. 188–202.

Buck, L.E. (1995) Agroforestry policy issues and research directions in the US and less developed countries: insights and challenges from recent experience. *Agroforestry Systems* 30(1–2), 57–73.

Buck, L.E. (2000) The social organization of agroforestry innovation: facilitating the emergence of a knowledge system in New York State and northeastern North America. PhD dissertation, Cornell University, Ithaca, New York.

Buck, L.E., Lassoie, J.P. and Fernandes, E.C.F. (eds) (1999) *Agroforestry in Sustainable Agricultural Systems*. CRC Press, St Lucie Division, Boca Raton, Florida.

Budd, W.W., Duchhart, I., Hardesty, L.H. and Steiner, F. (eds) (1990) *Planning for Agroforestry*. Elsevier, Amsterdam.

Bultena, G.L. and Hoiberg, E.O. (1983) Factors affecting farmers' adoption of conservation tillage. *Journal of Soil and Water Conservation* 38(3), 281–284.

Bumb, B. and Baanante, C. (1996) *The Role of Fertilizer in Sustaining Food Security and Protecting the Environment to 2020*. Food, Agriculture and the Environment Discussion Paper 17, International Food Policy Research Institute (IFPRI), Washington, DC.

Bumb, B.L., Teboh, J.F., Atta, J.K. and Asenso-Okeyre, W.K. (1996) Policy environment and fertilizer sector development in Ghana. Paper presented at the National Workshop on Soil Fertility Management Action Plan for Ghana, Efficient Soil Resources Management: A Challenge for the 21st Century, Cape Coast, Ghana, 2–5 July.

Bunch, R. and López, G. (1996) *Soil Recuperation in Central America: Sustaining Innovation after Intervention*. Gatekeeper Series SA 55, Sustainable Agriculture Programme, International Institute for Environment and Development, London.

Buol, S. and Stokes, M. (1997) Soil profile alteration under long-term high input agriculture. In: Buresh, R., Mokwunye, A.U., de Jager, A. and Smaling, E.M.A. (eds) *Replenishing Soil Fertility in Africa*. SSSA Special Publication No. 51, Soil Science Society of America, Madison, Wisconsin, pp. 97–110.

Bureau of Agriculture and Natural Resource Development (BoANRD) (1995) Agriculture in Tigray. Paper presented at Symposium on Agricultural Development in Humera Area of Tigray, Mekelle.

Bureau of Agriculture and Natural Resource Development (BoANRD) (1999) *Livestock Census Analysis Results*. Unpublished report, Mekelle.

Byiringiro, F.U. (1995) Determinants of farm productivity and the size–productivity relationship under land constraints: the case of Rwanda. MS thesis, Department of Agricultural Economics, Michigan State University, East Lansing, Michigan.

Byiringiro, F.U. and Reardon, T. (1996) Farm productivity in Rwanda: effects of farm size, erosion, and soil conservation investments. *Agricultural Economics* 15(2), 127–136.

Calder, I.R. (1994) *Eucalyptus, Water and Sustainability. A Summary Report*. ODA Forestry Series No. 6, ODA, London, 14 pp.

Cameron, L. (1999) The importance of learning in the adoption of high-yielding variety seeds. *American Journal of Agricultural Economics* 81(1), 83–94.

Campbell, J. (1991) Land or peasants? The dilemma confronting Ethiopian resource conservation. *African Affairs* 90(385), 5–21.

Carney, D. (1998) *Sustainable Rural Livelihoods*. Department for International Development, London.

Carr, S.J. (1996) Comments on Smale and Heisey paper. (Letter given to author).

Carsky, R.J., Tarawali, S.A., Becker, M., Chikoye, D., Tian, G. and Sanginga, N. (1998) *Mucuna – Herbaceous Cover Legume with Potential for Multiple Uses*. Resource and Crop Management Research Monograph No. 25, IITA, Ibadan, Nigeria, 52 pp.

Carter, J. (1995) *Alley Cropping: Have Resource Poor Farmers Benefited?* ODI Natural Resource Perspectives No. 3, ODI, London.

Carter, M.R. (1988) Equilibrium credit rationing of small farm agriculture. *Journal of Development Economics* 28(1), 83–103.

Carter, M.R. and Olinto, P. (forthcoming) Getting institution 'right' for whom: credit constraints and the impact of property rights on the quantity and composition of investment. *American Journal of Agricultural Economics* (in press).

Celis, R., Milimo, J. and Wanmali, S. (1991) *Adopting Improved Farm Technology: a Study of Smallholder Farmers in Eastern Province, Zambia*. IFPRI, Washington, DC.

Cernea, M.M. (ed.) (1985) *Putting People First: Sociological Variables in Rural Development*. Oxford University Press for the World Bank, New York.

Cernea, M.M. (1987) Farmer organisations and institution building for sustainable development. *Regional Development Dialogue* 8, 1–24.

Cernea, M.M. (1991) *Putting People First*, 2nd edn. Oxford University Press, Oxford.

Cernea, M.M. (1993) The sociologist's approach to sustainable development. *Finance and Development*, 11–13 December.

CGIAR (1997) Chairman's opening statement 'Meeting the challenges of a changing world', 27 October 1997. Electronic publication at: www.worldbank.org/html/cgiar/publications/icw97/

Chambers, R. (1983) *Rural Development: Putting the Last First*. Longman, London.

Chambers, R. (1997) *Whose Reality Counts? Putting the First Last*. Intermediate Technology, London.

Chambers, R. and Jiggins, J. (1986) *Agricultural Research for Resource-poor Farmers: A Parsimonious Paradigm*. Discussion Paper 220, Institute of Development Studies, University of Sussex, Brighton, UK.

Chambers, R. and Moris, J. (eds) (1973) *Mwea: an Irrigated Rice Settlement in Kenya*. IFO Afrika Studien No. 83, Weltforum Verlag, Munich.

Chapin, F.S., Vitousek, P.M. and Van Cleve, K. (1985) The nature of nutrient limitation in plant communities. *The American Naturalist* 127, 48–58.

Chapin, G. and Wasserstrom, R. (1981) Agricultural production and malaria resurgence in Central America and India. *Nature* 293, 181–185.

Chavas, J.P. (1994) Production and investment decisions under sunk costs and temporal uncertainty. *American Journal of Agricultural Economics* 76(1), 114–127.

Chavas, J.P., Kristjanson, P. and Matlon, P. (1991) On the role of information in decision making: the case of sorgum yield in Burkina Faso. *Journal of Development Economics* 35, 261–280.

Chianu, J., Kormowa, P. and Adesina, A. (2000) Re-invention: farmers modification of alley cropping in Nigeria. Research paper draft, International Institute of Tropical Agriculture, Ibadan, Nigeria.

Chipika, S. (1988) The coordinated agricultural and rural development crop project in Gutu district: extension, resource base and other important socio-economic aspects. Unpublished working paper, Monitoring and Evaluation Section, Agritex, Zimbabwe.

Chitedze Research Station (1987) *Local Maize Manure (Khola) and Fertiliser Trial*. Chitedze Research Station, Lilongwe.

Chiwona-Karltun, L. (2000) The winds of change: disparities of livelihoods and food security in cassava farming systems in Malawi. Paper presented to the Conference on Historical and Social Science Research in Malawi: Problems and Prospects, 26–29 June, Zomba, Malawi.

Chopra, K. and Hanummantha Rao, C.H. (1997) Institutional and technological perspectives on the links between agricultural sustainability and poverty: illustrations from India. In: Vosti, S.A. and Reardon, T. (eds) *Sustainability, Growth, and Poverty Alleviation: A Policy and Agroecological Perspective.* Johns Hopkins University Press, Baltimore, Maryland.

Christensen, G.N. (1989) Determinants of private investment in rural Burkina Faso. Unpublished PhD dissertation, Department of Agricultural Economics, Cornell University, Ithaca, New York.

Christiansson, C., Mbegu, A.C. and Yrgård, A. (1993) *The Hand of Man. Soil Conservation in Kondoa Eroded Area, Tanzania.* Sida's Regional Soil Conservation Unit (RSCU), Nairobi, 55 pp.

Clay, D.C. and Lewis, L.L. (1990) Land use, soil loss and sustainable agriculture in Rwanda. *Human Ecology* 18(2), 147–161.

Clay, D.C., Byiringiro, F., Kangasniemi, J., Reardon, T., Sibomana, B., Uwamariya, L. and Tardif-Douglin, D. (1995) *Promoting Food Security in Rwanda through Sustainable Agricultural Productivity: Meeting the Challenges of Population Pressure, Land Degradation, and Poverty.* MSU International Development Paper no. 17, Department of Agricultural Economics, Michigan State University, East Lansing, Michigan.

Clay, D.C., Reardon, T. and Kangasniemi, J. (1998) Sustainable intensification in the highland tropics: Rwandan farmers' investments in land conservation and soil fertility. *Economic Development and Cultural Change* 46(2), 351–378.

CNA (Commission Nationale d'Agriculture) (1991) *Rapport de synthèse: rapport préliminaire.* Government of Rwanda, Kigali.

Cohen, J. and Uphoff, N. (1980) Participation's place in rural development: seeking clarity through specificity. *World Development* 8(3), 213–236.

Coleman, J. (1988) Social capital and the creation of human capital. *American Journal of Sociology* 94 (Suppl.), S95–S120.

Coleman, J. (1990) *Foundations of Social Theory.* Harvard University Press, Cambridge, Massachusetts.

Collins, C.J. and Chippendale, P.J. (1991) *New Wisdom: the Nature of Social Reality.* Acorn Publications, Sunnybank, Queensland.

Conroy, A.C. and Kumwenda, J.D. (1994) Risks associated with the adoption of hybrid seed and fertiliser by smallholder farmers in Malawi. In: Jewell, D.C., Waddington, S.R., Ransom, J.K. and Pixley, K.V. (eds) *Proceedings of the 4th Eastern and Southern Africa Regional Maize Conference held in Harare, Zimbabwe, 28 March–1 April 1994.* CIMMYT, Mexico, DF, 279–284.

Conway, G. (1997) *The Doubly Green Revolution: Food for All in the Twenty-first Century.* Penguin Books, London.

Cooper, P.J., Leakey, R.R.B., Rao, M.R. and Reynolds, L. (1996) Agroforestry and the mitigation of land degradation in the humid and sub-humid tropics of Africa. *Experimental Agriculture* 32, 235–290.

Coppock, D.L. (1994) *The Borana Plateau of Southern Ethiopia: Synthesis of Pastoral Research, Development and Change, 1980–91.* Systems Study no. 5, International Livestock Centre for Africa, Addis Ababa.

Costanza, R., d'Arge, R., de Groot, R., Farber, S., Grasso, M., Hannon, B., Limburg, K., Naeem, S., O'Neil, R.V., Parvelo, J., Raskin, R.G., Sutton, P. and van den Belt, M. (1997, 1999) The value of the world's ecosystem services and natural capital. *Nature* 387, 253–260; also in *Ecological Economics* 25(1), 3–15.

Coulibaly, O.N. (1995) Devaluation, new technologies and agricultural policies in the Sudanian and Sudano-Guinean zones of Mali. Unpublished PhD dissertation, Department of Agricultural Economics, Purdue University, West Lafayette, Indiana.

Cragg, J.G. (1971) Some statistical models for limited dependent variables with application to the demand for durable goods. *Econometrica* 39(5), 829–844.

Critchley, W. (ed.) (1999) *Promoting Farmer's Innovation. Harnessing Local Environmental Knowledge in East Africa.* Workshop Report No. 2. UNDP and Sida RELMA.

Critchley, W.R.S., Reij, C. and Seznec, K. (1994) *Water Harvesting and Plant Protection.* Technical Paper 157, World Bank, Washington, DC.

Crowder, L.V. and Chheda, H.R. (1982) Herbage quality and nutritive value. In: *Tropical Grassland Husbandry.* Longman, London, pp. 346–384.

Crush, J. (ed.) (1995) *Power of Development.* Routledge, London.

Dalton, T. (1996) Soil degradation and technical change in southern Mali. PhD dissertation, Purdue University, West Lafayette, Indiana.

D'Arcy, R. (1998) *Gender and Soil Fertility Project Report: Dowa, Malawi.* Report to the University of Florida Soils Collaborative Research Support Project, Gainesville, Florida.

Dasgupta, P. and Serageldin, I. (eds) (2000) *Social Capital: a Multiperspective Approach.* World Bank, Washington, DC.

David, S. (1992) Open the door and see all the people: intra-household processes and the adoption of hedgerow intercropping. Paper presented at the Rockefeller Foundation Social Science Fellows' Meeting, CIMMYT, Mexico, 9–13 November.

Davidson, R. and MacKinnon, J.G. (1993) *Estimation and Inference in Econometrics.* Oxford University Press, New York.

Davis-Case, D. (1990) *The Community's Toolbox: The Idea, Methods and Tools for Participatory Assessment, Monitoring and Evaluation in Community Forestry.* Community Forestry Field Manual 2, FAO, Rome.

Deeg, A. and Freudenberger, K. (2000) *Analyse des Impacts du Système Ferroviaire FCE sur l'Economie Régionale: Résultats des Recherches Qualitatives,* Vol. III. Project PAGE (USAID), Antananarivo, Madagascar.

de Haan, C., Steinfeld, H. and Blackburn, H. (1997) *Livestock and the Environment. Finding a Balance.* Report of a study coordinated by FAO, USAID and the World Bank, FAO, Rome.

Deichmann, U. (1994) A medium resolution population database for Africa. National Center for Geographic Information and Analysis, Department of Geography, University of California, Santa Barbara, California.

deJanvry, A. (1981) *The Agrarian Question and Reformism in Latin America.* Johns Hopkins University Press, Baltimore, Maryland, 306 pp.

deJanvry, A., Fafchamps, M. and Sadoulet, E. (1991) Peasant household behaviour with missing markets: some paradoxes explained. *The Economic Journal* 101, 1400–1417.

de los Reyes, R. and Jopillo, S.G. (1986) *An Evaluation of the Philippines Participatory Communal Irrigation Program.* Institute of Philippine Culture, Quezon City.

Desai, G.M. (2001) Sustainable rapid growth of fertilizer use in Rwanda: a strategy and an action plan. Draft report from the Fertilizer Use and Marketing Policy Workshop, Sponsored by MINAGRI, Abt Associates, and Michigan State University (Food Security Research Project), Kigali, Rwanda, February.

Deshler, D. and Selener, D. (1991) Transformative research: in search of a definition. *Convergence* 24(3), 9–22.

DeWolf, J., Rommelse, R. and Pisanelli, A. (2000) *Improved Fallow Technology in Western Kenya: Potential and Reception by Farmers.* International Centre for Research in Agroforestry, Nairobi.

Dixit, A.K. and Pindyck, R.S. (1994) *Investment Under Uncertainty.* Princeton University Press, Princeton, New Jersey.

Dlamini, S. (1993) Factors associated with the adoption of basal fertilizers among Swazi nation land farmers. In: Mwangi, W., Rohrbach, D. and Heisey, D. (eds) *Cereal Grain Policy Analysis in the National Agriculture Research Systems of Eastern and Southern Africa.* CIMMYT, SADC and ICRISAT, Addis Ababa.

Dobbs, T.L. and Pretty, J. (2001) *Future Directions for Joint Agricultural-Environmental Policies: Implications of the United Kingdom Experience for Europe and the United States.* South Dakota State University, Brookings and University of Essex, Colchester, 117 pp.

Dommen, A.J. (1988) *Innovation in African Agriculture.* Westview Press, Boulder, Colorado.

Dommen, A.J. (1989) A rationale for African low-resource agriculture in terms of economic theory. Indigenous knowledge systems: implications for agriculture and international development. *Studies in Technology and Change* 11, 33–40. Technology and Social Change Program, Iowa State University, Ames, Iowa.

Douthwaite, B., Keatinge, J.D.H. and Park, J.R. (2001) Why promising technologies: the neglected role of user innovation during adoption. *Research Policy* (in press).

Dove, M.R. (1996) Center, periphery, and biodiversity: a paradox of governance and a developmental challenge. In: Brush, S.B. and Stabinsky, D. (eds) *Valuing Local Knowledge: Indigenous People and Intellectual Property Rights.* Island Press, Washington, DC, pp. 41–67.

Dregne, H.E. and Chou, N.T. (1992) Global desertification's dimensions and costs. In: Dregne, H.E. (ed.) *Degradation and Restoration of Arid Lands.* Texas Tech University, Lubbock, Texas.

Dube, J.S. and Ndlovu, L.R. (1995) Feed intake, chemical composition of faeces and nitrogen relution in goats consuming single browse species or browse mixtures. *Zimbabwe Journal of Agricultural Research* 33, 133–141.

Due, J.M. (1991) Policies to overcome the negative effects of structural adjustment programs on African female-headed households. In: Gladwin, C.H. (ed.) *Structural Adjustment and African Women Farmers.* University of Florida Press, Gainesville, Florida, pp. 103–127.

Due, J.M. and Gladwin, C.H. (1991) Impacts of structural adjustment programs on African female-headed households. *American Journal of Agricultural Economics* 73, 1431–1439.

Duff, S.N., Stonehouse, D.P., Brown, D.R., Baker, K.M., Blackburn, D.J., Coyle, D.O. and Hilts, S.G. (1990) *Understanding Soil Conservation Behaviour: a Critical Review.* Technical Publication 90-1, Centre for Soil and Water Conservation, University of Guelph, Guelph, Canada.

Dupriez, H. (1982) *Paysans d'Afrique Noire.* L'Harmattan, Paris.

Dvorak, K.A. (1993) Characterizing system dynamics for soil fertility management research in the humid tropics: studies in southeastern Nigeria. In: Dvorak, K.A. (ed.) *Social Science Research for Agricultural Technology Development. Spatial and Temporal Dimensions.* CAB International, Wallingford, UK, pp. 167–191.

Dvorak, K.A. (1996) *Adoption Potential of Alley Cropping. Final Project Report.* RCMD Research Monograph No. 23, International Institute of Tropical Agriculture, Ibadan, Nigeria.

Dzowela, B.H., Hove, L., Maasdorp, B.V. and Mafongoya, P.L. (1997) Recent work on the establishment of multipurpose trees as feed resource in Zimbabwe. *Animal Feed Science and Technology* 69, 1–5.

Easterbrook, G. (1997) Forgotten benefactor of humanity. *The Atlantic Monthly*, January, 75–82.

Ehui, S.K. (1987) Deforestation, soil dynamics and agricultural development in the tropics. PhD thesis submitted to Purdue University, West Lafayette, Indiana.

Ehui, S.K. (1995) Economic factors and policies encouraging environmentally detrimental land use practices in sub-Saharan Africa. In: Peters, G.H. and Wedley, D. (eds) *Proceedings of the 22nd International Conference of Agricultural Economists.* Dartmouth.

Ehui, S.K., Kang, B.T. and Dunstan, S.C. (1990) Economic analysis of soil erosion effects in alley cropping, no-till and bush-fallow systems in southwestern Nigeria. *Agricultural Systems* 34, 349–368.

Elabor-Idemudia, P. (1991) The impact of structural adjustment programs on women and their households in Bendel and Ogun States, Nigeria. In: Gladwin, C.H. (ed.) *Structural Adjustment and African Women Farmers.* University of Florida Press, Gainesville, Florida, pp. 128–150.

Elbasha, E., Thornton, P.K. and Tarawali, G. (1999) *An* ex post *Economic Impact Assessment of Planted Forages in West Africa.* ILRI Impact Assessment Series 2, International Livestock Research Institute (ILRI), Nairobi, Kenya, 61 pp.

Elliott, R.C. and Topps, J.H. (1963) Voluntary intake of low protein diets in sheep. *Animal Production* 5, 269–276.

Ellis, F. (2000) *Rural Livelihoods and Diversity in Developing Countries.* Oxford University Press, Oxford.

Elster, J. (1989) *The Cement of Society: A Study of Social Order.* Cambridge University Press, Cambridge.

Enoh, M.B., Yonkeu, S., Pingpoh, D.P., Messine, O. and Maadjou, N. (1999) Yield and composition of fodder banks on the Adamawa plateau of Cameroon. *Revue d'Elevage et Médecines Vétérinaires en Pays Tropicaux* 52(1), 55–62.

Equipe Ralaivao (2000) *Rapport d'Etude: Commune Rurale de Maromiandra.* CTA-AGERAS, Fianarantsoa, Madagascar.

Ervin, C.A. and Ervin, D.E. (1982) Factors affecting the use of conservation practices: hypotheses, evidence, and policy implications. *Land Economics* 58(3), 277–292.

Etzioni, A. (1995) *The Spirit of Community: Rights, Responsibilities and the Communitarian Agenda.* Fontana Press, London.

Eveleens, K.G., Chisholm, R., van de Fliert, E., Kato, M., Thi Nhat, P. and Schmidt, P. (1996) *Mid Term Review of Phase III Report – The FAO Intercountry Programme for the Development and Application of Integrated Pest Management Control in Rice in South and South East Asia.* GCP/RAS/145-147/NET-AUL-SWI, FAO, Rome

Fabiyi, Y.L., Idowu, E.O. and Oguntade, A.E. (1991) Land tenure and management constraints to the adoption of alley farming by women in Oyo State of Nigeria. *The Nigerian Journal of Agricultural Extension* 6(1 and 2), 40–46.

Fafchamps, M. (1993) Sequential labor decisions under uncertainty: an estimable household model of West African farmers. *Econometrica* 61(5), 1173–1197.

Fairhead, J. (1993) Representing knowledge: the 'new farmer' in research fashions. In: Pottier, J. (ed.) *Practicing Development: Social Science Perspectives.* Routledge, New York, pp. 187–204.

Fairhead, J. and Leach, M. (1996) *Misreading the African Landscape*. Cambridge University Press, Cambridge.

Fairhead, J. and Leach, M. (1998) *Reframing Deforestation: Global Analyses and Local Realities*. Macmillan, New York.

Falcon, W. (1970) The green revolution: generations of problems. *American Journal of Agricultural Economics* 52(5), 698–710.

FAO (1985) *Changes in Shifting Cultivation in Africa: Seven Case Studies*. FAO Forestry Paper 50/1, FAO, Rome.

FAO (1996) *FAO Fertilizer Yearbook 1995*. Food and Agriculture Organization, Rome.

FAO (2000) FAOSTAT Database, apps.fao.org, Food and Agriculture Organization of the United Nations, Rome.

Farming Systems Research Unit (1991) *Structural Adjustment and Communal Area Agriculture in Zimbabwe: The Case Studies from Mangwende and Chivi*.

Feder, G. and Umali, D.L. (1993) The adoption of agricultural innovations: a review. *Technological Forecasting and Social Change* 43, 215–239.

Feder, G., Just, R.E. and Zilberman, D. (1985) Adoption of agricultural innovations in developing countries: a survey. *Economic Development and Cultural Change* 33(2), 255–294.

Feder, G., Lau, L.J., Lin, J.Y. and Luo, X. (1992) The determinants of farm investment and residential construction in post-reform China. *Economic Development and Cultural Change* 41(1), 1–26.

Feldstein, H. and Poats, S. (eds) (1989) *Working Together: Gender Analysis in Agriculture*. Kumarian Press, West Hartford, Connecticut.

Feldstein, H., Rocheleau, D. and Buck, L. (1989) Kenya agroforestry extension and research: a case study from Siaya District. In: Feldstein, H. and Poats, S. (eds) *Working Together: Gender Analysis in Agriculture*. Kumarian Press, West Hartford, pp. 167–208.

Ferguson, A.E. (1991) So the grandparents may survive: farmer participation in bean improvement in Malawi. In: *Progress in Improvement in Common Bean in Eastern and Southern Africa*, Workshop Proceedings of the Ninth Sokoine University Bean-Cowpea CRSP and Second SADCC/CIAT Bean Research Workshop, held at Sokoine University of Agriculture, Morogoro, Tanzania, 17–22 September, 1990. *Bean Research* 5, 379–392.

Ferguson, A. (1994) Gendered science: a critique of agricultural development. *American Anthropologist* 96(3), 540–552.

Ferguson, A. and Derman, W. (2000) Writing against hegemony: development encounters in Zimbabwe and Malawi. In: Peters, P.E. (ed.) *Development Encounters: Sites of Knowledge and Participation*. Harvard University Press, Cambridge, Massachusetts.

Fernandes, E.C.M. (1998) Integrated farming systems to increase and sustain food production in the tropics. In: Fairclough, A.J. (ed.) *Sustainable Agriculture Solutions*. Novello Press, London.

Fernandez-Rivera, S., Williams, T.O., Hiernaux, P. and Powell, J.M. (1995) Faecal excretion by ruminants and manure availability for crop production in semi-arid West Africa. In: Powell, J.M., Fernandez-Rivera, S., Williams, T.O. and Renard, C. (eds) *Livestock and Sustainable Nutrient Cycling in Mixed Farming Systems of SubSaharan Africa*, Vol. II: *Technical Papers*. International Livestock Centre for Africa, Addis Adaba, pp. 149–170.

Finkelshtain, I. and Chalfant, J. (1991) Marketed surplus under risk: do peasants agree with Sandmo? *American Journal of Agricultural Economics* 73, 557–567.

Flora, J.L. (1998) Social capital and communities of place. *Rural Sociology* 63(4), 481–506.

Foster, A. and Rosenzweig, M. (1995) Learning by doing and learning from others: human capital and technical change in agriculture. *Journal of Political Economy* 103(6), 1176–1209.

Francis, P. (1987) Land tenure systems and the adoption of alley farming in Nigeria. In: *Land, Trees and Tenure. Proceedings of an International Workshop on Tenure Issues in Agroforestry, Nairobi, 27–31 May 1985*. International Centre for Research in Agroforestry, Nairobi, pp. 175–180.

Frank, A.B., Tanaka, D.L., Hoffmann, L. and Follett, R.F. (1995) Soil carbon and nitrogen of Northern Great Plains Grasslands as influenced by long term grazing. *Journal of Range Management* 48, 470–474.

Franzel, S. (1999) Socioeconomic factors affecting the adoption potential of improved tree fallows in Africa. *Agroforestry Systems* 47, 305–321.

Franzel, S., Phiri, D. and Kwesiga, F. (1997) Participatory on-farm research on improved fallows in Eastern Province, Zambia. Paper presented at Southern Africa Agroforestry Network Planning Workshop, 1–5 July, Harare, Zimbabwe.

Franzel, S., Phiri, D. and Kwesiga, F. (1999) *Assessing the Adoption Potential of Improved Fallows in Eastern Zambia*. AFRENA Report 124, International Centre for Research in Agroforestry, Nairobi.

Franzel, S., Phiri, D., Place, F. and Mwangi, M. (2000) *The Uptake of Improved Fallows in Eastern Zambia.* International Centre for Research in Agroforestry, Nairobi.

Franzel, S., Coe, R., Cooper, P., Place, F. and Scherr, S. (2001) Assessing the adoption potential of agroforestry practices in sub-Saharan Africa. *Agricultural Systems* 69, 37–62.

Freeman, H.A. (2001) Fertilizer market reforms and private traders' entry and investment decisions: evidence from semi-arid areas of Kenya. Unpublished manuscript, ICRISAT, Nairobi, Kenya.

Freeman, H.A. and Kaguongo, W. (2001) Impact of fertilizer market reforms on private traders behavior in Kenya. Unpublished report, ICRISAT, Nairobi, Kenya.

Freeman, H.A. and Omiti, J.M. (2001) Fertilizer use in semi-arid cropping systems of Kenya: analysis of smallholder farmers adoption behavior following market reforms. Unpublished manuscript, ICRISAT, Nairobi, Kenya.

Freire, P. (1973) *Education for Critical Consciousness.* Seabury Press, New York.

Freudenberger, K. (1998) *Livelihoods Without Livestock: A Study of Community and Household Resource Management in the Village of Andaladranovao, Madagascar.* Landscape Development Interventions, Fianarantsoa, Madagascar.

Freudenberger, K. (1999) *Flight to the Forest: A Study of Community and Household Resource Management in the Commune of Ikongo, Madagascar.* Landscape Development Interventions, Fianarantsoa, Madagascar.

Freudenberger, K., Ravelonahina, J. and Whyner, D. (1999) *Le Corridor Coincé: Une Etude sur l'Économie Familiale et la Gestion de Ressources Naturelles dans la Commune d'Alatsinainy Ialamarina, Madagascar.* Landscape Development Interventions, Fianarantsoa, Madagascar.

Fujisaka, S. (1994) Learning from six reasons why farmers do not adopt innovations intended to improve sustainability of upland agriculture. *Agricultural Systems* 46, 409–425.

Fukuyama, F. (1995) *Trust: The Social Values and the Creation of Prosperity.* Free Press, New York.

Galiba, M., Vissoh, P., Dagbenonbakin, G. and Fagbohoun, F. (1998) The reactions and fears of farmers utilizing velvet bean (*Mucuna pruriens*). In: Buckes, D. (ed.) *Cover Crops in West Africa: Contributions to Sustainable Agriculture, 1–3 October 1996, Cotonou, Benin.* International Development Research Center, Ottawa.

Gambara, P., Machemedze, T.E. and Mwenye, D. (2000) Chihota Soil Fertility Project, Annual Report 1998 to 1999. Mimeo, AGRITEX, Marondera, Zimbabwe.

Gambetta, D. (ed.) (1988) *Trust: Making and Breaking Cooperative Relations.* Blackwell, Oxford.

Gatsi, T. (1999) Preliminary incremental cost analysis of some 'best-bet' soil fertility options being promoted in Chihota, Zimbabwe. *Target (The Newsletter of the Soil Fertility Research Network for Maize-Based Cropping Systems in Malawi and Zimbabwe)* 20, 7.

Gatsi, T. and Mekuria, M. (2000) An economic analysis of productivity enhancing soil management technologies in smallholder agriculture in Murewa District of Zimbabwe. Draft EPWG Report, Harare, Zimbabwe.

Gatsi, T., Bellon, M.R. and Gambara, P. (2000) *The Adoption of Soil Fertility Technologies in Chihota, Zimbabwe: Potential and Constraints.* Soil Fert Net Research Results Working Paper Number 6, CIMMYT, Harare, Zimbabwe, 13pp.

Gavian, S. and Fafchamps, M. (1996) Land tenure and allocative efficiency in Niger. *American Journal of Agricultural Economics* 78(2), 460–471.

Gebremedhin, B. (1998) The economics of soil conservation investments in the Tigray region of Ethiopia. Unpublished PhD dissertation, Department of Agricultural Economics, Michigan State University, East Lansing, Michigan.

Gebremedhin, B. and Swinton, S.M. (2000) Investment in soil conservation in Ethiopia: the role of land tenure security and public programs. Unpublished manuscript, Department of Agricultural Economics, Michigan State University, East Lansing, Michigan.

Gebremedhin, B., Swinton, S.M. and Tilahun, Y. (1999) Effects of stone terraces on crop yields and farm profitability: results of on-farm research in Tigray, northern Ethiopia. *Journal of Soil and Water Conservation* 54(3), 568–573.

Gebremedhin, B., Pender, J. and Tesfay, G. (2000) Community resource management: the case of grazing lands in Tigray, northern Ehiopia. Paper presented at National Seminar on Policies for Sustainable Land Management in the Highlands of Ethiopia, 23–24 May 2000, International Livestock Research Institute, Addis Ababa.

Gebremedhin, B., Pender, J. and Tesfay, G. (2002) Community natural resource management: the case of woodlots in Tigray, northern Ethiopia. *Environment and Development Economics* (in press).

Gebremichael, Y. (1992) The effects of conservation on production in the Andit-Tid area, Ethiopia. In: Tato, K. and Hurni, H. (eds) *Soil Conservation for Survival*. Soil and Water Conservation Society, Ankeny, Iowa, pp. 239–250.

Gerdemann, M.M., Kreuzer, M., Fossard, W. and Kreuzer, M. (1999) Effect of different pig feeding strategies on the nitrogen fertilizing value of slurry for *Lolium multiflorum*. *Journal of Plant Nutrition and Soil Science* 162, 401–408.

Ghimire, K. and Pimbert, M. (1997) *Social Change and Conservation*. Earthscan Publications, London.

Giller, K. (1999) Summary of the 'best bets' discussions at Zomba. *Target (The Newsletter of the Soil Fertility Research Network for Maize-Based Cropping Systems in Malawi and Zimbabwe)* 20, 5.

Giller, K.E., McDonagh, J.F. and Cadisch, G. (1994) Can biological nitrogen fixation sustain agriculture in the tropics? In: Syers, J.K. and Rimmer, D.L. (eds) *Soil Science and Sustainable Land Management in the Tropics*. CAB International, Wallingford, UK, pp. 173–191.

Giller, K.E., Gilbert, R., Mugwira, L.M., Muza, L., Patel, B.K. and Waddington, S.R. (1998) Practical approaches to soil organic matter management for smallholder maize production in southern Africa. In: Waddington, S.R., Murwira, H.K., Kumwenda, J.D.T., Hikwa, D. and Tagwira, F. (eds) *Soil Fertility Research for Maize-Based Farming Systems in Malawi and Zimbabwe*. SoilFertNet and CIMMYT-Zimbabwe, Harare, Zimbabwe, pp. 139–153.

Giller, K.E., Mpepereki, S., Mapfumo, P., Kasasa, P., Sakala, W., Phombeya, H., Itimu, O., Cadisch, G., Gilbert, R.A. and Waddington, S.R. (2000) Putting legume N$_2$-fixation to work in cropping systems of southern Africa. In: Pedrosa, F.O., Hungria, M., Yates, H.G. and Newton, W.E. (eds) *Nitrogen Fixation: From Molecules to Crop Productivity*. Kluwer Academic, Dordrecht, The Netherlands, pp. 525–530.

Gjertsen, H. and Barrett, C.B. (2001) Context-dependent biodiversity conservation management regimes. Unpublished working paper, Cornell University, Ithaca, New York.

Gladwin, C.H. (1975) A model of the supply of smoked fish from Cape Coast to Kumasi. In: Plattner, S. (ed.) *Formal Methods in Economic Anthropology*. No. 4, American Anthropological Association, Washington, DC, pp. 77–127.

Gladwin, C.H. (1979) Cognitive strategies and adoption decisions: a case study of non-adoption of an agronomic recommendation. *Economic Development and Cultural Change* 28(1), 155–173.

Gladwin, C.H. (1989) *Ethnographic Decision Tree Modeling*. Sage Publications, Newbury Park, California, 95 pp.

Gladwin, C.H. (1991) Fertilizer subsidy removal programs and their potential impacts on women farmers in Malawi and Cameroon. In: Gladwin, C.H. (ed.) *Structural Adjustment and African Women Farmers*. University of Florida Press, Gainesville, Florida, pp. 191–216.

Gladwin, C.H. (1992) Gendered impacts of fertilizer subsidy removal programs in Malawi and Cameroon. *Agricultural Economics* 7, 141–153.

Gladwin, C.H. (1996) Gender in research design: old debates and new issues. In: Breth, S. (ed.) *Achieving Greater Impact from Research Investments in Africa*. Sasakawa Africa Association, Mexico City, pp. 127–149.

Gladwin, C.H. (1997) Targeting women farmers to increase food production in Africa. In: Breth, S. (ed.) *Women, Agricultural Intensification, and Household Food Security*. Sasakawa Africa Association, Mexico City, pp. 61–81.

Gladwin, C., Zabawa, R. and Zimet, D. (1984) Using ethnoscientific tools to understand farmers' plans goals, decisions. In: Malton, P.F., Cantrell, R., King, D. and Benoit-Cattin, M. (eds) *Coming Full Circle: Farmers' Participation in the Development of Technology*. IDRC, Ottawa, Canada, pp. 27–40.

Gladwin, C., Goldman, A., Randall, A., Schmitz, A. and Schuh, E. (1997a) *Are There Public Benefits to Private Use of Fertilizer in Africa?* International Working Paper Series IW97-15, International Agricultural Trade and Development Center, University of Florida, Gainesville, Florida.

Gladwin, C.H., Buhr, K.L., Goldman, A., Hiebsch, C.K., Hildebrand, P.E., Kidder, G., Langham, M., Lee, D., Nkedi-Kizza, P. and Williams, D. (1997b) Gender and soil fertility in Africa. In: Sanchez, P. and Buresh, R. (eds) *Replenishing Soil Fertility in Africa*. Special Publication no. 51, Soil Science Society of America, Madison, Wisconsin, pp. 219–236.

Gladwin, C.H., Peterson, J.S. and Mwale, A. (2000) The quality of science in participatory research: a case study from E. Zambia. Paper presented at the Uniting Science and Participatory Methods Workshop, III Workshop of CGIAR Program on Gender and Participatory Methods, ICRAF, Nairobi, Kenya, 6 November.

Gladwin, C.H., Thomson, A.M., Peterson, J.S. and Anderson, A.S. (2001) Improving food security in Africa via women farmers' multiple live-lihood strategies. *Food Policy*, March.

Goheen, M. (1991) The ideology and political economy of gender: women and land in Nso, Cameroon. In: Gladwin, C.H. (ed.) *Structural Adjustment and African Women Farmers*. University of Florida Press, Gainesville, Florida, pp. 239–256.

Goldstein, M. and Udry, C. (1999) Agricultural innovation and resource management in Ghana. Mimeo, Yale University, New Haven, Connecticut.

Gorse, J.E. and Steeds, D.R. (1988) *Desertification in the Sahelian and Sudanian Zones of West Africa*. Technical Paper No. 11, World Bank, Washington, DC.

Gould, B.W., Saupe, W.E. and Klemme, R.M. (1989) Conservation tillage: the role of farm and operator characteristics and the perception of erosion. *Land Economics* 65, 167–182.

Govereh, J. and Jayne, T. (1999) *Effects of Cash Crop Production of Food Crop Productivity in Zimbabwe: Synergies or Trade-offs?* MSU International Development Working Paper No. 74, Department of Agricultural Economics, Michigan State University, East Lansing, Michigan.

Government of Kenya (1999) *Statistical Abstract*. Ministry of Finance and Planning, Nairobi, Kenya.

Government of Kenya (2000) *Provisional Results of the 1999 Population and Housing Census*. Ministry of Finance and Planning, Nairobi, Kenya.

Government of Nigeria (1992) *Nigerian Livestock Resources*. Vol. 1: *Executive Summary and Atlas*. Federal Department of Livestock and Pest Control Services and Resource Inventory and Management Limited, St Helier, Jersey, UK.

Grandi, J.C. (ed.) (1996) *L'Evolution des systèmes de production agropastorale par rapport au développement rural durable dans les pays d'Afrique soudano-sahélienne*. Collection FAO, Gestions des exploitations agricoles, No. 11, FAO, Rome.

Gray, L. and Kevane, M. (2001) Evolving tenure rights and agricultural intensification in south-western Burkina Faso. *World Development* 29(4), 573–587.

Grillo, R. and Rew, A. (eds) (1985) *Social Anthropology and Development Policy*. ASA Monographs 23, Tavistock, London.

Grootaert, C. (1998) *Social Capital: the Missing Link*. Social Capital Initiative Working Paper No. 5, World Bank, Washington, DC.

Gryseels, G. (1988) Role of livestock on mixed small holder farms in the Ethiopian highlands: a case study from the Baso and Worenna District near Debre Berhan. Unpublished PhD thesis, University of Wageningen, Wageningen, The Netherlands.

Gryseels, G. and Anderson, F.M. (1983) *Research on Farm and Livestock Productivity in the Central Ethiopian Highlands: Initial Results, 1977–1980*. ILCA, Addis Ababa, Ethiopia.

Guyer, J.I. (1997) *An African Niche Economy: Farming to Feed Ibadan, 1968–88*. Edinburgh University Press, Edinburgh.

Habermas, J. (1987) *Theory of Communicative Action: Critique of Functionalist Reason*, Vol. II. Polity Press, Oxford.

Hagen, R. (1999) *Biodiversity Conservation Priorities, Key Challenges and Strategic Options for the Landscape Development Program*. Landscape Development Interventions, Fianarantsoa, Madagascar.

Hagmann, J. and Murwirwa, K. (1996) Indigenous SWC in Southern Zimbabwe: a study of techniques, historical changes and recent developments under participatory research and extension. In: Reij, C., Scoones, I. and Toulmin, C. (eds) *Sustaining the Soil: Indigenous Soil and Water Conservation in Africa*. Earthscan, London, pp. 97–106.

Hagos, F., Pender, J. and Gebreselassie, N. (1999) *Land Degradation in the Highlands of Tigray and Strategies for Sustainable Land Management*. Socioeconomic and Policy Research Working Paper No. 25, International Livestock Research Institute, Addis Ababa, Ethiopia.

Hamilton, N.A. (1995) Learning to learn with farmers. PhD thesis, Wageningen Agricultural University, Wageningen, The Netherlands.

Hannah, L., Rakotosamimana, B., Ganzhorn, J., Mittermeier, R., Olivieri, S., Iyer, L., Rajaobelina, S., Hough, J., Andriamialisoa, F., Bowles, I. and Tilkin, G. (1998) Participatory planning, scientific priorities, and landscape conservation in Madagascar. *Environmental Conservation* 25, 30–36.

Hassan, R.M. (ed.) (1998) *Maize Technology Development and Transfer. A GIS Application for Research Planning in Kenya*. CAB International, Wallingford, UK.

Hatibu, N. and Mtenga, N.A. (1996) *Smallholders Technological Constraints in Shinyanga District, Tanzania*. Consultancy report for FAO, Maswa, Tanzania, 101 pp.

Hayami, Y. and Ruttan, V. (1985) *Agricultural Development: An International Perspective*, 2nd edn., Johns Hopkins University Press, Baltimore, Maryland.

Hazell, P. (1998) IFPRI Draft proposal for development of bio-economic models to evaluate technology, policy and institutional options for ecoregional development, submitted to International Service for National Research (ISNER).

Hazell, P. and Norton, R.D. (1986) *Mathematical Programming for Economic Analysis in Agriculture*. Macmillan Publishing Company, New York.

Heisey, P.W. and Mwangi, W. (1997) Fertiliser use and maize production in sub-Saharan Africa. In: Byerlee, D. and Eicher, C.K. (eds) *Africa's Emerging Maize Revolution*. Lynne Reinner, Boulder, Colorado.

Henao, J. and Baanante, C. (1999) *Estimating Rates of Nutrient Depletion in Soils of Agricultural Lands of Africa*. International Fertilizer Development Center, Muscle Shoals, Alabama.

Henao, J., Brink, J., Coulibaly, B. and Traoré, A. (1992) *Fertilizer Policy Research Program for Tropical Africa: Agronomic Potential of Fertilizer Use in Mali*. International Fertilizer Development Center and Institut d'Economie Rurale, Muscle Shoals, Alabama.

Herdt, R.W. and Steiner, R.A. (1995) Agricultural sustainability: concepts and conundrums? In: Barnett, V., Payne, R. and Steiner, R. (eds) *Agricultural Sustainability: Economic and Statistical Considerations*. John Wiley & Sons, Chichester, UK.

Hijkoop, J., van der Poel, P. and Kaya, B. (1991) *Une lutte de longue haleine*. Institut d'Economie Rurale, Bamako, and Institut Royal des Tropiques, Amsterdam.

Hikwa, D. and Waddington, S.R. (1998) Annual legumes for improving soil fertility in the smallholder maize-based systems of Zimbabwe. *Transactions of the Zimbabwe Scientific Association* 72 (Suppl.), 15–26.

Hima Amadou, O. (2000) *L'analyse de l'impact de la fertilisation des sols sur le rendement et le revenu des paysans dans les zones rurales du Niger, cas de Kara-Bedji*. Mémoire de fin d'études, Institut Technologique Agricole Mostaganem, Algeria.

Hoff, K., Braverman, A. and Stiglitz, J.E. (eds) (1993) *The Economics of Rural Organisation. Theory, Practice, and Policy*. World Bank, Oxford University Press, Oxford.

Holden, S.T. and Binswanger, H. (1998) Small farmers, market imperfections, and natural resource management. In: Lutz, E., Binswanger, H.P., Hazell, P. and McCalla, A. (eds) *Agriculture and the Environment. Perspectives on Sustainable Rural Development*. World Bank, Washington, DC.

Holden, S.T. and Shanmugaratnam, N. (1995) Structural adjustment, production subsidies, and sustainable land-use. *Forum for Development Studies* 2, 247–266.

Holden, S.T. and Yohannes, H. (2000) *Land Tenure, Farm Input Intensity and Food Security: the Case of Farm Households in Southern Ethiopia*. Working Paper, Department of Economics and Social Sciences, Agricultural University of Norway, Ås.

Holden, S.T., Shiferaw, B. and Wik, M. (1998) Poverty, market imperfections and time preferences: of relevance for environmental policy? *Environment and Development Economics* 3, 105–130.

Holling, C.S. (1978) *Adaptive Environmental Assessment and Management*. John Wiley & Sons, London.

Honlonkou, A.N., Manyong, V.M. and Tchetche, N. (1999) Farmers' perceptions and the dynamics of adoption of a resource management technology: the case of *Mucuna* fallow in southern Benin, West Africa. *International Forestry Review* 1, 228–235.

Hope, A. and Timmel, S. (1984) *Training for Transformation*. Mambo Press, Gweru, Zimbabwe.

Hopkins, J. and Berry, P. (1994) *Determinants of Land and Labor Productivity in Crop Production in Niger*. Report to USAID, Niamey, Niger, International Food Policy Research Institute, Washington, DC.

Horowitz, M.M. (1979) *The Sociology of Pastoralism and African Livestock Projects*. AID Program Evaluation Discussion Paper no. 6, AID, Washington, DC.

Houndékon, V., Manyong, V.M., Gogan, C.A. and Versteeg, M. (1998) Déterminants de l'adoption de *Mucuna* dans le département du Mono au Bénin. In: Buckles, D., Etèka, A., Osiname, O., Galiba, M. and Galiano, G. (eds) *Cover Crops in West Africa: Contributing to Sustainable Agriculture*. International Development Research Centre (IDRC), Ottawa, International Institute of Tropical Agriculture (IITA), Ibadan, Nigeria, and Sasakawa Global 2000, Cotonou, Benin, pp. 45–54.

Hove, L., Topps, J.H., Sibanda, S. and Ndlovu, L.R. (2001) Nutrient intake and utilisation by goats fed dried leaves of the shrub legumes *Acacia angustissima*, *Calliandra calothyrsus* and *Leucaena leucocephala* as supplement to native pasture hay. *Animal Feed Science and Technology* 91, 95–106.

Howard, J.A., Jeje, J.J., Tschirley, D., Strasberg, P., Crawford, E.W. and Weber, M.T. (1998) *Is Agricultural Intensification Profitable for*

Mozambican Smallholders? An Appraisal of the Inputs Subsector and the 1996/97 DNER/SG2000 Program, Vol. 1, *Summary*. Research Report No. 31, Ministry of Agriculture and Fisheries, Government of Mozambique, Maputo, Mozambique, 17 pp.

Howard, J.A., Kelly, V., Stepanek, J., Crawford, E.W., Demeke, M. and Maredia, M. (1999) *Green Revolution Technology Takes Root in Africa*. MSU International Development Working Paper No. 76, Department of Agricultural Economics, Michigan State University, East Lansing, Michigan, 64 pp.

Hulugalle, N.R., Lal, R. and ter Kuile, C.H.H. (1986) Amelioration of soil physical properties by *Mucuna* after mechanized land clearing of a tropical rain forest. *Soil Science* 141, 219–224.

Hulugalle, N.R., Koning, J.D. and Matlon, P.J. (1990) Effect of rock bunds and tied ridges on soil water content and soil properties in the Sudan savannah of Burkina Faso. *Tropical Agriculture* 67(1), 149–153.

Hurni, H. (1985) *Soil Conservation Manual for Ethiopia. A Field Guide for Conservation Implementation*. Soil Conservation Research Project, Addis Ababa, Ethiopia.

Huxley, E. (1960) *A New Earth: An Experiment in Colonialism*. Chatto and Windus, London.

ICRA (1991) *An Analysis of Agricultural and Livestock Production Systems in Usagara Division in Tanzania*. Working Document Series 11, ICRA, Wageningen, The Netherlands, 132 pp.

ICRAF (1996) *Annual Report, 1996*. International Centre for Research in Agroforestry, Nairobi, Kenya.

IGAD/INTSORMIL/USAID/REDSO (2001) *Agricultural Technology for the Semi-arid African Horn, Country Study: Ethiopia*. INTSORMIL, University of Nebraska, Lincoln, Nebraska.

IITA (1996) *Annual Report*. International Institute of Tropical Agriculture, Ibadan, Nigeria.

IITA (1999) *IITA Annual Report for 1999*. International Institute of Tropical Agriculture, Ibadan, Nigeria, 82 pp.

Ikpe, F.N., Powell, J.M., Isirimal, N.O., Wahua, T.A.T. and Ngodingha, E.M. (1999) Effects of primary tillage and soil amendment practices on pearl millet yield and nutrient uptake in the Sahel of West Africa. *Experimental Agriculture* 35, 437–448.

Ikwuegbu, O.A. and Ofodile, S. (1992) Wet season supplementation of West African dwarf goats raised under traditional management in the subhumid zone of Nigeria. In: Ayeni, O.A. and Bosman, H.G. (eds) *Goat Production Systems*

in the Humid Tropics. Pudoc Scientific Publishers, Wageningen, The Netherlands, pp. 195–201.

ILCA (1991) *ILCA 1990: Annual Report and Programme Highlights*. International Livestock Centre for Africa (ILCA), Addis Ababa, Ethiopia, 84 pp.

Inaizumi, H., Singh, B.B., Sanginga, P.C., Manyong, V.M., Adesina, A.A. and Tarawali, S. (1999) *Adoption and Impact of Dry-season Dual-purpose Cowpea in the Semiarid Zone of Nigeria*. IMPACT Report, International Institute of Tropical Agriculture (IITA), Ibadan, Nigeria.

Innis, D.Q. (1997) *Intercropping and the Scientific Basis of Traditional Agriculture*. Intermediate Technology Publications, London.

Isnard, H. (1971) *Géographie de la décolonisation*. Presses Universitaires de France, Paris.

Izac, A.-M.N. and Swift, M.J. (1994) On agricultural sustainability and its measurement in small-scale farming in sub-Saharan Africa. *Ecological Economics* (11)2, 105–125.

Jabbar, M.A., Cobbina, J. and Reynolds, L. (1992) Optimum fodder-mulch allocation of tree foliage under alley farming in southwest Nigeria. *Agroforestry Systems* 20, 187–198.

Jacobs, J. (1961) *The Life and Death of Great American Cities*. Random House, London.

Jager, A., Nandwa, S.M. and Okoth, P.F. (1998) Monitoring nutrient flows and economic performance in African farming systems (NUTMON)1. Concepts and methodologies. *Agricultural Ecosystems and Environment* 71, 39–50.

Jagtap, S.S. (1995) Environmental characterisation of the moist lowland savanna of Africa. In: Kang, B.T., Akobundu, I.O., Manyong, V.M., Carsky, R.J., Sanginga, N. and Kuenenman, E.A. (eds) *Moist Savannas of Africa: Potential and Constraints for Crop Production. Proceedings of International Workshop held at Cotonou, Republic of Benin, 19–23 September, 1994*. International Institute of Tropical Agriculture (IITA), Ibadan, Nigeria, and FAO, Rome, pp. 9–30.

Jama, B., Swinkels, R. and Buresh, R.J. (1997) Agronomic and economic evaluation of organic and inorganic sources of phosphorus in western Kenya. *Agronomy Journal* 89, 597–604.

Jere, P.M. (1985) *Manure and Fertiliser Trial in 1984/85*. Chitedze Research Station, Lilongwe, Malawi.

Jere, P.M. (1986) *Manure × Fertiliser Trial 1985/86*. Chitedze Agricultural Research Station, Lilongwe, Malawi.

Jigawa State Agricultural and Rural Development Authority (JARDA) (1999) *Annual Report.* Dutse, Nigeria.

Jiggens, J., Reijntjes, C. and Lightfoot, C. (1996) Mobilising science and technology to get agriculture moving in Africa: a response to Borlaug and Dowswell. *Development Policy Review* 14, 89–103.

Jiggins, J. (1989) An examination of the impact of colonialism in establishing negative values and attitudes towards indigenous agricultural knowledge. Indigenous knowledge systems: implications for agriculture and international development. *Studies in Technology and Change* 11, 68–78. Technology and Social Change Program, Iowa State University, Ames, Iowa.

Jodha, N.S. (1990) Common property resources and rural poor in dry regions of India. *Economic and Political Weekly* 21, 1169–1181.

Jomini, P.A. (1990) The economic viability of phosphorus fertilization in southwestern Niger: a dynamic approach incorporating agronomic principles. Unpublished PhD dissertation, Department of Agricultural Economics, Purdue University, West Lafayette, Indiana.

Jones, R.B. and Heisey, P.W. (1993) Preliminary results from the MOA/UNDP/FAO Fertilizer Programme 1989–1992. Maize Commodity Team and MOA/UNDP/FAO Draft Working Paper, Ministry of Agriculture, Lilongwe, Malawi.

Jones, R.B., Freeman, H.A., Walls, S. and Londner, S.I. (2000) *Improving the Access of Small Farmers in Africa to Global Markets through the Development of Quality Standards for Pigeon Peas.* ICRISAT/ICRAF, Nairobi, Kenya.

Jutzi, S.C. (1987) *Animal Power for Improved Management of Deep Black Clay Soils in Sub-Saharan Africa (Phase 1: Ethiopia).* Joint ILCA/ICRISAT/MOA/Alemaya University of Agriculture/Addis Ababa University Research, Training and Outreach Project Report.

Kabore, J., Kambou, N.F., Dickey, J. and Lowenberg-Deboer, J. (1994) Economics of rock bunds, mulching and *zaï* in the Northern Central Plateau of Burkina Faso. In: Lowenberg-Deboer, J., Boffa, J.M., Dickey, J. and Robins, E. (eds) *Integrated Research in Agricultural Production and Natural Resource Management.* Purdue University, INERA and Winrock International, West Lafayette, Indiana, pp. 67–82.

Kambou, N.F., Taonda, S., Zougmoré, J.-B., Zougmoré, R., Keboré, B. and Dickey, J. (1994) Evolution of sedimentation, surface micro-morphology, and millet production in response to soil conservation practices on an eroded site at Yilou, Burkina Faso. In: Lowenberg-Deboer, J., Boffa, J.M., Dickey, J. and Robins, E. (eds) *Integrated Research in Agricultural Production and Natural Resource Management.* Purdue University, INERA and Winrock International, West Lafayette, Indiana, pp. 43–53.

Kang, B.T., Wilson, G.F. and Lawson, T.L. (1984) *Alley Cropping: a Stable Alternative to Shifting Cultivation.* International Institute of Tropical Agriculture, Ibadan.

Kang, B.T., Reynolds, L. and Atta-Krah, A.A. (1990) Alley farming. *Advances in Agronomy* 43, 315–359.

Kang, B.T., Hauser, S., Vanlauwe, B., Sanginga, N. and Attah-Krah, A.N. (1995) Alley farming research on high base status soils. In: *Alley Farming Research and Development. Proceedings of an International Conference on Alley Farming.* International Institute of Tropical Agriculture, Ibadan, Nigeria, pp. 25–39.

Kano State Agricultural and Rural Development Authority (KNARDA) (1999) *Annual Report.* KNARDA, Kano, Nigeria.

KARI (Kenya Agricultural Research Institute) (1995) *National Dryland Farming Research Centre – Katumani, Regional Research Programme.* KARI, Katumani, Kenya.

Kassie, M. (1997) Economics of food crops–forage legumes integration in mixed farms in the Ethiopian highlands. MSc thesis, Alemaya University of Agriculture, Ethiopia.

Kelly, V. and Murekezi, A. (2000) *Fertilizer Response and Profitability in Rwanda: A Synthesis of Findings from Minagri Studies Conducted by the Food Security Research Project and the FAO Soil Fertility Initiative.* Food Security Research Project, MINAGRI, Kigali, Rwanda.

Kelly, V., Mpyisi, E., Murekezi, A. and Niven, D. (2001) Fertilizer consumption in Rwanda: past trends, future potential, and determinants. Paper prepared for the Policy Workshop on Fertilizer Use and Marketing, Ministry of Agriculture, Kigali, Rwanda, 22–23 February.

Kenmore, P.E. (1999) IPM and farmer field schools in Asia. Paper for Conference on Sustainable Agriculture: New Paradigms and Old Practices? Bellagio Conference Centre, Italy, 26–30 April.

Kenmore, P.E., Carino, F.O., Perez, C.A., Dyck, V.A. and Gutierrez, A.P. (1984) Population regulation of the brown planthopper within rice fields in the Philippines. *Journal of Plant Protection in the Tropics* 1(1), 19–37.

Kerkhof, P. (1990) *Agroforestry in Africa. A Survey of Project Experience.* Panos Publications, London, 216 pp.

Kerr, J. and Sanghi, N.K. (1992) *Indigenous Soil and Water Conservation in India's Semi-arid Tropics.* Gatekeeper Series No. 34, Sustainable Agriculture Programme, International Institute for Environment and Development, London.

Kidane, G. and Rezene, F. (1989) *Dryland Agricultural Research Priorities to Increase Crop Productivity.* Ethiopian Agricultural Research Organization, Addis Ababa, Ethiopia.

Kim, T.-K., Hayes, D.J. and Hallam, A. (1992) Technology adoption under price uncertainty. *Journal of Development Economics* 38(1), 245–253.

Knight, J. (1992) *Institutions and Social Conflict.* Cambridge University Press, Cambridge.

Korten, D.C. (1980) Community organization and rural development: a learning process approach. *Public Administration Review* 40, 480–510.

Korten, D.C. (1984) Rural development programming: the learning process approach. In: Korten, D.C. and Klauss, R. (eds) *People Centered Development: Contributions toward Theory and Planning Frameworks.* Kumarian Press, West Hartford, Connecticut.

Kothari, A., Pathak, N., Anuradha, R.V. and Taneja, B. (1998) *Communities and Conservation: Natural Resource Management in South and Central Asia.* Sage, New Delhi.

Koudokpon, V., Versteeg, Adegbola, A. and Budelman, A. (1992) The adoption of hedgerow intercropping by the farmers in Adja plateau, south Benin Republic. In: Kang, B.T., Osiname, A.O. and Larbi, A. (eds) *Alley Farming Research and Development. Proceedings of the International Conference on Alley Farming, 14–18 September 1992.* International Institute of Tropical Agriculture, Ibadan, Nigeria, pp. 483–498.

Kristjanson, P., Tarawali, S., Okike, I. and Singh, B.B. (1999) *Farmer Impact Workshops for Evaluating Impact of Improved Food and Fodder Cowpea Varieties in Northern Nigeria.* A Report for the System-Wide Livestock Programme, International Livestock Research Institute, Nairobi, Kenya.

Kristjanson, P., Place, F., Franzel, S. and Thornton, P.K. (2001) Assessing research impact on poverty: the importance of farmers' perspectives. *Agricultural Systems* (in press).

Krueger, A.O. (1990) Government failures in development. *Journal of Economic Perspectives* 4(3), 9–23.

Krueger, A.O., Schiff, M. and Valdes, A. (1988) Agricultural incentives in developing countries: measuring the effect of sectoral and economywide policies. *World Bank Economic Review* 2(3), 255–271.

Kumwenda, A.S. (1998) Observations on the Malawi maize productivity task force. In: Waddington, S.R., Murwira, H.K., Kumwenda, J.D.T., Hikwa, D. and Tagwira, F. (eds) *Soil Fertility Research for Maize-Based Farming Systems in Malawi and Zimbabwe. Proceedings of the Soil Fert Net Results and Planning Workshop held from 7 to 11 July 1997 at Africa University, Mutare, Zimbabwe.* Soil Fert Net and CIMMYT-Zimbabwe, Harare, Zimbabwe, pp. 263–269.

Kumwenda, J.D. and Benson, T. (1996) *An Agronomic and Economic Analysis of Basal Fertiliser Types on Yield of Hybrid Maize, 1993–95.* Chitedze Research Station, Lilongwe, Malawi.

Kumwenda, J.D.T., Waddington, S.R., Snapp, S.S., Jones, R.B. and Blackie, M.J. (1996) *Soil Fertility Management Research for the Maize Cropping Systems of Smallholders in Southern Africa: A Review.* Natural Resources Group Paper 96-02, CIMMYT, Mexico, DF, 35 pp.

Kumwenda, J.D.T., Waddington, S.R., Snapp, S.S., Jones, R.B. and Blackie, M.J. (1997) Soil fertility management in southern Africa. In: Byerlee, D. and Eicher, C.K. (eds) *Africa's Emerging Maize Revolution.* Lynne Rienner, Boulder, Colorado, pp. 157–172.

Kupfuma, B. (1993) Fertiliser use in the semi arid communal area of Zimbabwe. In: Mwangi, W., Rohrbach, D. and Haisey, D.P. (eds) *Cereal Grain Policy Analysis in the National Agriculture Research Systems of Eastern and Southern Africa.* CIMMYT, SADC and ICRISAT, Addis Ababa.

Kwesiga, F. (1998) Improved fallows for sustainable food security in eastern Zambia. *Transactions of the Zimbabwe Scientific Association* 72 (Suppl.), 72–83.

Kwesiga, F.R. and Beniest, J. (1998) *Sesbania Improved Fallows for Eastern Zambia: an Extension Guideline.* International Centre for Research in Agroforestry, Nairobi.

Kwesiga, F. and Coe, R.S. (1994) The effect of short rotation *Sesbania sesban* planted fallows on maize yield. *Forest Ecology and Management* 64, 199–208.

Kwesiga, F.R., Franzel, S., Place, F., Phiri, D. and Simwanza, C.P. (1997) *Sesbania Improved Fallows in Eastern Zambia: their Inception, Development, and Farmer Enthusiasm.* Zambia/ICRAF Agroforestry Research Project, Chipata, Zambia.

Kwesiga, F., Franzel, S., Place, F., Phiri, D. and Simwanza, C.P. (1999) *Sesbania sesban* improved fallows in eastern Zambia: their inception, development and farmer enthusiasm. *Agroforestry Systems* 47, 49–66.

Kydd, J. (1989) Maize research in Malawi: lessons from failure. *Journal of International Development* 1, 112–144.

Kydd, J. and Christiansen, R. (1982) Structural change in Malawi since independence: consequences of a development strategy based on large-scale agriculture. *World Development* 10(5), 355–375.

Lal, R. (1989) Agroforestry systems and soil surface management of a tropical Alfisol. Soil moisture and crop yields. *Agroforestry Systems* 8, 7–29.

Lapar, A. and Pandey, S. (1999) Adoption of soil conservation: the case of Philippine uplands. *Agricultural Economics* 21, 241–256.

Larson, B. and Frisvold, G.B. (1996) Fertilizers to support agricultural development in sub-Saharan Africa: what is needed and why. *Food Policy* 21(6), 509–525.

Larson, P., Freudenberger, M. and Wyckoff-Baird, B. (1998) *WWF Integrated Conservation and Development Projects: Ten Lessons from the Field 1985–1996.* World Wildlife Fund, Washington, DC.

Lavado, R.S., Sierra, J.O. and Hashimoto, P.N. (1996) Impact of grazing on soil nutrients in a Pampean grassland. *Journal of Range Management* 49, 452–457.

Lawry, S., Stienbarger, D. and Jabbar, M.A. (1995) Land tenure and the adoption of alley farming in West Africa. In: Kang, B.T., Osiname, A.O. and Larbi, A. (eds) *Alley Farming Research and Development. Proceedings of the International Conference on Alley Farming, 14–18 September 1992.* International Institute of Tropical Agriculture, Ibadan, Nigeria, pp. 464–471.

LDI (2000) *Evaluation des activités de LDI Fianarantsoa à Alatsinainy-Ialamarina et Sendrisoa.* Landscape Development Interventions, Fianarantsoa, Madagascar.

Leach, M. and Mearns, R. (1996) *The Lie of the Land: Challenging Received Wisdom on the African Environment.* James Currey, Oxford.

Lee, D.R. and Barrett, C.B. (eds) (2000) *Tradeoffs or Synergies? Agricultural Intensification, Economic Development and the Environment in Developing Countries.* CAB International, Wallingford, UK.

Legislative Council of Tanganyika (1953) *Forest Policy.* Sessional Paper No. 1, Government Printers, Dar es Salaam, 5 pp.

Levin, S.A. (2000) *Fragile Dominion: Complexity and the Commons.* Perseus Books Group, Reading, Massachusetts.

Liang, K.-Y. and Zeger, S.L. (1986) Longitudinal data analysis using generalized linear models. *Biometrika* 73, 13–22.

Lindner, R., Pardey, P. and Jarrett, F.G. (1992) Distance to information source and the time lag to early adoption of trace element fertilizers. *Australian Journal of Agricultural Economics* 26(2), 98–113.

Little, P.D. and Watts, M.J. (eds) (1994) *Living Under Contract: Contract Farming and Agarian Transformation in Sub-Saharan Africa.* University of Wisconsin Press, Madison, Wisconsin.

Lohr, L. and Park, T.A. (1994) Discrete/continuous choices in contingent valuation surveys: soil conservation decisions in Michigan. *Review of Agricultural Economics* 16 January, 1–15.

Lohr, L. and Park, T.A. (1995) Utility-consistent discrete–continuous choices in soil conservation. *Land Economics* 71(4), 474–490.

Long, N. and Long, A. (eds) (1992) *The Battlefields of Knowledge: the Interlocking of Theory and Practice in Social Research and Development.* Routledge, London.

Lutz, E. and Pagiola, S. (1994) The costs and benefits of soil conservation: the farmers' viewpoint. *The World Research Observer* 9(2), 273–295.

Lynam, J.K., Nandwa, S.M. and Smaling, E.M.A. (1998) Introduction: nutrient balances as indicators of productivity and sustainability in sub-Saharan African agriculture. *Agricultural Ecosystems and Environment* 71, 1–4.

Lynne, G.D., Shonkwiler, J.S. and Rola, L.R. (1988) Attitudes and farmer conservation behavior. *American Journal of Agricultural Economics*, February, 12–19.

Maatman, A., Sawadogo, H., Schweigman, C. and Ouedraogo, A. (1998) Application of zaï and rock bunds in the northwest region of Burkina Faso: study of its impact on houselevel level by using a stochastic linear programming model. *Netherlands Journal of Agricultural Science* 46, 123–136.

McIntire, J., Bourzat, D. and Pingali, P. (1992) *Crop–Livestock Interactions in Sub-Sahara Africa.* World Bank, Washington, DC.

McLoughlin, P.F.M. (1967) Some aspects of land reorganization in Malawi (Nyasaland), 1950–60. *Ekistics* 24(141), 193–195.

McNamara, N. and Morse, S. (1992) *Developing On-farm Research – the Broad Picture.* On Stream Publication, Cloghroa, Blarney, County Cork, Ireland, 172 pp.

Maddala, G.S. (1983) *Limited Dependent and Qualitative Variables in Econometrics.* Cambridge University Press, London.

Mafongoya, P.L., Giller, K.E. and Palm, C. (1998) Decomposition and nitrogen release patterns of tree prunings and litter. *Agroforestry Systems* 38, 77–97.

Mafongoya, P.L., Barak, P. and Reed, J.D. (2000) Carbon, nitrogen and phosphorus mineralization of tree leaves and manure. *Biology and Fertility of Soils* 30, 298–305.

Malawi Maize Productivity Task Force (1996) *Organic Manure Technologies to Improve and Sustain Soil Fertility and Crop Yields: Progress Report.* Malawi Ministry of Agriculture and Irrigation, Lilongwe, Malawi.

Mandala, E.C. (1990) *Work and Control in a Peasant Economy.* University of Wisconsin Press, Madison, Wisconsin.

Mangan, J. and Mangan, M.S. (1998) A comparison of two IPM training strategies in China: the importance of concepts of the rice ecosystem for sustainable pest management. *Agric. Human Values* 15, 209–221.

Mann, C. (1998) *Higher Yields for All Smallholders through 'Best Bet' Technology: The Surest Way to Restart Economic Growth in Malawi.* Soil Fert Net Research Results Working Paper 3, CIMMYT, Harare, Zimbabwe, 14 pp.

Manson, S.C., Leighner, D.E. and Vorst, J.J. (1986) Cassava–cowpea and cassava–peanut intercropping. III. Nutrient concentration and removal. *Agronomy Journal* 78, 441–444.

Manyong, V.M., Smith, J., Weber, G.K., Jagtap, S.S. and Oyewole, B. (1996) *Macrocharacterization of Agricultural Systems in West Africa: An Overview.* Resource and Crop Management Monograph No. 21, International Institute of Tropical Agriculture (IITA), Ibadan, Nigeria.

Mashaka, R.I., Anania, J.B., Busungu, D.G. and Braun, P.M. (1992) *Farming System Baseline Survey for Shinyanga Region, Tanzania.* Tanzania–German Project for Integrated Pest Management, Shinyanga, 45 pp.

Matabwa, C.J. and Wendt, J.W. (1993) Soil fertility management: present knowledge and prospects. In: Munthali, D.C., Kumwenda, J.D.T. and Kisyombe, F. (eds) *Proceedings of the Conference on Agricultural Research for Development.* Department of Agricultural Research, Chancellor College, Zomba, Malawi, pp. 107–123.

Mataya, C., Simotowe, F. and Mekuria, M. (2000) Choice between compost manure and inorganic fertilizer: a post liberalization dilemma of smallholder maize producers in Zomba rural development program, Malawi. EPWG Draft Report.

Maturana, H.R. and Varela, F.J. (1982) *The Tree of Knowledge: The Biological Roots of Human Understanding.* Shabhala Publications, Boston.

Meena, R. (1991) The impact of structural adjustment programs on rural women in Tanzania. In: Gladwin, C.H. (ed.) *Structural Adjustment and African Women Farmers.* University of Florida Press, Gainesville, Florida, pp. 169–190.

Meertens, H.C.C. and Lupeja, P.M. (1996) *A Collection of Agricultural Background Information for Mwanza Region.* Field Note, Kilimo/FAO Plant Nutrition Programme, URT/FAO, Dar es Salaam, 74 pp.

Mehra, R. (1991) Can structural adjustment work for women farmers? *American Journal of Agricultural Economics* 73(5), 1440–1447.

Meinzen-Dick, R.S., Brown, L.R., Feldstein, H.S. and Quisumbing, A.R. (1997) Gender, property rights, and natural resources. *World Development* 25(8), 1303–1315.

Mercer, D.E. and Miller, R.P. (1998) Socio-economic research in agroforestry: progress, prospects and priorities. *Agroforestry Systems* 38, 177–193.

Mezirow, J. (1991) *Transformative Dimensions of Adult Learning.* Jossey-Bass, San Francisco, California.

Migot-Adholla, S.E., Place, F. and Oluoch-Kosura, W. (1990) *Tenure Security and Productivity in Kenya.* World Bank, Washington, DC.

Migot-Adholla, S.E., Hazell, P.B.R., Blarel, B. and Place, F. (1991) Indigenous land rights systems in sub-Saharan Africa: a constraint on productivity? *World Bank Economic Review* 5, 155–175.

Miller, F.R. and Kebede, Y. (1984) Genetic contributions to yield gains in sorghum, 1950 to 1980. In: Fehr, W.R. (ed.) *Genetic Contributions to Yield Gains of Five Major Crop Plants. Proceedings of Symposium Sponsored by Crop Science Society of America, 21 December 1981, Atlanta, Georgia.* Special Publication No. 7, Crop Science Society of America, Madison, Wisconsin, pp. 1–44.

Mitchell, R. and Carson, R. (1989) *Using Surveys to Value Public Goods: the Contingent Valuation Method.* Resources for the Future, Washington, DC.

MoA (Ministry of Agriculture) (1982) *Tanzania National Agricultural Policy. Final Report.* Government Printer, Dar es Salaam.

MoA (Ministry of Agriculture) (1983a) *The Agricultural Policy of Tanzania.* Government Printer, Dar es Salaam.

MoA (Ministry of Agriculture) (1983b) *The Livestock Policy of Tanzania.* Government Printer, Dar es Salaam.

MoAC (Ministry of Agriculture and Cooperatives) (1997) *National Agriculture and Livestock Policy.* Government Printer, Dar es Salaam.

MoAC (Ministry of Agriculture and Cooperatives) (1998) *Basic Data, Agriculture and Livestock Sector, 1991–1998.* Government Printer, Dar es Salaam.

Mohamed-Saleem, M.A. (1985) Effect of sowing time on the grain yield and fodder potential of sorghum undersown with stylo in the sub-humid zone of Nigeria. *Tropical Agriculture (Trinidad)* 62, 151–153.

Mohamed-Saleem, M.A. (1998) Nutrient balance patterns in African livestock systems. *Agriculture, Ecosystems and Environment* 71, 241–254.

Mohamed-Saleem, M.A. and Suleiman, H. (1986) Fodder banks: dry season feed supplementation for traditionally managed cattle in the subhumid zone. *World Animal Review* 59, 11–17.

Mokwunye, A. and Vlek, P. (eds) (1986) *Management of Nitrogen and Phosphorus Fertilizers in Sub-Saharan Africa.* Martinus Nijhoff, Dordrecht.

Mokwunye, A.U. and Hammond, L.L. (1992) Myths and science of fertilizer use in the tropics. In: Lal, R. and Sanchez, P.A. (eds) *Myths and Science of Soils in the Tropics.* Special Publication No. 29, Soil Science of America and American Society of Agronomy, Madison, Wisconsin.

Mose, L.O. (1998) Factors affecting the distribution and use of fertilizers in Kenya: a preliminary assessment. Paper presented at Conference on Strategies for Raising Productivity in Kenya, Tegemeo Institute of Agricultural Policy and Development, Egerton University, Kenya.

Mose, L.O., Nyangito, H.O. and Mugunieri, L.G. (1997) An analysis of socio-economic factors that influence chemical fertilizer use among small holder maize producers in western Kenya. In: Ransom, J.K., Palmer, A.F.E., Zambezi, B.T., Mduruma, Z.O., Waddington, S.R., Pixley, K.V. and Jewell, D.C. (eds) *Maize Productivity Gains Through Research and Technology Dissemination: Proceedings of the Fifth Eastern and Southern Africa Regional Maize Conference, held in Arusha, Tanzania,* 3–7 June 1996. CIMMYT, Addis Ababa, Ethiopia, pp. 43–46.

Moser, C.M. (2001) Technology adoption decisions of farmers facing seasonal liquidity constraints: a case study of the system of rice intensification in Madagascar. MS thesis, Department of Applied Economics and Management, Cornell University, Ithaca, New York.

Mpofu, I.D.T., Ndlovu, L.R. and Casey, N.H. (1999) The copper, cobalt, iron, selenium and zinc status of cattle in the Sanyati and Chinamhora smallholder grazing areas of Zimbabwe. *Asian–Australian Journal of Animal Science* 12, 579–584.

MPTF (Maize Productivity Task Force) (1998) *Malawi: Soil Fertility Issues and Options.* Discussion Paper. The Rockfeller Foundation, Lilongwe, Malawi.

MTNR (Ministry of Tourism and Natural Resources) (1998) *The National Forestry Policy.* Government Printer, Dar es Salaam.

MTNRE (Ministry of Tourism, Natural Resources and Environment) and Sida (Swedish International Development Agency) (1995) *Dodoma Region Soil Conservation Project. Final Evaluation Report.* MTNR, Dar es Salaam.

Muhammad, L.W. and Parton, K.A. (1992) Smallholder farming in semi-arid eastern Kenya – basic issues relating to the modelling of adoption. In: Probert, M.E. (ed.) *A Search for Strategies for Sustainable Dryland Cropping in Semi-arid Eastern Kenya.* Proceedings No. 41, ACIAR, Nairobi, Kenya, pp. 119–123.

Murray, S., Burke, L., Tunstall, D. and Gilruth, P. (1999) Drylands population assessment II. Draft, World Resources Institute and UNDP Office to Combat Drought and Desertification, New York.

Murwira, H.K. (1995) Ammonia losses from Zimbabwean cattle manure before and after incorporation into soil. *Tropical Agriculture* 72, 269–273.

Murwira, H.K., Swift, M.J. and Frost, P.G.H. (1995) Manure as a key resource in sustainable agriculture. In: Powell, J.M., Fernandez-Rivera, S., Williams, T.O. and Renard, C. (eds) *Livestock and Sustainable Nutrient Cycling in Mixed Farming systems of SubSaharan Africa,* Vol. II, *Technical Papers.* International Livestock Centre for Africa, Addis Adaba, pp. 131–148.

Murwira, H.K., Tagwira, F., Chikowo, R. and Waddington, S. (1998) An evaluation of the agronomic effectiveness of low rates of cattle manure and combinations of inorganic N in Zimbabwe. In: Waddington, S., Murwira, H.K.,

Kumwenda, J.D.T., Hikwa, D. and Tagwira, F. (eds) *Soil Fertility Research for Maize-Based Farming Systems in Malawi and Zimbabwe. Proceedings of the Soil Fert Net Results and Planning Workshop held from 7 to 11 July at Africa University, Mutare, Zimbabwe.* Soil Fert Net and CIMMYT-Zimbabwe, Harare, Zimbabwe, pp. 179–182.

Mushayi, P.T., Waddington, S.R. and Chiduza, C. (1999) Low efficiency of nitrogen use by maize on smallholder farms in sub-humid Zimbabwe. In: *Maize Production Technology for the Future: Challenges and Opportunities. Proceedings of the Sixth Eastern and Southern Africa Regional Maize Conference, 21–25 September 1998, Addis Ababa, Ethiopia.* CIMMYT and EARO, Addis Ababa, Ethiopia, pp. 278–281.

Mwangi, W. (1997) Low use of fertilizers and low productivity in sub-Saharan Africa. *Nutrient Cycling in Agroecosystems* 47, 135–147.

Mwaura, F.M. and Woomer, P.L. (1999) Fertilizer retailing in the Kenyan Highlands. *Nutrient Cycling in Agroecosystems* 55, 107–116.

MWEM (Ministry of Water, Energy and Minerals) (1991) *Water Policy.* Government Printer, Dar es Salaam.

Mworia, J.K., Mnene, W.N., Musembi, D.K. and Reid, R.S. (1997) Resilience of soils and vegetation subjected to different grazing intensities in a semi arid rangeland of Kenya. *African Journal of Range and Forage Science* 14, 26–31.

Nandwa, S.M. and Bekunda, M.A. (1998) Research on nutrient flows and balances in East and southern Africa: state-of-the-art. *Agriculture, Ecosystems and Environment* 71, 5–18.

Napier, T.L. (1991) Factors affecting acceptance and continued use of soil conservation practices in developing countries: a diffusion perspective. *Agriculture, Ecosystems and Environment* 36, 127–140.

Napier, T.L. (1996) Socioeconomic factors affecting adoption of soil and water conservation practices in lesser-scale societies. In: Buerkert, F., Allison, B. and Von Oppen, M. (eds) *Wind Erosion in West Africa.* Margraf Verlag, Weikersheim.

Narayan, D. and Pritchett, L. (1996) *Cents and Sociability: Household Income and Social Capital in Rural Tanzania.* Policy Research Working Paper 1796, World Bank, Washington, DC.

Nash, J.A. (1991) *Loving Nature: Ecological Integrity and Christian Responsibility.* Abingdon Press, Nashville, Tennessee.

Ndiaye, S. and Sofranko, A. (1994) Farmers' perceptions of resource problems and adoption of conservation practices in a densely populated area. *Agriculture, Ecosystems and Environment* 48(1), 35–47.

Ndjeunga, J. and Bationo, A. (2000) *Stochastic Dominance Analysis of Soil Fertility Restoration Options on Sandy Sahelian Soils in Southwest Niger.* Working Paper, ICRISAT-IFDC, 14 pp.

Ndlovu, L.R., Francis, J. and Hove, E. (1996) Performance of draught cattle in communal farming areas in Zimbabwe after dry season supplementation. *Tropical Animal Health and Production* 28, 298–306.

Neef, A., Haigis, J. and Heidhues, F. (1996) L'impact du pluralisme international et juridique sur l'introduction des mesures anti-erosives. Le cas du Niger. In: Buerkert, F., Allison, B. and Von Oppen, M. (eds) *Wind Erosion in West Africa.* Margraf Verlag, Weikersheim.

Negatu, W. and Parikh, A. (1999) The impact of perception and other factors on the adoption of agricultural technology in the Moret and Jiru Woreda (district) of Ethiopia. *Agricultural Economics* 21(1), 205–216.

Neill, S.P. and Lee, D.R. (2001) Explaining the adoption and disadoption of sustainable agriculture: the case of cover crops in northern Honduras. *Economic Development and Cultural Change* 49(4), 793–820.

Ngambeki, D.S. (1985) Economic evaluation of alley cropping *Leucaena* with maize–maize and maize–cowpea in southern Nigeria. *Agricultural Systems* 17, 243–258.

Nherera, F.V., Ndlovu, L.R. and Dzowela, B.H. (1998) Utilisation of *Leucaena diversifolia, Leucaena esculenta, Leucaena pallida,* and *Calliandra calothyrsus* as nitrogen supplements for growing goats fed maize stover. *Animal Feed Science and Technology* 74, 15–28.

Nicou, R. and Charreau, C. (1985) Soil tillage and water conservation in semi-arid West Africa. In: Ohm, W. and Nagy, J.C. (eds) *Appropriate Technologies for Farmers in Semi-arid West Africa.* Purdue University, West Lafayette, Indiana, pp. 9–37.

Nieuwolt, S. (1973) *Rainfall and Evaporation in Tanzania.* BRALUP Research Paper No. 24, University of Dar es Salaam, Dar es Salaam.

Norman, D.W. and Baker, D.C. (1986) Components in farming systems research, FSR credibility and experiences in Botswana. In: Moock, J.L. (ed.) *Understanding Africa's Rural*

Households and Farming Systems. Westview Press, Boulder, Colorado, pp. 36–57.

Norris, P.E. and Batie, S.S. (1987) Virginia farmers' soil conservation decisions: an application of Tobit analysis. *Southern Journal of Agricultural Economics* 19, 79–89.

North, D. (1990) *Institutions, Institutional Change and Economic Performance*. Cambridge University Press, Cambridge.

Noss, R. (n.d.) *The Wildlands Project Conservation Strategy*.

Nsahlai, I.V., Umunna, N.N. and Osuji, P.O. (1998) Complementarity of bird resistant and non-bird-resistant varieties of sorghum stover with cottonseed cake and noug (*Guizotia abyssinica*) cake when fed to sheep. *Journal of Agricultural Science, Cambridge* 130, 229–239.

Obonyo, E. (2000) Adoption of biomass transfer agroforestry systems in Western Kenya. MSc thesis, University of Ghana, Accra.

Ohlsson, E., Shepherd, K.D. and David, S. (1998) *A Study of Farmers' Soil Fertility Management Practices on Small-Scale Mixed Farms in Western Kenya*. Interna Publikationer 25, Institutionen for Vaxtodlingslara, Swedish University of Agricultural Sciences, Uppsala.

OHVN (1999) *Septième Session du Conseil d'Administration, Plan de Campagne 1999–2000*. Office de la Haute Vallée du Niger, Bamako.

Okike, I. (1999a) Crop–livestock interactions and economic efficiency of mixed farmers in the savanna zones of Nigeria. Unpublished PhD thesis, University of Ibadan, Nigeria.

Okike, I. (1999b) *Phase 1 Cowpea Impact Assessment Survey, Kano and Jigawa States, Nigeria*. A Consultant's Report to the International Livestock Research Institute (ILRI), Nairobi, Kenya.

Okike, I., Jabbar, M.A., Manyong, V.M., Smith, J.W., Akinwumi, J.A. and Ehui, S.K. (2001) *Agricultural Intensification and Efficiency in the West African Savannahs: Evidence from Northern Nigeria*. Socio-Economics and Policy Research Working Paper 33. International Livestock Research Institute, Nairobi, Kenya, 54 pp.

Okumu, B.N. (2000) Bio-economic modeling of watershed resources in Ethiopia. PhD thesis, University of Manchester, Manchester, UK.

Okumu, B.N., Jabbar, M.A., Colman, D., Russell, N., Mohamed Saleem, M.A. and Pender, J. (2000) *Technology and Policy Impacts on Economic Performance, Nutrient Flows and Soil Erosion at Watershed Level: The Case of Ginchi in Ethiopia*. Unpublished report to the World Bank.

Oldeman, L.R. (1998) *Soil Degradation: A Threat to Food Security?* International Soil Reference and Information Centre, Wageningen, The Netherlands.

Olson, D. and Dinerstein, E. (1998) *The Global 2000: A Representation Approach to Conserving the Earth's Distinctive Ecoregions*. Conservation Science Programme, World Wildlife Fund, Washington, DC.

Olson, M. (1965) *The Logic of Collective Action: Public Goods and the Theory of Groups*. Harvard Press, London.

Omamo, S.W. (1996) *Fertilizer Trade Under Market Liberalization: The Case of Njoro Division, Nakuru District, Kenya*. Rockefeller Foundation, Nairobi, Kenya.

Omamo, S.W. (2001) Policy research on African agriculture: trends, gaps, challenges. Unpublished report to the Rockefeller Foundation.

Omamo, S.W. and Mose, L.O. (2001) Fertilizer trade under market liberalization: preliminary evidence from Kenya. *Food Policy* 26, 1–10.

Omiti, J. (1995) Economic analysis of crop livestock integration: the case of the Ethiopian highlands. Unpublished PhD thesis submitted to University of New England, Armidale, Australia.

Omiti, J.M., Freeman, H.A., Kaguongo, W. and Bett, C. (1999) *Soil Fertility Maintenance in Eastern Kenya: Current Practices, Constraints, and Opportunities*. CARMASAK Working Paper No. 1, KARI/ICRISAT Nairobi, Kenya.

Ong, C. (1994) Alley cropping: ecological pie in the sky? *Agroforestry Today* 6(3), 8–10.

O'Reagan, P.J. and Schwartz, J. (1995) Dietary selection and foraging strategies of animals on rangeland: coping with spatial and temporal variability. In: Journet, M., Grenet, Farce, M.-H., Theriez, M. and Demarguilly, C. (eds) *Recent Developments in the Nutrition of Herbivores. Proceedings of the TV International Symposium on the Nutrition of Herbivores, 11–15 September 1995, Clermont-Ferrand, France*. Institut National de la Recherche Agronomique (INRA) Editions, Paris, pp. 407–423.

Orskov, E.R. (1982) *Protein Nutrition in Ruminants*. Academic Press, London, 160 pp.

Ostrom, E. (1990) *Governing the Commons: The Evolution of Institutions for Collective Action*. Cambridge University Press, New York.

Ostrom, E. (1998) *Social Capital: a Fad or Fundamental Concept?* Center for the Study of

Institutions, Population and Environmental Change, Indiana University, Bloomington, Indiana.

Otchere, E.O. (1986) The effects of supplementary feeding of Bunaji cattle in the subhumid zone of Nigeria. In: von Kaufmann, R., Chater, S. and Blench, R. (eds) *Livestock Systems Research in Nigerias' Subhumid Zone.* International Livestock Centre for Africa (ILCA), Addis Ababa, Ethiopia, pp. 204–212.

Otsuka, K. and Place, F. (eds) (1999) *Land Tenure and Natural Resource Management: Forest, Agroforest, and Crop land Management in Asia and Africa.* Unpublished manuscript, International Food Policy Research Institute, Washington, DC.

Otsyina, R.M., von Kaufmann, R., Mohamed-Saleem, M.A. and Suleiman, H. (1987) *Manual on Fodder Bank Establishment and Management.* International Livestock Centre for Africa (ILCA), Addis Ababa, Ethiopia, 27 pp.

PAGE (2000) *Le Rôle de la Ligne Ferroviare FCE dans l'Economie Régionale: Synthèse de l'Analyse Économique Quantitative et Qualitative du FCE,* Vol. I. Antananarivo, Madagascar.

Page, S.L. and Page, H.E. (1992) *Western Hegemony over African Agriculture in Southern Rhodesia and its Continuing Threat to Food Security in Independent Zimbabwe.* Working Paper No. 7, The Project on African Agriculture, Joint Committee on African Studies, SSRC/ACLS, New York.

Pagiola, S. (1995) The effect of subsistence requirements on sustainable land use practices. Paper presented at the American Agricultural Economics Association Annual Meeting, 6–9 August, Indianapolis, Indiana.

Pagiola, S. (1996) Price policy and returns to soil conservation in semi-arid Kenya. *Environmental and Resource Economics* 8, 255–271.

Pala-Okeyo, A. (1980) Daughters of the lakes and rivers: colonization and land rights of Luo women. In: Davison, J. (ed.) *Land and Women's Agricultural Production: the African Experience.* Westview Press, Boulder, Colorado, pp. 185–213.

Palm, C.A., Myers, R.J.K. and Nandwa, S.M. (1997) Combined use of organic and inorganic nutrient sources for soil fertility maintenance and replenishment. In: Sanchez, P. and Buresh, R. (eds) *Replenishing Soil Fertility in Africa.* Special Publication no. 51, Soil Science Society of America, Madison, Wisconsin, pp. 193–217.

Palmer, I. (1976) *The New Rice in Asia: Conclusions from Four Country Studies.* UNRISD, Geneva.

Pell, A. (1999) Integrated crop–livestock management systems in sub-Saharan Africa. Mimeo, Cornell University, Ithaca, New York.

Pender, J. (1999) *Rural Population Growth, Agricultural Change and Natural Resource Management in Developing countries: a Review of Hypotheses and Some Evidence from Honduras.* Discussion Paper No. 48, Environment and Production Technology Division, International Food Policy Research Institute, Washington, DC.

Pender, J. and Kerr, J. (1998) Determinants of farmers' indigenous soil and water conservation investments in semi-arid India. *Agricultural Economics* 19(1), 113–125.

Pender, J. and Scherr, S. (1999) *Organizational Development and Natural Resource Management: Evidence from Central Honduras.* Discussion Paper No. 49, Environment and Production Technology Division, International Food Policy Research Institute, Washington, DC.

Pender, J., Place, F. and Ehui, S. (1999) *Strategies for Sustainable Agricultural Development in the East African Highlands.* Discussion Paper No. 41, Environment and Production Technology Division, International Food Policy Research Institute, Washington, DC.

Penning de Vries, F.W.T. and Djiteye, M.A. (eds) (1991) *La productivité des pâturages sahéliens: Une étude des sols, des végétations et de l'exploitation de cette ressource naturelle.* Pudoc, Wageningen, The Netherlands.

Peters, P.E. (1997) Against the odds: matriliny, land and gender in the Shire Highlands of Malawi. *Critique of Anthropology* 17(2), 189–210.

Peterson, J.S. (1998) *Kubweleza Nthaka: Gender and Soil Fertility in Africa.* Report to the University of Florida's 'Gender and Soil Fertility in Africa' Soils Management Collaborative Research Support Program (CRSP), United States Agency for International Development, and the International Centre for Research on Agroforestry.

Peterson, J.S. (1999) *Kubweletza Nthaka: Ethnographic Decision Trees and Improved Fallows in the Eastern Province of Zambia.* Report to the University of Florida's 'Gender and Soil Fertility in Africa' Soils Management Collaborative Research Support Program (CRSP) and the International Centre for Research on Agroforestry.

Peterson, J.S., Tembo, L., Kawimbe, C. and Mwang'amba, E. (1999) *The Zambia Integrated Agroforestry Project Baseline Survey: Lessons Learned in Chadiza, Chipata, Katete and*

Mambwe Districts, Eastern Province, Zambia. Report to the World Vision International Zambia Integrated Agroforestry Project staff, the University of Florida's 'Gender and Soil Fertility in Africa' Soils Management Collaborative Research Support Program, the International Centre for Research on Agroforestry, and the Zambian Ministry of Agriculture, Food, and Fisheries.

Phiri, D., Franzel, S., Mafongoya, P., Jere, I., Katanga, R. and Phiri, S. (1999) *Who is Using the New Technology? A Case Study of the Association of Wealth Status and Gender with the Planting of Improved Tree Fallows in Eastern Province, Zambia.* International Centre for Research in Agroforestry, Nairobi.

Pieri, C. (1992) *Fertility of Soils: A Future for Farming in the West African Savannah.* Springer-Verlag, Berlin. [Note: This is essentially the English translation of Pieri's 1989 publication *Fertilité des terres de savanes.*]

Pinstrup-Andersen, P., Pandya-Lorch, R. and Rosegrant, M. (1997) *The World Food Situation: Recent Developments, Emerging Issues, and Long-term Prospects.* 2020 Food Policy Report, International Food Policy Research Institute, Washington, DC.

Pisanelli, A., Franzel, S., DeWolf, J., Rommelse, R. and Poole, J. (2000) *Adoption of Improved Tree Fallows in Western Kenya: Farmer Practices, Knowledge and Perception.* International Centre for Research in Agroforestry, Nairobi.

Place, F. (1999) *Village Impact Assessment Workshop for Agroforestry-based Soil Fertility Replenishment Practices in Western Kenya.* International Centre for Research in Agroforestry, Nairobi.

Place, F. and Dewees, P. (1999) Policies and incentives for the adoption of improved fallows. *Agroforestry Systems* 47(1/3), 323–343.

Place, F.P. and Hazell, P. (1993) Productivity effects of indigenous land tenure systems in sub-Saharan Africa. *American Journal of Agricultural Economics* 75(1), 10–19.

Place, F. and Swallow, B. (2000) Assessing the relationships between property rights and technology adoption in smallholder agriculture: a review of issues and empirical methods. In: Meinzen-Dick, R., Knox, A., Place, F. and Swallow, B. (eds) *Property Rights, Collective Action, and Technology Adoption.*

Place, F., Roth, M. and Hazell, P. (1994) Land tenure security and agricultural performance in Africa: overview of research methodology. In: Bruce, J.W. and Migot-Adholla, S. (eds) *Searching for Land Tenure Security in Africa.* Kendall/Hunt, Ames, Iowa, pp. 15–39.

Planning Commission (1996) *The Rolling Plan and Forward Budget for Tanzania for the Period 1996/97–1998/99,* Vol. I. Government Printer, Dar es Salaam.

Platteau, J.-P. (1994a) Behind the market stage where real societies exist – Part I: the role of public and private order institutions. *Journal of Development Studies* 30(3), 533–577.

Platteau, J.-P. (1994b) Behind the market stage where real societies exist – Part II: the role of moral norms. *Journal of Development Studies* 30(4), 753–817.

Platteau, J.-P. (1997) Mutual insurance as an elusive concept in traditional communities. *Journal of Development Studies* 33(6), 764–796.

Polson, R. and Spencer, D.S.C. (1991) The technology adoption process in subsistence agriculture: the case of cassava in Southwestern Nigeria. *Agricultural Systems* 36, 65–77.

Pompi, I., Mpepereki, S. and Gwata, E. (1998) Zimbabwe soybean promotion taskforce: objectives, achievements and agenda for the future. In: Waddington, S.R., Murwira, H.K., Kumwenda, J.D.T., Hikwa, D. and Tagwira, F. (eds) *Soil Fertility Research for Maize-Based Farming Systems in Malawi and Zimbabwe. Proceedings of the Soil Fert Net Results and Planning Workshop held from 7 to 11 July 1997 at Africa University, Mutare, Zimbabwe.* Soil Fert Net and CIMMYT-Zimbabwe, Harare, Zimbabwe, pp. 271–273.

Portes, A. and Landolt, P. (1996) The downside of social capital. *The American Prospect* 26, 18–21.

Pottier, J. (ed.) (1993) *Practicing Development: Social Science Perspectives.* Routledge, New York.

Powell, J.M. (1986) Manure for cropping: a case study from central Nigeria. *Experimental Agriculture* 22, 15–24.

Powell, J.M. and Williams, T.O. (1993) *Livestock, Nutrient Cycling and Sustainable Agriculture in the West African Sahel.* Gatekeeper Series No. 37, Sustainable Agriculture Programme, International Institute for Environment and Development (IIED), London.

Powell, J.M., Fernandez-Rivera, S. and Hofs, S. (1994) Sheep diet effects on nutrient cycling in semi-arid mixed farming systems of West Africa. *Agriculture, Ecosystems and Environment* 48, 263–271.

Powell, J.M., Fernandez-Rivera, S., Hiernaux, P. and Turne, N.D. (1996) Nutrient cycling in integrated rangeland/cropland systems of the Sahel. *Agricultural Systems* 52, 143–170.

Powell, J.M., Ikpe, F.N., Sonda, Z.C. and Fernandez-Rivera, S. (1998a) Urine effects on soil chemical properties and the impacts of urine and dung on pearl millet yield. *Experimental Agriculture* 34, 259–276.

Powell, J.M., Pearson, R.A. and Hopkins, J.C. (1998b) Impacts of livestock on crop production. In: Gill, M., Smith, T., Pollock, G.E., Owen, E. and Lawrence, T.L.J. (eds) *Food, Land, and Livelihoods – Setting Research Agendas for Animal Science, Proceedings of an International Symposium, 3–5 January 1998, Nairobi, Kenya.* Occasional Publication 21, British Society of Animal Science, Edinburgh, pp. 53–66.

Preston, T.R. and Leng, R.A. (1987) *Matching Ruminant Production Systems with Available Resources in the Tropics and Sub-Tropics.* Penambul Books, Armidale, Australia, 245 pp.

Pretty, J.N. (1995a) *Regenerating Agriculture: Policies and Practice for Sustainability and Self-Reliance.* Earthscan Publications, London, National Academy Press, Washington, DC, and ActionAid, Bangalore.

Pretty, J.N. (1995b) Participatory learning for sustainable agriculture. *World Development* 23(8), 1247–1263.

Pretty, J.N. (1998) *The Living Land: Agriculture, Food and Community Regeneration in Rural Europe.* Earthscan Publications, London.

Pretty, J. and Hine, R. (2000) The promising spread of sustainable agriculture in Asia. *Natural Resources Forum (UN)* 24, 107–121.

Pretty, J.N. and Pimbert, M. (1995) Beyond conservation ideology and the wilderness myth. *Natural Resources Forum* 19(1), 5–14.

Pretty, J.N. and Shah, P. (1997) Making soil and water conservation sustainable: from coercion and control to partnerships and participation. *Land Degradation and Development* 8, 39–58.

Pretty, J. and Ward, H. (2001) Social capital and the environment. *World Development* 29(2).

Pretty, J.N., Thompson, J. and Kiara, J.K. (1995) Agricultural regeneration in Kenya: the catchment approach to soil and water conservation. *Ambio* 24(1), 7–15

Pretty, J., Brett, C., Gee, D., Hine, R., Mason, C.F., Morison, J.I.L., Raven, H., Rayment, M. and van der Bijl, G. (2000) An assessment of the total external costs of UK agriculture. *Agricultural Systems* 65(2), 113–136.

Provenza, F.D. (1995) Postingestive feedback as an elementary determinant of food selection and intake in ruminants. *Journal of Range Management* 48, 2–17.

Putnam, R. (1995) Bowling alone: America's declining social capital. *Journal of Democracy* 6(1), 65–78.

Putnam, R.D. with Leonardi, R. and Nanetti, R.Y. (1993) *Making Democracy Work: Civic Traditions in Modern Italy.* Princeton University Press, Princeton, New Jersey.

Quin, F.M. (1997) Introduction. In: Singh, B.B., Mohan Raj, D.R., Dashiell, K.E. and Jackai, L.E.N. (eds) *Advances in Cowpea Research.* International Institute of Tropical Agriculture (IITA) and Japan International Research Centre for Agricultural Sciences (JIRCAS), Ibadan, Nigeria, pp. ix–xv.

Quiñones, M.A., Borlaug, N.E. and Dowswell, C.R. (1997) A fertilizer-based green revolution for Africa. In: Buresh, R., Sanchez, P. and Calhoun, F. (eds) *Replenishing Soil Fertility in Africa.* SSSA Special Publication No. 51, American Society of Agronomy and Soil Science Society of America, Madison, Wisconsin, pp. 81–96.

Quisumbing, A.R. (1996) Gender differences in agricultural productivity: methodological issues and empirical evidence. *Economic Development and Cultural Change* 24, 1579–1596.

Rachlin, H. (1990) *Judgment, Decision, and Choice.* W.H. Freeman, New York, 288 pp.

Rahm, M.R. and Huffman, W.E. (1984) The adoption of reduced tillage: the role of human capital and other variables. *American Journal of Agricultural Economics*, November, 405–413.

Raiffa, H. (1968) *Decision Tree Analysis.* McGraw Hill, New York, 300 pp.

Railovy, R., Josianne, and Andriamihaja, N. (2000) *Analyse Coût-Bénéfice Economique et Environnemental de la Privatisation du Chemin de Fer FCE,* Vol. II. Projet PAGE (USAID), Antananarivo, Madagascar.

Raintree, J.B. (1986) Agroforestry pathways: land tenure, shifting cultivation and sustainable agriculture. *Unasylva* 38, 2–15.

Raintree, J.B. (1987a) The state of the art of agroforestry diagnosis and design. *Agroforestry Systems* 5(3), 219–250.

Raintree, J.B. (1987b) *D&D User's Manual: An Introduction to Agroforestry Diagnosis and Design.* ICRAF, Nairobi.

Randrianarijoana, P. (1983) The erosion of Madagascar. *Ambio* 12, 308–311.

Rao, M.R. and Mathuva, M.N. (2000) Legumes for improved maize yields and income in semi-arid Kenya. *Agriculture, Ecosystems and Environment* 78, 123–137.

Rasmussen, L.N. and Meinzen-Dick, R. (1995) *Local Organizations for Natural Resource Management: Lessons from Theoretical and Empirical Literature.* Discussion Paper No. 11, Environment and Production Technology

Division, International Food Policy Research Institute, Washington, DC.

Razafindralambo, N. (1999) *La Ligne Ferroviare Fianarantsoa–Côte Est.* LDI, Fianarantsoa, Madagascar.

Reardon, T. (1997a) Using evidence of household income diversification to inform study of the rural nonfarm labor market in Africa. *World Development* 25(5), 735–747.

Reardon, T. (1997b) African agriculture: productivity and sustainability issues. In: Eicher, C. and Staatz, J. (eds) *Agricultural Development in the Third World*, 3rd edn. Johns Hopkins University Press, Baltimore, Maryland, pp. 444–457.

Reardon, T. and Vosti, S.A. (1992) Issues in the analysis of effects of policy on conservation and productivity at the household level in developing countries. *Quarterly Journal of International Agriculture* 31(4), 380–396.

Reardon, T. and Vosti, S.A. (1995) Links between rural poverty and the environment in developing countries: asset categories and investment poverty. *World Development* 23(9), 1495–1506.

Reardon, T. and Vosti, S.A. (1997) Policy analysis of conservation investments: extensions of traditional technology adoption research. In: Vosti, S.A. and Reardon, T. (eds) *Sustainability, Growth and Poverty Alleviation: A Policy and Agroecological Perspective.* Johns Hopkins University Press, Baltimore, Maryland, pp. 135–145.

Reardon, T., Kelly, V., Crawford, E., Jayne, T., Savadogo, K. and Clay, D. (1996) *Determinants of Farm Productivity in Africa: A Synthesis of Four Case Studies.* MSU International Development Paper No. 22, Department of Agricultural Economics, Michigan State University, East Lansing, Michigan.

Reardon, T., Stamoulis, K., Balisacan, A., Cruz, M.E., Berdegue, J. and Banks, B. (1998) Rural nonfarm income in developing countries. In: *The State of Food and Agriculture 1998.* Food and Agricultural Organization of the United Nations, Rome.

Reardon, T., Barrett, C., Kelly, V. and Savadogo, K. (1999) Policy reforms and sustainable agricultural intensification in Africa. *Development Policy Review* 17(4), 375–395.

Reed, J.D. (1995) Nutritional toxicology of tannins and related polyphenols in forage legumes. *Journal of Animal Science* 73, 1516–1528.

Reed, J.D., Soller, H. and Woodward, A. (1990) Fodder tree and straw diets for sheep: intake, growth and digestibility and the effects of phenolics on nitrogen utilization. *Animal Feed Science and Technology* 30, 39–50.

Reij, C. (1983) *Evolution de la lutte-anti-érosive en Haute-Volta depuis l'indépendance. Vers une plus grande participation de la population.* Amsterdam Institute of Environmental Studies, Free University, Amsterdam.

Reij, C., Mulder, P. and Begemann, L. (1988) *Water Harvesting for Plant Production.* Technical Paper No. 91, World Bank, Washington, DC.

Reij, C., Scoones, I. and Toulmin, C. (eds) (1996) *Sustaining the Soil: Indigenous Soil and Water Conservation in Africa.* Earthscan, London.

Reijntjes, C., Haverkrot, B. and Waters-Bayer, A. (1992) *Farming for the Future. An Introduction to Low External Input and Sustainable Agriculture.* Macmillan, London.

Reynolds, L. and de Leeuw, P.N. (1995) Myth and manure in nitrogen cycling: a case study of Kaloleni Division in Coast Province, Kenya. In: Powell, J.M., Fernandez-Rivera, S., Williams, T.O. and Renard, C. (eds) *Livestock and Sustainable Nutrient Cycling in Mixed Farming Systems of SubSaharan Africa*, Vol. II: *Technical Papers.* International Livestock Centre for Africa, Addis Adaba, pp. 509–522.

Reynolds, L. and Jabbar, M.A. (1995) Livestock in alley farming systems. In: Kang, B.T., Osimane, O. and Larbi, A. (eds) *Alley Farming Research and Development. Proceedings of the International Conference on Alley Farming, 14–18 September 1992.* International Institute of Tropical Agriculture, Ibadan, Nigeria, pp. 52–69.

Richards, P. (1983) Ecological change and the politics of African land use. *African Studies Review* 26(2), 1–72.

Richards, P. (1985) *Indigenous Agricultural Revolution: Ecology and Food Production in West Africa.* Hutchinson, London.

Richards, P. (1993) Cultivation: knowledge or performance? In: Hobart, M. (ed.) *An Anthropological Critique of Development: The Growth of Ignorance.* Routledge, New York, pp. 61–78.

Riklefs, R., Naveh, Z. and Turner, R. (1984) *Conservation of Ecological Processes.* The International Union for the Conservation of Nature and Natural Resources, Gland, Switzerland.

RIM (1992) *Nigerian Livestock Resources*, Vol. 1: *Executive Summary and Atlas.* Federal Government of Nigeria and Resource Inventory and Management Limited, Jersey, UK.

Robertson, A.F. (1984) *The People and the State: an Anthropology of Planned Development.* Cambridge University Press, Cambridge.

Rocheleau, D. (1995) Women, men, and trees: gender, power, and property in forest and agrarian landscapes. Paper prepared for GENDER-PROP, an International E-mail Conference on Gender and Property Rights.

Roderick, R. (1986) *Habermas and the Foundations of Critical Theory*. St Martins Press, New York.

Rodriguez, M.S. (1987) Agronomic practices for reducing drought stress and improving maize yield in the semi-arid tropics of West Africa. In: Menyogo, J., Bezuneh, T. and Youdeowei, A. (eds) *Food Grain Production in Semi-arid Africa*. OAU/STRC/SAFGRAD, Ouagadougou, Burkina Faso, pp. 493–509.

Rogers, E.M. (1995) *Diffusion of Innovations*. The Free Press, New York.

Rohrbach, D. (1986) *Sources of Growth in Smallholder Maize Production in Zimbabwe*. Working Paper.

Röling, N.G. (1988) *Extension Science: Information Systems in Agricultural Development*. Cambridge University Press, Cambridge.

Röling, N.R. and Wagemakers, M.A. (eds) (1998) *Social Learning for Sustainable Agriculture*. Cambridge University Press, Cambridge.

Rommelse, R. (2000) *Economic Analyses of On-farm Biomass Transfer and Improved Fallow Trials in Western Kenya*. International Centre for Research in Agroforestry, Nairobi.

Romney, D.L., Thorne, P.J. and Thomas, D. (1994) Some animal related factors influencing the cycling of nitrogen in mixed farming systems in sub-Saharan Africa. *Agriculture, Ecosystems and Environments* 49, 163–172.

Roose, E. (1990) Méthodes traditionnelles de gestion de l'eau et des sols en Afrique de l'Ouest. Définitions, fonctionnement limités et améliorations possibles. *Bulletin Erosion* 10, 98–107.

Rosegrant, M., Agcaoili-Sombilla, M. and Perez, N. (1995) *Global Food Projections to 2020: Implications for Investment*. Food, Agriculture, and the Environment Discussion Paper 5, International Food Policy Research Institute, Washington, DC.

Rowley, J. (1999) *Working with Social Capital*. Report for DFID, London.

Rukandema, M., Mavua, J.K. and Audi, P.O. (1981) *The Farming System of Lowland Machakos District, Kenya – Report on Farm Survey Results from Mwala Location*. Technical Report No. 1. NDFRS, Katumani, Kenya.

Runge-Metzger, A. (1995) Closing the cycle: obstacles to efficient P management for improved global food security. In: Tiessen, H. (ed.) *Phosphorus in the Global Environment*. John Wiley & Sons, New York, pp. 27–42.

Rusike, J., Sukume, C., Dorward, A., Mpepereki, S. and Giller, K. (2000) *The Economic Potential for Smallholder Soyabean Production in Zimbabwe*. Soil Fert Net Special Publication, CIMMYT-Zimbabwe, Harare, Zimbabwe, 64 pp.

Ruthenberg, H. (1985) *Innovation Policy for Small Farmers in the Tropics. The Economics of Technical Innovations for Agricultural Development*. Clarendon Press, Oxford, 176 pp.

Ruttan, V.W. and Hayami, Y. (1991) Rapid population growth and technical and institutional change. In: *United Nations Expert Group Meeting, Consequences of Rapid Population Growth in Developing Countries*. Taylor and Francis, New York.

Saliba, B. and Bromley, D. (1984) Empirical analysis of the relationship between soil conservation and farmland characteristics. Paper presented at the AAEA Annual Meeting, Ithaca, New York.

Samaké, O., Kodio, A., Traoré, B. and Ouattara, M. (1999) *Caractérisation multi-échelle pour améliorer la gestion des ressources naturelles dans les zones en marge du désert en Afrique de l'Ouest: Le cas du district de Bankass et du village de Lagassagou au Mali*. Projet DMP/ORU Rapport M1, IER-CRRA, Mopti, Mali.

Sanchez, P.A. (1995) Science in agroforestry. *Agroforestry Systems* 30, 5–55.

Sanchez, P.A. (1996) Science in agroforestry. *Agroforestry Systems* 9, 259–274.

Sanchez, P.A., Buresh, R.J. and Leakey, R.R.B. (1997a) Trees, soils and food security. Presented at the discussion meeting on Land Resources: on the Edge of the Malthusian Precipice? *Philosophical Transactions of the Royal Society, Series B London* 352, 949–961.

Sanchez, P.A., Shepherd, K.D., Soule, M.J., Place, F.M., Mokwunye, A.U., Buresh, R.J., Kwesiga, F.R., Izac, A.-M.N., Ndiritu, C.G. and Woomer, P.L. (1997b) Soil fertility replenishment in Africa: an investment in natural resource capital. In: Buresh, R.J., Sanchez, P.A. and Calhoun, F. (eds) *Replenishing Soil Fertility in Africa*. Soil Science Society of America Special Publication 51, SSSA and ASA, Madison, Wisconsin, pp. 1–46.

Sanchez, P., Jama, B., Niang, A.I. and Palm, C.A. (2000) Soil fertility management in sub-Saharan Africa: implications for development and the environment. In: Lee, D.R. and Barrett, C.B. (eds) *Tradeoffs or Synergies?*

Agricultural Intensification, Economic Development and the Environment. CAB International, Wallingford, UK, pp. 325–344.

Sanders, J.H. (1989) Agricultural research and cereal technology introduction in Burkina Faso and Niger. *Agricultural Systems* 30, 139–154.

Sanders, J.H. and Ahmed, M.M. (2001) Developing a fertilizer strategy for sub-Saharan Africa. In: Payne, W.A., Keeney, D.R. and Rao, S.C. (eds) *Sustainability of Agricultural Systems in Transition*. American Society of Agronomy Special Publication No. 64, Madison, Wisconsin, pp. 173–181.

Sanders, J.H., Shapiro, B.I. and Ramaswamy, S. (1996) *The Economics of Agricultural Technology in Semi-arid Sub-Saharan Africa*. Johns Hopkins University Press, Baltimore, Maryland.

Saunier, R. and Meganck, R. (1995) *Conservation of Biodiversity and New Regional Planning*. Department of Regional Development and Environment, Organization of American States, Washington, DC.

Savadogo, K., Reardon, T. and Pietola, K. (1998) Adoption of improved land-use technologies to increase food security in Burkina Faso: relating animal traction, productivity, and nonfarm income. *Agricultural Systems* 58(3), 441–464.

Scherr, S.J. (1995) Economic factors in farmer adoption of agroforestry: patterns observed in Western Kenya. *World Development* 23(5), 787–804.

Scherr, S.J. (1999) *Soil Degradation: A Threat to Developing-Country Food Security by 2020?* International Food Policy Research Institute, Washington, DC.

Scherr, S.J. and Alitsi, E. (1991) *The Development Impact of the CARE Agroforestry Extension Project in Siaya and South Nyanza Districts, Kenya*. Working Paper No. 64, ICRAF, Nairobi.

Scherr, S.J. and Hazell, P.B.R. (1994) *Sustainable Agricultural Development Strategies in Fragile Lands*. Discussion Paper no. 1, Environment and Production Technology Division, International Food Policy Research Institute, Washington, DC.

Scherr, S.J. and Oduol, P.A. (1989) *Farmer Adoption of Alley Cropping and Border Planting in the CARE Agroforestry Project*, Parts 1–3. ICRAF-CARE Project Reports Nos 15–17, ICRAF, Nairobi.

Schimel, D.S., Parton, W.J., Adamsen, F.J., Woodmasee, R.G., Senft, R.L. and Stillwell, M.A. (1986) *Biogeochemistry* 2, 39–52.

Schleik, K. (1986) The use of cattle dung in agriculture. *Natural Resources and Resource Development* 23, 53–87.

Schultz, T.W. (1975) The value of the ability to deal with disequilibria. *Journal of Economic Literature* 13(2), 827–846.

Scoones, I. (1993) Why are there so many animals? Cattle population dynamics in the communal areas of Zimbabwe. In: Behnke, R., Scoones, I. and Kerven, C. (eds) *Range Ecology at Disequilibrium: New Models of Natural Variability and Pastoral Adaptation in African Savannas*. ODI, London, pp. 334–356.

Scoones, I. (1994) *Living with Uncertainty: New Directions in Pastoral Development in Africa*. IT Publications, London.

Scoones, I. (1995) Exploiting heterogeneity: habitat use by cattle in the communal areas of Zimbabwe. *Journal of Arid Environments* 29, 221–237.

Scoones, I. (1998) *Sustainable Rural Livelihoods: a Framework for Analysis*. IDS Discussion Paper 72, University of Sussex, Brighton.

Scoones, I., Reij, C. and Toulmin, C. (1996) Sustaining the soil: indigenous soil and water conservation in Africa. In: Reij, C., Scoones, I. and Toulmin, C. (eds) *Sustaining the Soil: Indigenous Soil and Water Conservation in Africa*. Earthscan Publications, London, pp. 1–27.

Sedogo, M. (1993) Evolution des sols ferrugineux lessives sous culture: Influence des modes de gestion de la fertilité des sols. Thèse de doctorat ès Sciences, Université Nationale de Côte d'Ivoire, Abidjan.

Shapiro, B.I. (1990) New technology adoption in two agricultural systems of the Niamey region of Niger. Unpublished PhD dissertation, Department of Agricultural Economics, Purdue University, West Lafayette, Indiana.

Shapiro, B.I. and Sanders, J.H. (1998) Fertilizer use in semi-arid West Africa: profitability and supporting policy. *Agricultural Systems* 56(4), 467–482.

Shaxon, L.J. (1990) Intercropping and diversity: an economic analysis of cropping patterns on smallholder farms in central Malawi. Master's thesis, Cornell University, Ithaca, New York.

Shepherd, A. (1989) Approaches to privatization of fertilizer marketing in Africa. *Food Policy* 14, 143–154.

Shepherd, G. (1991) The communal management of forests in the semi-arid and sub-humid regions of Africa: past practice and prospects for the future. *Development Policy Review* 9(2), 151–176.

Shepherd, K.D. and Soule, M. (1998) Soil fertility management in west Kenya: dynamic simulation of productivity, profitability, and sustainability at different resource endowment levels. *Agriculture, Ecosystems, and Environment* 71, 131–145.

Shepherd, K.D., Ndufa, J.K., Ohlsson, E., Sjogren, H. and Swinkels, R. (1997) Adoption potential of hedgerow intercropping in maize-based cropping systems in the Highlands of Western Kenya: I. Background and agronomic evaluation. *Experimental Agriculture* 33, 197–209.

Sherlund, S.M., Barrett, C.B. and Adesina, A.A. (2000) *Smallholder Technical Efficiency with Stochastic Exogenous Production Conditions.* Working Paper WP 98-15, Department of Agricultural, Resource and Managerial Economics, Cornell University, Ithaca, New York.

Shiferaw, B. and Holden, S. (1997) Peasant agriculture and land degradation in Ethiopia: reflections on constraints and incentives for soil conservation and food security. *Forum for Development Studies* 2, 277–306.

Shiferaw, B. and Holden, S. (1998) Resource degradation and adoption of land conservation technologies in the Ethiopian highlands: a case study in Andit Tid North Shewa. *Agricultural Economics* 18, 233–247.

Shiferaw, B. and Holden, S.T. (1999) Soil erosion and smallholders' conservation decisions in the Ethiopian highlands. *World Development* 24(4), 739–752.

Shiferaw, B. and Holden, S. (2000) Policy instruments for sustainable land management: the case of highland smallholders in Ethiopia. *Agricultural Economics* 22(3), 217–232.

Shively, G.E. (1997) Consumption risk, farm characteristics and soil conservation among low-income farmers in the Philippines. *Agricultural Economics* 17, 165–177.

Silanikove, N., Gillboa, N. and Nitsan, Z. (1997) Interactions among tannins, supplementation and polyethylene glycol in goats fed oak leaves. *Animal Science* 64, 479–483.

Simon, H.A. (1979) Information processing models of cognition. *Annual Reviews of Psychology* 30, 363–396.

Singh, B.B. and Tarawali, S.A. (1997) Cowpea: an integral component of sustainable mixed crop/livestock farming systems in West Africa and strategies to improve its productivity. In: Renard, C. (ed.) *Crop Residues in Sustainable Mixed Crop–Livestock Farming System.* CAB International in association with the International Crops Research Institute for the Semi-Arid Tropics (ICRISAT) and the International Livestock Research Institute (ILRI), Wallingford, UK, pp. 79–100.

Singh, B.B., Sharma, B.M. and Terao, T. (1995) A simple screening method for drought tolerance in cowpea. In: *Agronomy Abstracts.* American Society of Agronomy, Madison, Wisconsin, p. 71.

Singh, I., Squire, L. and Strauss, J. (eds) (1986) *Agricultural Household Models. Extensions, Applications, and Policy.* Johns Hopkins University Press, Baltimore, Maryland.

Singh, K. and Ballabh, V. (1997) *Cooperative Management of Natural Resources.* Sage, New Delhi.

Sivakumar, M.V.K. and Salaam, S.A. (1999) Effect of year and fertilizer on water-use efficiency of pearl millet (*Pennisetum glaucum*) in Niger. *Journal of Agricultural Science* 132, 139–148.

Sjaastad, E. and Bromley, D.W. (1997) Indigenous land rights in sub-Saharan Africa: appropriation, security and investment demand. *World Development* 25(4), 549–562.

Smale, M. and Heisey, P.W. (1994) Maize research in Malawi revisited: an emerging success story? *Journal of International Development* 6 (Nov.–Dec.), 689–706.

Smale, M. with Kaunda, Z.H.W., Makina, H.L., Mkandawire, M.M.M.K., Msowoya, M.N.S., Mwale, D.J.E.K. and Heisey, P.W. (1991) *Chimanga cha Makolo, Hybrids, and Composites: an Analysis of Farmers' Adoption of Maize Technology in Malawi, 1989–91.* CIMMYT, Lilongwe, Malawi.

Smaling, E.M.A., Fresco, L.O. and de Jager, A. (1996) Classifying monitoring and improving soil nutrient stocks and flows in African agriculture. *Ambio* 25, 492–496.

Smaling, E.M., Nandwa, S.M. and Janssen, B.H. (1997) Soil fertility in Africa is at stake. In: Sanchez, P. and Buresh, R. (eds) *Replenishing Soil Fertility in Africa.* Special Publication no. 51, Soil Science Society of America, Madison, Wisconsin, pp. 47–62.

Smith, J.W., Naazie, A., Larbi, A., Agyemang, K. and Tarawali, S. (1997) Integrated crop–livestock systems in sub-Sahara Africa: an option or imperative? *Outlook on Agriculture* 26(4), 237–346.

Smith, V.K. (1996) Can contingent valuation distinguish economic values for different public goods? *Land Economics* 72(2), 139–151.

Snapp, S. (1999) Initial report on best bet legume-based soil fertility technologies. Prepared for Soil Fertility Workshop participants, 25–26 March 1999, ICRISAT, Lilongwe, Malawi.

Snapp, S.S. and Silim, S. (2002) Farmer preference and legume intensification for low nutrients environments. *Plants and Soil* (in press).

Snapp, S.S., Mafongoya, P.L. and Waddington, S. (1998) Organic matter technologies for integrated nutrient management in smallholder cropping systems of southern Africa. *Agriculture, Ecosystems and Environment* 71, 185–200.

Sogbedji, J. (2000) *Green Manure Cover Crops in West Africa: Constraints and Research Needs.* MOIST/CAWG Seminar Series, Cornell University, Ithaca, New York.

Soil Fert Net (1999) *'Best Bet' Soil Fertility Technologies Revisited for Malawi and Zimbabwe. Report on a Workshop held in Zomba, Malawi, 26 to 28 August 1999.* CIMMYT-Zimbabwe and ICRISAT-Malawi, Harare, Zimbabwe, 21 pp.

Soko, H.N. (1998) Pigeonpeas: issues and options for research, production, marketing and utilisation in Malawi. A discussion paper presented at the Groundnut and Pigeonpea Workshop, Mangochi, Malawi.

Somda, Z.C., Powell, J.M., Fernandez-Rivera, S. and Reed, J.D. (1995) Feed factors affecting nutrient excretion by ruminants and the fate of nutrients when applied to soil. In: Powell, J.M., Fernandez-Rivera, S., Williams, T.O. and Renard, C. (eds) *Livestock and Sustainable Nutrient Cycling in Mixed Farming Systems of SubSaharan Africa, Vol. II: Technical Papers.* International Livestock Centre for Africa, Addis Adaba, pp. 149–170.

Soule, M. and Shepherd, K.D. (2000) An ecological and economic analysis of phosphorus replenishment for Vihiga Division, western Kenya. *Agricultural Systems* 64, 83–98.

Southgate, D., Hitzhusen, F. and Macgregor, R. (1984) Remedying Third World soil erosion problems. *American Journal of Agricultural Economics* 66(5), 879–884.

Spencer, D.S. (1996) Infrastructure and technology constraints to agricultural development in the humid and subhumid tropics of Africa. *African Development Review* 8(2), 68–93.

Spradley, J.P. (1979) *The Ethnographic Interview.* Harcourt Brace Jovanovich College Publishers, Orlando, Florida, 200 pp.

Spradley, J.P. (1980) *Participant Observation.* Holt, Rinehart and Winston, New York, 196 pp.

Srivastava, K.L., Abebe, M., Astatke, A., Haile, M. and Regassa, H. (1993) Distribution and importance of Ethiopian vertisols and location of study sites. In: Mamo, T., Astatke, A., Srivastava, K.L. and Dibabe, A. (eds) *Improved Management of Vertisols for Sustainable Crop–Livestock Production in the Ethiopian Highlands: Synthesis Report 1986–92.* Technical Committee of the Joint Vertisol Project, Addis Ababa.

Staal, S.J., Kruska, R., Baltenweck, I., Kenyanjui, M., Wokabi, A., Njubi, D., Thornton, P.K. and Thorpe, W. (1999) Combined household and GIS analysis of smallholder production systems: an application to intensifying smallholder dairy systems in Central Kenya. Paper presented at the Third International Symposium on Systems Approaches for Agricultural Development (SAAD-III), 8–10 November, National Agrarian University, La Molina, Lima, Peru.

Steer, A. and Lutz, E. (1993) Measuring environmentally sustainable development. *Finance and Development*, December, 20–23.

Stiglitz, J.E. (1990) Peer monitoring and credit markets. *World Bank Economic Review* 4(2), 271–297.

Stocking, M.A. (1988) Socio-economics of soil conservation in developing countries. *Journal of Soil Conservation* 43, 381–385.

Stoorvogel, J. and Smaling, E. (1990) *Assessment of Soil Nutrient Depletion in Sub-Saharan Africa: 1983–2000*, Vols 1–4. Report No. 28, Winand Staring Centre, Wageningen, The Netherlands.

Strasberg, P., Jayne, T., Yamano, T., Nyoro, J., Karanja, D. and Strauss, J. (1999) *Effects of Agricultural Commercialization on Food Crop Input Use and Productivity in Kenya.* MSU International Development Working Paper No. 71, Department of Agricultural Economics, Michigan State University, East Lansing, Michigan.

Sumberg, J.E., McIntire, J., Okali, C. and Attah-Krah, A.N. (1987) Economic analysis of alley farming with small ruminants. *ILCA Bulletin* 28, 2–6.

Sureshwaran, S., Londhe, S.R. and Frazier, P. (1996) Factors influencing soil conservation decisions in developing countries: a case study of upland farmers in the Phillipines. *Journal of Agribusiness* 14(1), 83–94.

Sussman, R.W., Green, G. and Sussman, L. (1994) Satellite imagery, human ecology, anthropology and deforestation in Madagascar. *Human Ecology* 22, 333–354.

Swallow, B.M. and Bromley, D.W. (1995) Institutions, covernance and incentives in common property regimes for African rangelands. *Environment and Resource Economics* 6, 99–118.

Swift, M.J. (1996) *Sustainable Management of the Soil Resource: Developing a Framework for Research and Development.* World Bank, Washington, DC.

Swift, M.J., Heal, O.W. and Anderson, J.M. (1979) *Decomposition in Terrestrial Eco-systems.*

Studies in Ecology 5, Blackwell, Oxford, 243 pp.

Swinkels, R. and Franzel, S. (1997) Adoption potential of hedgerow intercropping systems in the highlands of Western Kenya: II. Economic and farmers' evaluation. *Experimental Agriculture* 33, 211–223.

Swinkels, R., Franzel, S., Shepherd, K.D., Ohlsson, E. and Ndufa, J.K. (1997) The economics of short rotation improved fallows: evidence from areas of high population density in western Kenya. *Agricultural Systems* 55, 99–121.

Tanner, J.C., Reed, J.D. and Owen, E. (1990) The nutritive value of fruits (pods with seeds) from four *Acacia* spp. compared with extracted noug (*Guizotia abyssinica*) meal as supplements to maize stover for Ethiopian highland sheep. *Animal Production* 51, 127–133.

Tarawali, G. (1991) Residual effect of *Stylosanthes* fodder banks on grain yield of maize. *Tropical Grasslands* 25, 26–31.

Tarawali, G. and McNamara, N. (1998) *Green Manure Cover Crops Systems for Smallholder Farmers of Igalaland, Kogi State, Nigeria.* A preliminary report submitted to Diocesan Development Services (DDS), Idah, Kogi State, 34 pp.

Tarawali, G., Manyong, V.M., Carsky, R.J., Vissoh, P., Osei-Bonsu, P. and Galiba, M. (1999) Adoption of improved fallows in West Africa: lessons learned from the velvet bean and stylo case studies. *Agroforestry Systems* 47, 93–122.

Tarawali, S.A., Smith, J.W., Hiernaux, P., Singh, B.B., Gupta, S.C., Tabo, R., Harris, F., Nokoe, S., Fernandez-Rivera, S. and Bationo, A. (2000a) Integrated natural resource management – putting livestock in the picture. Paper presented at the CGIAR workshop on Integrated Natural Resource Management held at Penang, Malaysia, 20–25 August 2000.

Tarawali, S.A., Singh, B.B., Gupta, S.C., Tabo, R., Harris, F., Nokoe, Fernandez-Rivera, S., Bationo, A., Manyong, V.M. and Odion, E.C. (2000b) Cowpea as a key factor for a new approach to integrated crop–livestock systems research in the dry savannas of West Africa. Presented at the World Cowpea Research Conference III held in International Institute of Tropical Agriculture (IITA), Ibadan, Nigeria, 4–8 September 2000.

Taylor, M. (1982) *Community, Anarchy and Liberty.* Cambridge University Press, Cambridge.

Tegegne and Bekele (1989) An overview of environmental degradation in Ethiopia and set priority program of the sector. In: Institute of Agricultural Research (IAR, now renamed Ethiopian Agricultural Research Organization (EARO)) (ed.) *Proceedings, First Natural Resource Conservation Conference, Natural Resources Degradation: A Challenge to Ethiopia.* IAR, Addis Ababa.

Tekalign, M., Abiye, A., Srivastava and Asgelil, D. (1993) *Improved Management of Vertisols for Sustainable Crop–Livestock Production in the Ethiopian Highlands. Synthesis Report 1986–92.* ILCA, Addis Ababa, Ethiopia.

Tekie, A. (1999) Land tenure and soil conservation: evidence from Ethiopia. Unpublished PhD dissertation, Department of Economics, Göteborg University, Göteborg.

Templeton, S. and Scherr, S.J. (1999) Effects of demographic and related microeconomic change on land quality in hills and mountains of developing countries. *World Development* 27(6), 903–918.

Tiffen, M., Moltimore, M. and Gichuki, F. (1994) *More People, Less Erosion: Environmental Recovery in Kenya.* ACTS Press, Nairobi.

Timmer, C.P. (1986) The agricultural transformation. In: Chenery, H. and Srinivasan, T.N. (eds) *Handbook of Development Economics,* Vol. 1. Elsevier, Amsterdam, pp. 275–331.

Tisdell, C. (1995) Economic indicators to assess the sustainability of conservation farming projects: an evaluation. *Agriculture, Ecosystems and Environment* 57(2/3), 117–131.

Tobin, J. (1958) Estimation of relationships for limited dependent variables. *Econometrica* 26, 24–36.

Tomich, T.P., Kilby, P. and Johnston, B.F. (1995) *Transforming Agrarian Economies: Opportunities Seized, Opportunities Missed.* Cornell University Press, Ithaca, New York.

Tonye, J., Meke-Me-Ze, C. and Titi-Nwel, P. (1993) Implications of national land legislation and customary land and tree tenure on the adoption of alley farming. *Agroforestry Systems* 22, 153–160.

Turner, B.L., Hyden, G. and Kates, R. (1993) *Population Growth and Agricultural Change in Africa.* University of Florida Press, Gainesville, Florida.

Turner, R.K., Pearce, D. and Bateman, I. (1994) *Environmental Economics: an Introduction.* Harvester Wheatsheaf, Hertfordshire.

Udry, C. (1996) Gender, agricultural production, and the theory of the household. *Journal of Political Economy* 104(5), 1010–1046.

Université AM de Niamey et Ecole Polytechnique Fédérale de Zurich (1995) *Mesures simples de protection anti-erosive des ecosystemes au Niger.* Bulletin No. 4, Conservation et Gestion des Eaux et des Sols au Niger, Tahoua, Niger.

Uphoff, N. (1992) *Learning from Gal Oya: Possibilities for Participatory Development and Post-Newtonian Science.* Cornell University Press, Ithaca, New York.

Uphoff, N. (1998) Understanding social capital: learning from the analysis and experience of participation. In: Dasgupta, P. and Serageldin, I. (eds) *Social Capital: A Multiperspective Approach.* World Bank, Washington, DC.

Uphoff, N. (1999) Agroecological implications of the system of rice intensification (SRI) in Madagascar. *Environment, Development and Sustainability* 1, 297–313.

Uphoff, N. (2000) Agroecological implications of the system of rice intesification (SRI) in Madagascar. *Environment, Development and Sustainability* 1(3).

Uphoff, N. (ed.) (2001) *Agroecological Innovations: Increasing Food Production with Participatory Development.* Earthscan, London.

Uphoff, N., Esman, M.J. and Krishna, A. (1998) *Reasons for Success: Learning from Instructive Experiences in Rural Development.* Kumarian Press, West Hartford, Connecticut.

Uttaro, R.P. (1998) *Diminishing Returns: Soil Fertility, Fertilizer, and the Strategies of Farmers in Zomba RDP in Southern Malawi.* Report to the University of Florida Soils Collaborative Research Support Project, Gainesville, Florida.

Vabi, M.B., Yamoah, C. and Tambi, E. (1995) On farm trials with *Calliandra* in the highland region of northwestern Cameroon. In: Kang, B.T., Osiname, A.O. and Larbi, A. (eds) *Alley Farming Research and Development. Proceedings of the International Conference on Alley Farming, 14–18 September 1992.* International Institute of Tropical Agriculture, Ibadan, Nigeria, pp. 449–454.

van de Fliert, E. (1997) From pest control to ecosystem management: how IPM training can help? Paper presented to International Conference on Ecological Agriculture, Chandigarh, India, November 1997.

Van den Brink, R., Bromley, D.W. and Chavas, J.P. (1995) The economics of Cain and Abel: agropastoral property rights in the Sahel. *Journal of Development Studies* 31(3), 373–399.

van der Pol, R. (1992) *Soil Mining: An Unseen Contributor to Farm Income in Southern Mali,* Bulletin 325. Royal Tropical Institute (KIT), Amsterdam.

van Reuler, H. and Prins, W.H. (eds) (1993) *The Role of Plant Nutrients for Sustainable Food Crop Production in Sub-Saharan Africa.* Part 1. *The Role of Plant Nutrients for Sustainable Food Crop Production in Sub-Saharan Africa.* Part 2. *Nutrient Supply and Distribution at Country Level. Case Studies of Malawi and Ethiopia.* VKP, Leidschendam, The Netherlands, 232 pp.

Van Soest, P.J. (1994) *Nutritional Ecology of the Ruminant,* 2nd edn. Cornell University Press, Ithaca, New York, 476 pp.

Vaughan, M. (1987) *The Story of an African Famine.* Cambridge University Press, New York.

Veit, P.G., Mascarenhas, A. and Ampadu-Agyei, O. (1995) *Lessons from the Ground Up: African Development that Works.* World Resources Institute, Washington, DC.

Versteeg, M. and Koudokpon, V. (1990) Participative farmer testing of four low external input technologies to address soil fertility decline in Mono Province (Benin). *Agricultural Systems* 42, 265–276.

Versteeg, M. and Koudokpon, V. (1991) *Mucuna* helps control *Imperata* in southern Benin. *West African Farming Systems Research Network Bulletin* 7, 7–8.

Versteeg, M.N. and Koudokpon, V. (1993) Participative farmer testing of four low external input technologies, to address soil fertility decline in Mono province (Benin). *Agricultural Systems* 42, 265–276.

Viederman, S. (1994) *Ecological Literacy: Can Colleges Save the World? Keynote address to Associated Colleges of the Midwest Conference on Ecological Education, 11 March, Beloit College.* Jessie Smith Noyes Foundation, New York.

Viets, F.G., Jr (1962) Fertilizers and the efficient use of water. *Advances in Agronomy* 14, 223–264.

Vissoh, P., Manyong, V.M., Carsky, R.J., Osei-Bonsu, P. and Galiba, M. (1998) Experiences with *Mucuna* in West Africa. In: Buckles, D., Eteka, A., Osiname, O., Galiba, M. and Galiano, N. (eds) *Cover Crops in West Africa: Contributing to Sustainable Agriculture.* International Development Research Centre (IDRC), Ottawa, Canada, International Institute of Tropical Agriculture (IITA), Ibadan, Nigeria, and Sasakawa Global 2000, Cotonou, Benin, pp. 1–32.

Vitale, J.D. (2001) The economic impacts of new sorghum and millet technologies in Mali. Unpublished PhD dissertation, Department of Agricultural Economics, Purdue University, West Lafayette, Indiana.

Vosti, S.A. and Reardon, T. (eds) (1997) *Sustainability, Growth and Poverty Alleviation.* Johns Hopkins University Press, Baltimore, Maryland.

Waddington, S.R. and Heisey, P.W. (1997) Meeting the nitrogen requirements of maize grown by resource-poor farmers in southern Africa by integrating varieties, fertilizer use, crop management and policies. In: Edmeades, G.O., Bänziger, M., Mickelson, H.R. and Peña-Valdivia, C.B. (eds) *Developing Drought- and Low N-Tolerant Maize. Proceedings of a Symposium, 25 to 29 March 1996, CIMMYT, El Batan, Texcoco, Mexico.* CIMMYT, Mexico DF, pp. 44–57.

Waddington, S.R., Murwira, H.K., Kumwenda, J.D.T., Hikwa, D. and Tagwira, F. (eds) (1998) *Soil Fertility Research for Maize-Based Farming Systems in Malawi and Zimbabwe. Proceedings of the Soil Fert Net Results and Planning Workshop held from 7 to 11 July 1997 at Africa University, Mutare, Zimbabwe.* Soil Fert Net and CIMMYT-Zimbabwe, Harare, Zimbabwe, 312 pp.

Waghorn, G.C., Reed, J.D. and Ndlovu, L.R. (1997) Nutritional effects of browse condensed tannins (proanthocyanidins) on ruminant digestion. In: Buchanan-Smith, J.G., Bailey, L.D. and McCaughey, P. (eds) *Proceedings of the XVIII International Grasslands Congress: Grasslands 2000, 8–19 June 1997, Winnipeg, Canada.* Association Management Centre, Calgary, Canada, pp. 153–166.

Walker, B.H., Matthew, D.A. and Dye, P.J. (1986) Management of grazing systems – existing versus an event-orientated approach. *South African Journal of Science* 82, 176–181.

Walker, D.H., Sinclair, F.L. and Thapa, B. (1995) Incorporation of indigenous knowledge and perspectives in agroforestry development. Part 1: Review of methods and their application. *Agroforestry Systems* 30(1–2), 235–248.

Wallace, M.B. (1997) *Fertilizer Use and Environmental Impacts – Positive and Negative: a Review with Emphasis upon Inorganic Fertilizers in Africa.* Winrock International Institute for Agricultural Development, Washington, DC.

Walters, B.B., Cadelina, A., Cardano, A. and Visitacion, E. (1999) Community history and rural development: why some farmers participate more readily than others. *Agricultural Systems* 59, 193–214.

Ward, H. (1998) State, association, and community in a sustainable democratic polity: towards a green associationalism. In: Coenen, F., Huitema, D. and O'Toole, L.J. (eds) *Participation and the Quality of Environmental Decision Making.* Kluwer Academic Publishers, Dordrecht.

Warren, A. and Khogali, M. (1992) *Assessment of Desertification and Drought in the Sudano-Sahelian Region, 1985–1991.* UNSO, New York.

Watson, C.J., Billingsley, P., Croft, D.J. and Huntsberger, D.V. (1990) *Statistics for Management and Economics.* Allyn and Bacon, Needham Heights, Massachusetts.

Weight, D. and Kelly, V. (1999) *Fertilizer Impacts on Soils and Crops of Sub-Saharan Africa.* MSU International Development Paper No. 21, Michigan State University, East Lansing, Michigan.

Wellard, K. and Copestake, J. (eds) (1993) *Non-Governmental Organizations and the State in Africa.* Routledge, New York.

Whitehead, D.C. (1990) Atmospheric ammonia in relation to grassland agriculture and livestock production. *Soil Use and Management* 6, 63–65.

Whittome, M.P.B., Spencer, D.S.C. and Bayliss-Smith, T. (1995) IITA and ILCA on-farm alley farming research: lessons for extension workers. In: Kang, B.T., Osiname, A.O. and Larbi, A. (eds) *Alley Farming Research and Development. Proceedings of the International Conference on Alley Farming, 14–18 September 1992.* International Institute of Tropical Agriculture, Ibadan, Nigeria, pp. 423–435.

Williams, D.E. (1997) Gender and integrated resource management: the case of Western Kenya. Masters' thesis, University of Florida, Gainesville, Florida.

Williams, T.O., Powell, J.M. and Fernandez-Rivera, S. (1993) Manure utilization, drought cycles and herd dynamics in the Sahel: implications for crop land productivity. In: *Proceedings of the Nutrient Cycling Workshop at ILCA, 26–30 November 1993.* ILCA, Addis Ababa, Ethiopia, pp. 89–113.

Williams, T.O., Hiernaux, P. and Fernandez-Rivera, S. (1999) Crop–livestock systems in sub-Saharan Africa: determinants and intensification pathways. In: McCarthy, N., Swallow, B., Kirk, M. and Hazell, P. (eds) *Property Rights, Risk and Livestock Development in Africa.* International Livestock Research Institute, Nairobi, Kenya, and International Food Policy Research Institute, Washington, DC, pp. 132–151.

Williamson, J. (1975) *Useful Plants of Malawi* (1st edn, 1955). University of Malawi, Limbe, Malawi.

Winrock (1992) *Assessment of Animal Agriculture in Sub-Saharan Africa.* Winrock International, Morrilton, Arkansas, 125 pp.

Wischmeier, W.H. and Smith, D.D. (1978) *Predicting Rainfall Erosion Losses, a Guide to*

Conservation Planning. Agricultural Handbook No. 537, USDA, Washington, DC, pp. 1–58.

Wood, S., Sebastian, K., Nachtergaele, F., Nielsen, D. and Dai, A. (1999) *Spatial Aspects of the Design and Targeting of Agricultural Development Strategies.* EPTD Discussion Paper No. 44, Environment and Production Technology Division, International Food Policy Research Institute, Washington, DC.

Woodward, A. and Reed, J.D. (1997) Nitrogen metabolism of sheep and goats consuming *Acacia brevispica* and *Sesbania sesban. Journal of Animal Science* 75, 1130–1139.

Woolcock, M. (1998) Social capital and economic development: towards a theoretical synthesis and policy framework. *Theory and Society* 27, 151–208.

World Bank (1995) *World Development Report, 1995. Workers in an Integrating World.* Oxford University Press for the World Bank.

World Bank (1996) *Towards Environmentally Sustainable Development in Sub-Saharan Africa.* World Bank, Washington, DC.

World Bank (1999) *World Development Report.* World Bank, Washington, DC.

World Bank (2000a) *World Development Report.* World Bank, Washington, DC.

World Bank (2000b) *Can Africa Claim the 21st Century?* World Bank, Washington, DC.

Wright, P. (1985) Water and soil conservation by farmers. In: Ohm, H.W. and Nagy, J.G. (eds) *Appropriate Technologies for Farmers in Semiarid West Africa.* Purdue University, West Lafayette, Indiana, pp. 54–60.

Wubenah, N. and Sanders, J.H. (2001) Diffusion of *Striga*-resistant cultivars in Tigray, Ethiopia: a preliminary report to INTSORMIL. Draft. ILRI, Addis Ababa, Ethiopia.

Wyatt, T.J. (1998) Investments in soil conservation by Malagasy farmers of the Haute Terre. PhD dissertation, Department of Agricultural and Resource Economics, University of California, Davis, California.

Yai, K. (1998) Experiénces du projet de développement de l'élévage du Borgou-est (PDEBE) sur les plantes de couvature. In: Buckles, D., Eteka, A., Osiname, O., Galiba, M. and Galiano, N. (eds) *Cover Crops in West Africa. Contributing to Sustainable Agriculture.* International Development Research Centre (IDRC), Ottawa, Canada, International Institute of Tropical Agriculture (IITA), Ibadan, Nigeria, and Sasakawa Global 2000, Cotonou, Benin, p. 239.

Yanggen, D., Kelly, V., Reardon, T. and Naseem, A. (1998) *Incentives for Fertilizer Use in Sub-Saharan Africa: a Review of Empirical Evidence on Fertilizer Response and Profitability.* MSU International Development Working Paper No. 70, Department of Agricultural Economics, Michigan State University, East Lansing, Michigan.

Yaron, J., Benjamin, M.P. and Piprek, G.L. (1997) *Rural Finance: Issues, Design, and Best Practices.* World Bank, Washington, DC.

Yoder, D. and Lown, J. (1995) The future of RUSLE: inside the new revised universal soil loss equation. *Journal of Soil and Water Conservation* 50(5), 484–489.

Yohannes, H. and Holden, S. (2000a) Land tenure, farm input intensity and food security: a study of farm households in southern Ethiopia. Unpublished manuscript.

Yohannes, H. and Holden, S. (2000b) Food self-sufficiency as a means of achieving food security: a case of farm households in southern Ethiopia. Unpublished manuscript.

Zeller, M., Schreider, G., von Braun, J. and Heidhueg, F. (1997) *Rural Finance for Food Security for the Poor.* International Food Policy Research Institute, Washington, DC.

Index

OCIAL